YOUNG PEOPLE
AND THE BELT AND ROAD

Opportunities And Challenges
In Central And Eastern Europe

U0106527

ACKNOWLEDGEMENTS

We, the editors in the Young Belt and Road Series Editorial Board and the authors who contributed to this publication, offer our sincere gratitude to the following institutions and individuals.

In 2017, this project commenced under the support of the Research Center for Regional Coordinated Development, Zhejiang University and the Academy of International Strategy and Law of Zhejiang University. The subsequent editing and publication processes were generously sponsored by the International Academy of the Belt and Road and Greater Bay Area & Belt and Road Center, Basic Law Foundation. We express our most sincere gratitude to these organisations that have enabled this publication.

We benefitted enormously from the edits and suggestions made by Donal Scully, who is a veteran (since 1987) print and digital media journalist with experience writing and editing in UK provincial newspapers, the prestigious South China Morning Post in Hong Kong and several online news platforms as well as assorted book proofreading credits. We would also like to thank Wang Xin, a postdoctoral researcher at Zhejiang University, who helped us in refining the academic aspects of the publication.

Last but not least, there is Anne Lee and the editorial team at Joint Publishing Hong Kong, without which this publication would not have been possible.

PREFACE

The Belt and Road Initiative (BRI) is an international, inter-generational vision that includes more than half the population around the globe. The young people of today will be the main beneficiaries as well as contributors to this defining initiative of the century. However, despite the common recognition of the importance of young people as part of the initiative, very little work has been done to study the opportunities, roles, and value of youth involvement as part of the Belt and Road Initiative. The Young Belt and Road publication series aims to shed some light into these unexplored yet significant areas of research.

The current book examines the Central and Eastern European (CEE) region and the '16+1' initiative. In total, 33 authors worked on this project. The teams are composed of undergraduate and graduate students, PhD candidates, and graduates from the region. After nearly six months of hard work, we invited youth representatives (1-4 per country) from each of the 16 countries to write English manuscripts that form part of this publication. Most authors were recommended to us by universities, research institutes and foreign ministries in the region. To ensure the quality of the English manuscripts, we established the Young Belt and Road Series Editorial Board, with 11 members made up of university students from Canada, Mainland China, Denmark, Hong Kong, India, Singapore, the United Kingdom, and the United States. They include Winnie Lam, Wang Hongxi, William P Campbell, Ujjwal Sharma, Charlie Parker, Goh Shu Li, Rasmus N Hogh, Alexandra Davis, Adrian Lopez, as well as Charlotte Chiu, who contributed to the substance and style of this publication.

The first chapter of this book is an article that provides an overview of the role of Central and Eastern Europe (CEE) in the Belt and Road Initiative. In the ensuing chapters, readers may expect each to cover one of the 16 countries in the CEE region. Each chapter is a report (or country paper) that, in most cases, is divided into five parts. The first part gives an overview of the political structure of a country, summarising its modern historical development and the formation of its political system. The second part examines the political direction of a country, analysing potential reforms (e.g. constitutional reforms) and other important social movements. The third part evaluates the current opportunities and challenges faced by young people in their respective countries, providing a youth's perspective of what their countries have done and need to do for the rising generation. The fourth part focuses on the Belt and Road Initiative, discussing the developmental possibilities and obstacles associated with the Initiative. In response to the obstacles, the final part suggests alternatives to circumvent the barriers involved in the BRI and ways to increase youth involvement in the Initiative.

The main idea of the publication is to cover the political realities of the present and the political trends of the future, as well as the opportunities and challenges that young people face in each of these countries. Through the work of our 33 authors, we wish to present the voice of young people in the region and to provide a gateway of insights into the future leaders of Central and Eastern Europe.

The Belt and Road Initiative is an inter-generational movement, akin to a long-term investment made into different societies in the world. Most of the benefits from current infrastructure projects may only be reaped over the next twenty to thirty years. Therefore, it is crucial to place attention onto the largest group of benefactors of the BRI – the younger generation. Nearly every author suggests that society needs to increase the opportunities for the youth for them to increase social mobility. Nearly every paper mentions the expansion of international cooperation and exchange as the way to maximise the untapped potential young people can offer. Taking this publication as an example, the difficulties experienced in the search process for authors itself illustrates the lack of an international platform for young people. Even within the Central and Eastern European region, there is an insufficient level of cross-border youth exchanges, highlighted by the fact that many authors do not know students or other contact points in neighbouring countries. This further points to the imminent need of constructing a strong youth network.

In response to the aforementioned reasons, the Academy of International Strategy and Law of Zhejiang University hosted the 1st International Youth Forum on the Belt and Road in December 2017. The event invited more than 100 students from nearly 30 countries across Asia, Africa, Europe, and North America. This included six of our authors from the Czech Republic, Hungary, Lithuania, Romania, Montenegro, and Slovenia. Over the course of the two-day conference, participants presented materials on a variety of BRI-related issues, such as political risk, economic corridors, innovative technology, education, international law, dispute resolution mechanisms, cultural exchange, etc. One of the main results of the Youth Forum was the signing of the Young Belt and Road Initiative, a document that urges an increase in youth-level participation in BRI-related projects. Inspired by the value of exchanges such as the Youth Forum, numerous students joined together to express support for similar events in the future.

Subsequently, in December 2018, the 2nd International Youth Forum on the Belt and Road was hosted in Zhejiang University. The Forum focused on Central and Eastern Europe and invited all authors and editors who contributed to this publication to present their research. In addition, through the authors, we also gathered eight universities and research institutes in the Central and Eastern Europe region to participate as Forum co-organisers. Student representatives from 12 Chinese universities were also present. Panel discussions covered the '16+1' Initiative, future development in the Balkan Peninsula, as well as technology and business opportunities. The Forum also included an 'open-floor colloquium', in which all participants engaged in a candid exchange discussing BRI-related issues and presented ideas on how to increase the impact of youth collaboration.

We hope this book would be the starting point in a greater project committed to connecting young people around the world and one of the first sprouts in the Young Belt and Road.

Wang Chenxi
12th February 2019
London

CHAPTER I

by WANG CHENXI
University of Oxford, Master of Public
Policy (Class of 2021).

INTRODUCTION:

On the Role of Central and Eastern Europe in the Belt and Road Initiative

This paper aims to provide a brief examination of the Belt and Road Initiative (BRI) in four separate parts. First, we provide an overview of BRI, outlining the different types of projects that are proposed and those that are now under way. Second, we analyse the main political and economic characteristics of the Central and Eastern Europe (CEE). Third, we evaluate the prospects of cooperation in CEE. Finally, we discuss some obstacles to closer China-CEE relationships that have emerged in recent years.

1. The Belt and Road Initiative

Since the announcement of its concept in Kazakhstan in 2013, the Belt and Road Initiative has undergone major changes and development. The Initiative alludes to the ancient Silk Road, a network of merchant routes that date back to the Chinese Han Dynasty (206 BCE). The BRI includes two main concepts: the 'Silk Road Economic *Belt*' and the '21ˢᵗ Century Maritime Silk *Road*', which involves upwards of 60 countries and more than 60% of the world's population[1]. In its conception, the 'belt' and the 'road' spans across most of Asia, Africa, and Europe. In a 2015 policy brief issued jointly by the National Development and Reform Commission, Ministry of Foreign Affairs, and Ministry of Commerce of the People's Republic of China[2], it was stated that the BRI remains 'open and inclusive' and welcomes all national, international, as well as regional organisations. Thus, it is most appropriate to interpret the two main horns of the BRI as a rough sketch of its implementation[3].

A main vision of the BRI is to increase connectivity and cooperation between the involved countries. This may be specified into five major goals: policy co-ordination, facilities

1 | Campbell, C. (2017, May 12). China: 5 Facts on Xi Jinping's Belt & Road Initiative Summit. *Time*. Retrieved September 01, 2018, from http://time.com/4776845/china-xi-jinping-belt-road-initiative-obor/.

2 | The National Development and Reform Commission, Ministry of Foreign Affairs, and Ministry of Commerce of the People's Republic of China. (2015, March). 推動共建絲綢之路經濟帶和 21 世紀海上絲綢之路的願景與行動。 *Vision and Actions on Jointly Building Silk Road Economic Belt and 21st-Century Maritime Silk Road.*

connectivity, unimpeded trade, financial integration, and people-to-people bonds. Of these goals, financial integration is considered the primary objective to be attained. Two major institutions have arisen in the international society since the announcement of the Initiative in 2013: the Silk Road Fund and the Asian Infrastructure Investment Bank (AIIB). The former is a US$40 billion medium-to-long-term investment fund that is the Chinese government's primary financial arm in support of BRI projects; the latter is a multilateral development bank with 87 country members that promises to address the severe shortage of infrastructure in Asia[4].

Another important area is policy coordination that enhances intergovernmental partnership. This is symbolised by the signing of Memorandums of Understanding (MoUs) between the BRI related parties. The MoUs are important tools in showing commitment to the Initiative and agreeing upon common standards of cooperation. Currently, the majority of BRI projects focus on increasing infrastructure hardware. The wide variety of projects can be observed through a survey of the China-Pakistan Economic Corridor, which includes the building of railways, seaports, gas pipes, highways, etc. Beyond Pakistan, more projects are under way, such as the freight train links connecting East Asia to Europe through Central Asia via the Khorgos Gateway, the Mombasa–Nairobi Standard Gauge Railway in Kenya, as well as energy transport systems such as the Power of Siberia project and the crude oil pipe projects in Myanmar.

3 | Ren. J. (2018, July 7). 正確定位 '一帶一路' 實施。 Zhengque dingwei 'yi dai yi lu' changyi (The Proper Positioning of the Belt and Road Initiative). Chinese Academy of Social Sciences. Retrieved September 1, 2018, from http://www.cssn.cn/index/xldg/201805/t20180507_4238444.shtml.

4 | Shepard, W. (2017, July 15). The Real Role of the AIIB in China's New Silk Road. Forbes. Retrieved September 1, 2018, from https://www.forbes.com/sites/wadeshepard/2017/07/15/the-real-role-of-the-aiib-in-chinas-new-silk-road/#dc6d00974727.

2. 'Central and Eastern Europe'

In this context, we turn to consider Central and Eastern Europe. Roughly, this is the region between Germany and Austria to the west, and Ukraine and Belarus to the east. In this book, we specify the region as the 16 countries composed of Central European states, the Baltic states, and the Southeastern European states. At first sight, this is a region that is easy to overlook, as it is not the first place one has in mind when thinking about infrastructure investment and international economic cooperation opportunities. In this section, we aim to present a Central and Eastern Europe (CEE) that is rich and diverse, in terms of its history, economy, and socio-political situation.

A unique historical trait of the CEE region is the communist legacy shared by the 16 countries. Each of these countries was either a former Soviet state or a satellite state in the Eastern Bloc that declared their independence, albeit via different routes, upon the collapse of the Soviet Union. Eight of the CEE countries restored their independence through referendums or through conducting democratic elections, whereas the other eight were borne from the dissolution of the former Czechoslovakia and former Yugoslavia. While some of the political shifts that brought an end to the Eastern Bloc and Czechoslovakia were non-peaceful,[5] none compared to the violence that ensued following the dissolution of the former Yugoslavia. The breakup of Yugoslavia was a process of political turmoil characterised by the Yugoslav wars and a long series of territories being drawn and redrawn, and states being formed and reformed. This is evident in the fact that Montenegro only declared its independence from the State Union of Serbia and Montenegro in 2006. This historical legacy plays a salient role in the modern social as well as political attitudes in the region. Besides, the fact that the independence wars in Croatia and, more severely so, in Bosnia & Herzegovina were marked by ethnic conflict entails crucial challenges to a multilateral project such as the BRI.

Before discussing the opportunities related to the BRI in the Central and Eastern Europe region, we need to have an overall understanding of the current socioeconomic status

5 | For instance, the independence movement in Romania in 1989, see Preface in Sztompka's *Society in Action: The Theory of Social Becoming* (1991).

6 | European Commission. (2018, February 6). European Commission - Fact Sheet Q&A: A Credible Enlargement Perspective for an Enhanced EU Engagement with the Western Balkans. *European Commission.* Retrieved September 1, 2018, from http://europa.eu/rapid/press-release_MEMO-18-562_en.htm.

7 | The frontrunners of the accession process are generally recognised to be Serbia and Montenegro. According to a European Commission press release, both countries could be accepted into the EU by 2025. Source: European Commission. (2018, February 6). *European Commission - Fact Sheet Q&A: A Credible Enlargement Perspective for an Enhanced EU Engagement with the Western Balkans.* Retrieved from http://europa.eu/rapid/press-release_MEMO-18-562_en.htm.

8 | Estonia, Latvia, Lithuania, Slovakia, and Slovenia.

9 | Bosnia & Herzegovina (Bosnia & Herzegovina convertible mark), Bulgaria (Bulgarian lev), and North Macedonia (Macedonian denar).

10 | Stone, J. (2018, February 6). Serbia and Montenegro Could Join EU by 2025, European Commission Says. *Independent*. Retrieved September 1, 2018, from https://www.independent.co.uk/news/world/europe/eu-enlargement-serbia-montenegro-macedonia-albania-kosovo-brexit-juncker-2025-a8197201.html.

11 | European Commission. (2018, February 6). European Commission - Fact Sheet Q&A: A Credible Enlargement Perspective for an Enhanced EU Engagement with the Western Balkans. *European Commission*. Retrieved September 1, 2018, from https://ec.europa.eu/commission/presscorner/detail/en/MEMO_18_562.

12 | Stone, J. (2018, February 6). Serbia and Montenegro Could Join EU by 2025, European Commission Says. *Independent*. Retrieved September 1, 2018, from https://www.independent.co.uk/news/world/europe/eu-enlargement-serbia-montenegro-macedonia-albania-kosovo-brexit-juncker-2025-a8197201.html.

13 | For instance, the naming dispute between Macedonia and Greece since the dissolution of the former Yugoslavia. A major, positive development has occurred upon the signing of the Prespa agreement in June 2018.

14 | Emmott, R. (2018, June 25). EU Divided Over Balkan Accession as NATO Says Macedonia Welcome. *Reuters*. Retrieved September 1, 2018, from https://www.reuters.com/article/us-eu-balkans/eu-divided-over-balkan-accession-as-nato-says-macedonia-welcome-idUSKBN1JL0OL.

15 | The HDI is a composite index based on a country's life expectancy, average educational level, and standard of living.

of the countries. Although all 16 countries share a common communist legacy, their post-independence economic development have greatly diverged. To begin with, there is the aspect of European integration. As of January 2018, 11 of the 16 countries are members of the European Union; the rest of them, the West Balkan states including Albania, Bosnia & Herzegovina, North Macedonia, Montenegro, and Serbia, are in an active process of seeking accession into the EU.[6,7] While only five of the 16 countries are in the Euro Zone,[8] several others have their currencies pegged to the Euro.[9] Evidently, the region maintains a close relationship with the rest of Europe. Generally, from the perspective of the countries seeking accession, the major benefits of joining the EU include the ability to access the common market as well as the vital reforms, in areas such as the rule of law, that will likely improve the post-war stability and maintain peace in the Balkan states. The major obstacles to successful accession may be characterised into two categories. First, there are internal obstacles, mainly manifesting as the gap between the EU standards of successful candidacy and the social conditions within the Balkan states. The main issues that raise attention are severe corruption and organised crime[10] as well as the lack of credible commitment in conducting the necessary reforms in areas such as the rule of law, independence of the judiciary, safeguarding private investment, and constitutional reforms[11,12]. Second, there are external obstacles catalysed between West Balkan states and other EU members.[13]

In addition, the more prosperous European members, such as France and Denmark[14], fear that opening accession talks would strengthen the rhetoric of the domestic far-right political parties that are anti-immigration. Despite the hurdles on multiple dimensions, given the support by major EU member states such as Germany, we consider the expansion of the EU's membership as a likely occurrence over the next decades. Thus, it is crucial for long-term BRI projects to factor in the European integration process in their planning.

According to the Human Development Index country ranking in 2018,[15] 12 of the 16 countries are considered to have 'very high human development', the other have 'high human development'. Of the 16 countries, Slovenia is ranked the highest, performing better in all sub-categories other than the 'mean years of schooling'. Overall, based on the HDI alone, we may conclude that the Central and Eastern Europe region constitutes a relatively well-developed area of the globe. When it comes to other economic indices, the region presents a nuanced picture. For instance, according to the World Bank data on unemployment rates around the world in 2017[16], six of the countries are above 10%,[17] two of these even exceed 20%.[18] In contrast, the developing economies in Asia have an average unemployment rate of around 5-6%. A similar predicament persists amongst the youth population, as indicated by the youth unemployment data published in the same year. This corresponds to numerous authors of this publication mentioning young people's lack of access to good labour opportunities in their countries. Combining this with the fact that a high percentage of the population is literate and educated, there remains a huge, untapped pool of quality labour in these markets.

On the whole, the region is more developed than many other countries in the world. Yet internally, it is an area that is unbalanced in the distribution of development. After the collapse of the USSR, economic development trends of the CEE countries have diverged. Some managed to prosper, while others lagged. To this day, there remains an evident developmental disparity between the Southeast European states, with the possible exception of Slovenia, and the rest of the CEE countries. For instance, the GDP per capita based on purchasing power parity (PPP) in the region ranges from US$11,714 to US$32,606 in 2017[19], indicating a near three times difference between the lowest ranked (Bosnia) and highest ranked (Czech Republic) countries. Furthermore, this gap is also apparent in the infrastructure of these countries. The World Economic Forum publishes an annual Global Competitiveness Index, of which road quality is one of the sub-components. Using this data as a proxy, we will find that the Southeast European states have been consistently outperformed by the neighbouring countries[20].

16 | World Bank. (2018). *Unemployment, Total (% of total labor force) (Modeled ILO Estimate)*. Retrieved September 1, 2018, from https://data.worldbank.org/indicator/sl.uem.totl.zs?year_high_desc=true.

17 | Albania (15%), Bosnia & Herzegovina (25.8%), Croatia (11.5%), North Macedonia (24.4%), Montenegro (17.7%), and Serbia (14.4%).

18 | North Macedonia and Bosnia & Herzegovina.

19 | World Bank. (n.d.). *GDP Per Capita, PPP (Constant 2011 International $)*. Retrieved September 1, 2018, from https://data.worldbank.org/indicator/NY.GDP.PCAP.PP.KD?end=2017&start=2017&view=bar&year_high_desc=true.

20 | World Economic Forum. (2018) *The Global Competitiveness Report 2017-2018, Executive Opinion Survey Appendix C*. Available at http://reports.weforum.org/pdf/gci-2017-2018-scorecard/WEF_GCI_2017_2018_Scorecard_EOSQ057.pdf.

The region's diversity in political and socioeconomic development has deep implications for realising the BRI vision in the region.[21] Varied political development over the past decade led to varied political institutions, which then impacted the laws and regulations of a political system – including rules of trade, foreign investment, and foreign funds. Thus, extra attention should be paid towards the EU, for it is an entity that involves multiple countries and has the necessary mechanisms and bargaining power in place to efficaciously enforce its rules. The success of BRI projects, especially in EU member states, will be contingent upon the compliance of EU laws and regulations. Varied economic development leads to varied types of cooperation. For instance, while Balkan states may demand constructing and upgrading their infrastructure hardware in the short term, other more developed regions may prefer expanding the portfolios of exports.

21 | Adding to the complexity of the region, the foreign policy positions of these countries also diverge, especially pertaining to Russia. For example, Serbia has a positive relationship with Russia whereas the Baltic states perceive Russia as a security threat. See: McLaughlin, D. (2018, February 22). Serbia Says It Will Not Sacrifice Russia Ties for EU Membership. *Irish Times*. Retrieved September 1, 2018, from https://www.irishtimes.com/news/world/europe/serbia-says-it-will-not-sacrifice-russia-ties-for-eu-membership-1.3401978; and Ivanauskas, V., Keršanskas, V., & Kasčiūnas, L. (2017). Kaliningrad Factor in Lithuanian - Russian Relations: Implications to the Security Issues of Lithuania. *Lithuanian Annual Strategic Review*, 15(1), pp. 119-149. doi:10.1515/lasr-2017-0006.

Country	Human Development Index (Hdi)	Life Expectancy At Birth (Years)	Expected Years Of Schooling (Years)	Mean Years Of Schooling (Years)	Gross National Income (GNI) Per Capita (2011 PPP$)	HDI Rank 2018	HDI Rank 2017
VERY HIGH HUMAN DEVELOPMENT							
SLOVENIA	0.902	81.2	17.4	12.3	32,143	24	24
CZECHIA (Czech Republic)	0.891	79.2	16.8	12.7	31,597	26	27
ESTONIA	0.882	78.6	16.1	13.0	30,379	30	30
POLAND	0.872	78.5	16.4	12.3	27,626	32	33
LITHUANIA	0.869	75.7	16.5	13.0	29,775	34	34
SLOVAKIA	0.857	77.4	14.5	12.6	30,672	36	37
LATVIA	0.854	75.2	16.0	12.8	26,301	39	39
HUNGARY	0.845	76.7	15.1	11.9	27,144	43	44
CROATIA	0.837	78.3	15.0	11.4	23,061	46	46
BULGARIA	0.816	74.9	14.8	11.8	19,646	52	51
MONTENEGRO	0.816	76.8	15.0	11.4	17,511	52	51
ROMANIA	0.816	75.9	14.3	11.0	23,906	52	51
HIGH HUMAN DEVELOPMENT							
SERBIA	0.799	75.8	14.8	11.2	15,218	63	65
ALBANIA	0.791	78.5	15.2	10.1	12,300	69	69
BOSNIA & HERZEGOVINA	0.769	77.3	13.8	9.7	12,690	75	75
NORTH MACEDONIA	0.759	75.7	13.5	9.7	12,874	82	81

Figure 1: HDI ranking of the 16 Central and Eastern European countries.
Source: Human Development Reports. (2018). *Human Development Index and Its Components*. UNDP. Retrieved from http://hdr.undp.org/en/content/table-1-human-development-index-and-its-components-1.

3. Prospects for Cooperation in the CEE

22 | Buckley, N. (2017, May 7). Opportunities and Risks for Investors in Central and East Europe. *Financial Times.* Retrieved September 1, 2018, from https://www.ft.com/content/4248a712-07da-11e7-ac5a-903b21361b43.

23 | HKTDC Research. (2016, October 5). Belt and Road Opportunities in Central and Eastern Europe. *Hong Kong Trade Development Council Research.* Retrieved September 1, 2018, from http://economists-pick-research.hktdc.com/business-news/article/Research-Articles/Belt-and-Road-Opportunities-in-Central-and-Eastern-Europe/rp/en/1/1X000000/1X0A7MSE.htm.

24 | The State Council of The People's Republic of China. (2016). *The Riga Guidelines for Cooperation between China and Central and Eastern European Countries.* Retrieved 1 September 2018, from http://english.gov.cn/news/international_exchanges/2016/11/06/content_281475484363051.htm.

25 | Ministry of Foreign Affairs of The People's Republic of China. (2017). *The Budapest Guidelines for Cooperation between China and Central and Eastern European Countries.*

26 | The State Council of The People's Republic of China. (2018). *The Sofia Guidelines for Cooperation between China and Central and Eastern European Countries.* Retrieved 1 September 2018, from http://english.gov.cn/news/international_exchanges/2018/07/16/content_281476224693086.htm.

We move on to evaluate the region's various prospects of cooperation. To begin, the Central and Eastern European region contains some of the fastest growing markets in the world. The 'core nations' in the region, including Poland, Hungary, the Czech Republic, Slovakia, and Romania have emerged to be second only to the Asia-Pacific nations in terms of economic growth[22]. The geographical location of the 16 countries also means they are a significant connection point between Europe, Central Asia, and East Asia. Given that one of the main goals of the BRI is the commitment to creating a transcontinental economic corridor, the CEE nations will be crucial transport links joining with Western Europe as well as the Scandinavian countries. This will be especially important because of expanding cooperation with the EU market. As discussed above, much of the CEE region's modern history is conflict-ridden alongside economic downturns.

Especially in the Balkan papers of this publication, one finds the sincere yearning of the people towards a better future for their countries. The willingness to embrace international engagement paves the way for successful cooperation between these countries and China, which is evident given the exponential rise of Chinese investment in the region[23]. To facilitate high-level governmental coordination between China and the CEE countries, communication platforms have been set up. Since 2011, there has been an annual Summit of China and Central and Eastern European Countries, where heads of states, senior officials, and other representatives (including those from the EU and neighbouring countries) gather to review the ongoing projects and agree on the direction for future cooperation. The main outcomes of the summit are the Guidelines for Cooperation – the Riga Guidelines (2016)[24], the Budapest Guidelines (2017)[25], and the Sofia Guidelines (2018)[26] – that lay down the central principles and scope of the '16+1' format.

In the CEE region, the main economic opportunities are infrastructure and investment. We have already noted the lack of rail links and roads between and within these countries. The importance of infrastructure investment is an issue emphasised in nearly every paper submitted for this publication

27 | Including the Trans-European Transport Networks (TEN-T), the Solidarity Fund, the European Regional Development Fund, the European Social Fund, European Agricultural Fund for Rural Development, and the European Maritime and Fisheries Fund.

28 | Kratz, A., & Pavlićević, D. (2016, November 21). Belgrade-Budapest Via Beijing: A Case Study of Chinese Investment in Europe. *European Council on Foreign Relations.* Retrieved September 1, 2018, from https://www.ecfr.eu/article/commentary_belgrade_budapest_via_beijing_a_case_study_of_chinese_7188.

29 | Chen, X. (2017, December 19). Europe Should Support China-CEE Cooperation. *Reconnecting Asia.* Retrieved September 1, 2018, from https://reconnectingasia.csis.org/analysis/entries/europe-should-support-china-cee-cooperation/.

30 | European Commission INEA. (n.d.) *TEN-T programme 2007-2013 30 Priority Projects.* Retrieved September 1, 2018, from https://ec.europa.eu/inea/ten-t/ten-t-projects/projects-by-priority-project.

31 | Railway axis Lyon-Trieste-Divača/Koper-Divača-Ljubljana-Budapest-Ukrainian border (Project 6), Motorway axis Igoumenitsa/Patra-Athina-Sofia-Budapest (Project 7), Railway axis Paris-Strasbourg-Stuttgart-Wien-Bratislava (Project 17), Railway axis Athina–Sofia–Budapest–Wien–Praha–Nürnberg/Dresden (Project 22), Railway axis Gdańsk–Warszawa–Brno/Bratislava-Wien (Project 23), Motorway axis Gdańsk–Brno/Bratislava-Vienna (Project 25), and 'Rail Baltica' axis: Warsaw-Kaunas-Riga-Tallinn-Helsinki (Project 27).

32 | Projects 6, 7, 22, and 27.

33 | Projects 22 and 27.

34 | Karnitschnig, M. (2017, July 18). Beijing's Balkan Backdoor. *Politico.* Retrieved September 1, 2018, from https://www.politico.eu/article/china-serbia-montenegro-europe-investment-trade-beijing-balkan-backdoor/.

project. Notwithstanding the existing European Union policies and institutions directed towards overhauling and upgrading outdated infrastructure in the region,[27] there are reasons to support the '16+1' framework. First and foremost, not all CEE countries are EU member states. The non-members, often those most in need of better infrastructure, are left with limited support from European institutions such as the European Bank for Reconstruction and Development. In these states, especially the 'cash-strapped, largely deindustrialised countries'[28], foreign investment can be vital in infrastructure development. Second, even for CEE countries that are EU members, there is an imbalance in funds allocated to them[29]. Priority is often given to bringing CEE countries closer to Western Europe, which leaves the inter-connectivity between the CEE countries a severely neglected issue. This is evident from evaluating the 30 Priority Projects[30] included in the EU's Trans-European Transport Networks (TEN-T). Of the 30 projects, seven concern the CEE members,[31] of which four connect more than one CEE nation.[32] Among the four 'inter-connecting' projects, only two run from north to south.[33] On top of these factors, the EU's bureaucracy and delays are also perceived as barriers for development in the region[34].

One of the key BRI projects in the region is the Budapest-Belgrade high speed railway, which could shorten travel time from eight hours to three hours between the capitals of Hungary and Serbia. The railway is part of a larger vision – the 'Budapest-Belgrade-Skopje-Athens railway' – that connects Hungary, Serbia, North Macedonia, and Greece. If completed, the future railway would serve as the one of the main transport pathways for goods that arrive at the Greek port of Piraeus to enter into the CEE market.

While the plan was first introduced by Hungary, Serbia, and China in 2013, the railway also resonates

35 | Miltiadou, M., Taxiltaris, C., Mintsis, G., & Basbas, S. (2012). Pan-European Corridor X Development: Case of Literal Implementation of the European Transport Strategy Itself or of Change of the General Environment in the Region? *Procedia-Social and Behavioral Sciences*, 48, pp. 2361-2373.

36 | Savary, G. (2013). *Priority Project 22 Annual Report of the Coordinator.* European Commission.

37 | *Ibid.*

38 | GCR. (2018, July 18). *Montenegro a Crossroads with Expensive Chinese Motorway.* Retrieved September 1, 2018, from http://www.globalconstructionreview.com/news/montenegro-crossroads-expensive-chinese-motorway/.

39 | Podgorica (AFP). (2010, December 27). Montenegro Drops Greek-Israeli Consortium to Build Highway. *Terra Daily.* Retrieved September 1, 2018, from http://www.terradaily.com/reports/Montenegro_drops_Greek-Israeli_consortium_to_build_highway_999.html.

40 | The project also did not gain much support from Western Europe, possibly due to the perceived lack of economic benefit as a result of Montenegro's small population size. As of 2018, Montenegro has a population of around 620,000. World Bank. (n.d.). GDP per capita, PPP (constant 2011 international $). Retrieved September 1, 2018, from https://data.worldbank.org/indicator/SP.POP.TOTL.

41 | Drobnjak, A. (2018, May 16). Bar-Boljare Highway to Improve Life Conditions for Northern Montenegro. *Total Montenegro News.* Retrieved September 1, 2018, from https://www.total-montenegro-news.com/business/1130-montenegrin-highway-will-improve-life-conditions-for-northern-montenegro.

with other, older European transport initiatives. For instance, it is part of Corridor X, one of the ten Pan-European Corridors that were defined in the Pan-European Transport Conferences in Crete (1994) and Helsinki (1997)[35]. In the 2013 Annual Report for the EU's TEN-T, the Coordinator of Priority Project 22 extended a brief consideration of the Balkan Route of Corridor X, which included the Budapest-Belgrade railway[36]. The report compared Priority Project 22 (one of the EU's aforementioned 30 Priority Projects) with the Balkan Route, stating several practical and technical advantages in favour of the latter's implementation, including the potential shorter distance of travel, higher proportions of electrified tracks, and higher travel speed, but cited the substantial upgrading costs as a factor that hampers government investment[37].

Thus, it appears that projects carried out as part of the '16+1' are not necessarily 'new ones' introduced by China. Rather, China picked up on the existing plans of regional governments that never received the necessary support for them to be carried through.

Another example is the Belgrade-Bar highway, a project that envisions creating a highway route between Belgrade, the inland capital of Serbia, and Bar, a seaport of Montenegro. Montenegro bears the most expensive part of this project estimated to cost at least €2 billion, a section known as the Bar-Boljare highway, due to the rough and mountainous terrains[38]. From early 2009 to 2010, the Montenegrin government actively sought collaboration with Croatian, Greek, and Israeli parties, but these companies 'failed to provide obligatory banking guarantees for the contract'[39,40].

Notwithstanding, the local government views the project as 'the most critical infrastructure project in Montenegro'[41], notably because Montenegro is the

42 | The China Road and Bridge Corporation as well as the Export-Import Bank of China. The latter offered a loan of near €690 million with a six-year grace period, covering 85% of the €809 million construction cost.

43 | Drobnjak, A. (2018, May 16). Bar-Boljare Highway to Improve Life Conditions for Northern Montenegro. *Total Montenegro News*. Retrieved September 1, 2018, from http://www.total-montenegro-news.com/business/1130-montenegrin-highway-will-improve-life-conditions-for-northern-montenegro.

44 | Davis, A. (2018, July 23). The Longest Tunnel of the Bar-Boljare Highway Project in Montenegro Breaks Through. *Highways Today*. Retrieved September 1, 2018, from https://www.highways.today/2018/07/23/the-longest-tunnel-of-the-bar-boljare-highway-project-in-montenegro-breaks-through/.

45 | The Baltic Review. (2016, October 11). Skype - Estonia's Greatest Contribution to the Global Telecommunication Industry. *The Baltic Review*. Retrieved September 1, 2018, from https://baltic-review.com/estonia-skype/.

46 | Heller, N. (2017, December 18). Estonia, the Digital Republic. *New Yorker*. Retrieved September 1, 2018, from https://www.newyorker.com/magazine/2017/12/18/estonia-the-digital-republic.

47 | McLean, A. (2018, August 13). E-Estonia: What is All the Fuss About? *ZDNet*. Retrieved September 1, 2018, from https://www.zdnet.com/article/e-estonia-what-is-all-the-fuss-about/.

48 | Olson, P. (2012, September 6). Why Estonia Has Started Teaching Its First-Graders to Code. *Forbes*. Retrieved September 1, 2018, from https://www.forbes.com/sites/parmyolson/2012/09/06/why-estonia-has-started-teaching-its-first-graders-to-code/#24d71241aa3d.

49 | GLOBSEC. (2017, December 1). The Danube Valley: From Manufacturers To Innovators Read. *GLOBSEC*. Retrieved September 1, 2018, from https://www.globsec.org/danube-valley-manufacturers-innovators/.

only European country that does not have a highway. Ultimately, an agreement was reached between the Montenegrin government and Chinese companies in 2014,[42] aiming to complete the first section by 2019[43]. In July 2018, one of the longest tunnels in the project, the Vjeternik Tunnel, was broken through the mountains[44]. From such case studies, it is clear that Chinese companies have much to offer in terms of infrastructure construction experience, technological know-how and financial support. On the other hand, Chinese involvement can be critical to enabling countries to carry out crucial investments, particularly for smaller nations.

Another opportunity for investment in the region is its blossoming high-tech industry, demonstrable by the increasing number of tech-related companies that have arisen in CEE countries. Estonia is a world-renowned start-up hub with a dynamic free market economy that encourages foreign direct investment and is most famed for its being the breeding ground of the global telecommunication company, Skype[45]. The focus on digital technology permeates Estonian society – 99% of its government services, from filing for taxes to seeking public health services[46,47], are digitalised, and, since 2012, children from the age of seven have been trained to code in schools[48]. In Slovakia, there is the 'Danube valley', which has developed advanced technologies in the mobility and energy industry[49]. AeroMobil, a Slovakian company, has surfaced as one of the major players in the flying car industry[50]. Polish computer engineers have consistently been ranked some of the best in the world, fostering world champions in international programming competitions such as The Topcoder Open (TCO)[51]. Romania has one of the highest quality IT infrastructures in the world and more than 15,000 specialised software companies, employed with skilled and diversified labour[52]. The Czech Republic is known for its cybersecurity companies, such as Avast. Upcoming plans to set up a '16+1 Smart City Coordination Center' in Romania & '16+1 Fintech Coordination Center' in Lithuania were both mentioned in the recently published Sofia Guidelines (2018)[53].

Digital and innovative technology is at the core of many of the CEE countries' future objectives. Countries such as Slovenia, Hungary, Poland, the Czech Republic, and Estonia have been allocating a significant proportion of their public expenditure, around 1-2% of their GDP, on research and development, matching the levels of western European countries[54]. The combination of government support as well as the inherent underlying talent pool has fostered a vibrant ecosystem for innovation that is emerging and expanding within the CEE region.

Beyond infrastructure and digital technology, the region also offers a range of goods and services for trade. To begin with, food production is an area with strong interest within the Chinese consumer market — most probably in light of the persistent food safety problems that have arisen in China over the past decade. As many consumers in China now turn to imported food produce for safety guarantees, this has opened opportunities for Central and Eastern European countries, many of which are traditional agricultural powerhouses. Recently, Croatian companies have begun exporting canned sardines into China and have been showing a keen interest to expand into the dairy industry[55]. Hungary is another intriguing case study to bring up related to food quality; it is one of the only countries in the world that has 'anti-Genetically-Modified-Organisms laws' written into its constitution[56]. Regarding future development in the region's agricultural industry, reports show that there is a huge potential for the expansion of regional crop yield that could come from the investment in improving land quality and training local farmers to

50 | O'Brien, C. (2018, March 21). AeroMobil Unveils New Concept for Flying Car that can Take Off Vertically and Drive on Roads. *Venture Beat*. Retrieved September 1, 2018, from https://venturebeat.com/2018/03/21/aeromobil-unveils-new-concept-for-flying-car-that-can-takeoff-vertically-and-drive-on-roads/.

51 | Cybercom Group. (n.d.). Polish Programmers Among Best in the World | Ranking of Developers. *Cybercom*. Retrieved September 1, 2018, from https://www.cybercom.com/pl/Poland/Software-House/blog/ranking-of-developers/.

52 | Gheorghe, G. (2017, December 11). Romanian IT sector Valued at EUR 5 Bln in 2016. *BR Business Review*. Retrieved September 1, 2018, from http://business-review.eu/news/romanian-it-sector-valued-at-eur-5-billion-in-2017-154177.

53 | The State Council of The People's Republic of China. (2018). *The Sofia Guidelines for Cooperation between China and Central and Eastern European Countries*. Retrieved 1 September 2018, from http://english.gov.cn/news/international_exchanges/2018/07/16/content_281476224693086.htm.

54 | Spisak, A. (2017, June 06). Central and Eastern Europe Unveils Its Tech Ambitions. *Financial Times*. Retrieved September 1, 2018, from https://www.ft.com/content/889422a8-09ad-11e7-ac5a-903b21361b43.

55 | HINA. (2018, June 17). Croatia to Export Dairy Products, Tuna, and Poultry to China. *Total Croatia News*. Retrieved September 1, 2018, from https://www.total-croatia-news.com/business/29147-croatia-to-export-dairy-products-tuna-and-poultry-to-china.

56 | The Fundamental Law of Hungary. (2018). Article XX, paragraph 2.

adopt modern agronomic practices[57].

Another area we should consider is tourism. The CEE region is known for its scenic beauty and rich cultural heritage, which has become the destination for an increasing number of Chinese tourists[58]. In recent years, more and more people are becoming aware of and interested in the CEE countries, an awareness which has been bolstered by increased exposure in international events as well as presence in popular TV-shows. For instance, after the 2018 World Cup semi-final (when Croatia beat England to reach the final), Qunar, a tourist booking website in China reported that flight ticket prices to Croatia had more than doubled, accompanied by a three-fold increase in online searches for hotels[59]. These trade and cultural aspects of the BRI have also received attention from governments, companies and the public. Since 2014, the annual China-CEEC Investment and Trade Expo has been held in Ningbo, China. In 2018, the 4th Expo featured hundreds of products from the CEE countries, alongside tourism and youth exchange sessions. Hence, beyond the infrastructure projects and digital-tech cooperation, there are other significant trade and cultural exchange opportunities that are arising in the region.

57 | Smit, H. (2015, January 21). How Can Farmers in Central and Eastern Europe Close Yield Gap with West? *The Guardian.* Retrieved September 1, 2018, from https://www.theguardian.com/sustainable-business/2015/jan/21/farmers-central-eastern-europe-grain-yield-gap.

58 | Nielson. (2017). *2017 Outbound Chinese Tourism and Consumption Trends.* Nielsen Holdings.

59 | Zhu, W. (2018, July 18). Soccer Success Boosts Croatia Tourism. *China Daily.* Retrieved September 1, 2018, from http://usa.chinadaily.com.cn/a/201807/18/WS5b4ead70a310796df4df727b.html.

4. Current Obstacles and Suggestions

At the same time, an international, long-term cooperation vision such as the Belt and Road Initiative, that aims to bring together diverse parties, is likely to face some obstacles. Here, we shall limit our discussion to the ones which we consider the most significant. First, critics fear that the BRI projects may exacerbate local conditions, by introducing financial instability

60 | Montenegro's GDP in 2017 is estimated to be US$4.774 billion, around €4.111 billion. See World Bank. (n.d.). *GDP (Current US$)*. Retrieved September 1, 2018, from https://data.worldbank.org/indicator/NY.GDP.MKTP.CD?locations=ME&view=chart; Global Construction Review. (2018, July 18). Montenegro At Crossroads with Expensive Chinese Motorway. *Global Construction Review*. Retrieved September 1, 2018, from http://www.globalconstructionreview.com/news/montenegro-crossroads-expensive-chinese-motorway/.

61 | Global Construction Review. (2018, July 18). Montenegro At Crossroads with Expensive Chinese Motorway. *Global Construction Review*. Retrieved September 1, 2018, from http://www.globalconstructionreview.com/news/montenegro-crossroads-expensive-chinese-motorway/; International Monetary Fund. (2017). *Montenegro Selected Issues*. International Monetary Fund, Washington D.C.

62 | International Monetary Fund. (2017). *Montenegro Selected Issues*. International Monetary Fund, Washington D.C.

63 | Ishikawa, J. (2018, May 25). China Unsettles EU with Belt and Road Moves in the Balkans. *Nikkei Asian Review*. Retrieved September 1, 2018, from https://asia.nikkei.com/Politics/International-Relations/China-unsettles-EU-with-Belt-and-Road-moves-in-the-Balkans.

64 | Ma, A. (2018, August 21). Malaysia Has Axed $22 Billion of Chinese-Backed Projects, in a Blow to China's Grand Plan to Dominate World Trade. *Business Insider*. Retrieved September 1, 2018, from http://uk.businessinsider.com/malaysia-axes-22-billion-of-belt-and-road-projects-blow-to-china-2018-8?r=US&IR=T.

65 | Ralev, R. (2018, June 11). Hungary Hopes Belgrade-Budapest Rail Project to be Completed in 2023. *See News*. Retrieved September 1, 2018, from https://seenews.com/news/hungary-hopes-belgrade-budapest-rail-project-to-be-completed-in-2023-615874.

in terms of debt risk as well as channeling domestic corruption. Major concerns have been raised over the Western Balkan states, particularly given their relatively high debt-to-GDP ratios. For example, returning to the aforementioned Bar-Belgrade motorway, the construction costs for phase 1 alone (estimated at €809 million) are equivalent to over 20% of Montenegro's GDP in 2017.[60] The dollar-denominated loan, covering 85% of the costs, that was financed by the Chinese Export and Import Bank in 2014 has a six-year grace period, with repayments set to begin in 2021. Accompanied by an appreciation of the dollar since 2014, the debt expanded to an estimated €1 billion[61].

Red flags have been raised by the IMF with regards to Montenegro's medium-to long-run fiscal sustainability. Especially in light of the prospect of Montenegro's debt-to-GDP ratio reaching near 90% in 2019[62]. Since the benefits of infrastructure projects are unlikely to redound in the short-term, questions have been raised as to whether the Bar-Belgrade motorway will, as the Montenegrin government sees it, be a boon to its economic development, or whether it will be a bane driving the country into bankruptcy. Other than public debt, allegations of abetting corruption and enabling public officials to embezzle project money for private and political gain have also arisen[63]. Such concerns have been heightened by responses from other regions such as Malaysia[64].

Second, in projects that involve EU countries, concerns over compliance with EU regulations have been raised. Take, for example, the Budapest-Belgrade high speed rail project. The agreement for the construction was reached between Hungary, Serbia, and China in 2013, a subsequent MoU between the three parties was signed in 2014[65], and the railway was originally planned

66 | The State Council of The People's Republic of China. (2015, December 23). 李克强致信祝賀匈塞鐵路項目塞爾維亞段正式啟動 Li Keqiang zhixin zhuhe aoxiongtieluxiangmu saierweiyaduan zhengshi qidong (Li Keqiang Sends Letter in Congratulation of the Official Launching of the Serbian Section of the Hungary-Serbia Railway). *Zhonguo zhengfu wang*. Retrieved 1 September 2018, from http://www.gov.cn/guowuyuan/2015-12/23/content_5027075.htm.

67 | Kynge, J. (2017, February 20). EU Sets Collision Course with China Over 'Silk Road' Rail Project. *Financial Times*. Retrieved September 1, 2018, from https://www.ft.com/content/003bad14-f52f-11e6-95ee-f14e55513608.

68 | Belt and Road Advisory. (2017, November 11). Legal Quagmire Blocks Belt and Road Initiative in CEE?. *Belt and Road Advisory*. Retrieved September 1, 2018, from https://beltandroad.ventures/beltandroadblog/2017/11/11/legal-quagmire-blocks-belt-and-road-initiative-in-cee.

69 | Kynge, J. (2017, February 20). EU Sets Collision Course with China Over 'Silk Road' Rail Project. *Financial Times*. Retrieved September 1, 2018, from https://www.ft.com/content/003bad14-f52f-11e6-95ee-f14e55513608.

70 | Magyar Távirati Iroda (MTI). (2017, November 27). Tender for Budapest-Belgrade Rail Upgrade Published. *Budapest Business Journal*. Retrieved September 1, 2018, from https://bbj.hu/economy/tender-for-budapest-belgrade-rail-upgrade-published_142131.

71 | Railway Pro Communication Platform. (2018, June 18). Two JVs Compete for Hungarian Section on Budapest-Belgrade Line. *Railway Pro Communication Platform*. Retrieved September 1, 2018, from https://www.railwaypro.com/wp/two-jvs-compete-for-hungarian-section-on-budapest-belgrade-line/.

72 | Budapest Business Journal. (2019, April 29). Mészáros Firm Among Winners of Massive Rail Contract. *Budapest Business Journal*. Retrieved 29 June, 2019, from https://bbj.hu/business/meszaros-firm-among-winners-of-massive-rail-contract_164893.

to begin operation by 2018[66]. Yet, at the time of writing, the construction of the Hungarian section of the railway still had not commenced. The stagnation stems from the fact that Hungary, as an EU member, is subject to the EU procurement regulations, which stipulate the requirement for public tenders to be offered for large-scale investment projects[67]. When China and Hungary agreed to partake in the construction of the railway, a few companies were selected to carry out the project, without subjecting the contract to public bidding, and thus arose the accusations of a breach of EU law[68].

In February 2017, the European Commission launched a probe into the project, evaluating its financial viability and its compliance with EU law[69]. Nine months later, Hungary called for a two-phase tender that specifies the scope of the contract as well as the necessary qualifications for companies to attain the contract[70]. By June 2018, two Hungarian-Chinese joint ventures submitted valid applications for the tender[71]. In May 2019, the contract was awarded to a consortium including RM International Zrt., China Tiejiuju Engineering & Construction Kft., and China Railway Electrification Engineering Group Kft.[72,73]. The new deadline for completion of the project is set at 2025[74] – a seven-year delay in one of China's key projects in the region. The initial non-compliance is probably fuelled by a combination of the want to accelerate '16+1' projects and the emphasis placed on the Budapest-Belgrade Railway as an integral part of the Eurasian Land-Sea Express Route[75]. To make matters worse, non-compliance cases also lend convenient ammunition for opponents to incite fear that the BRI is a Chinese stratagem to corrode existing orders and pacts between countries.

This leads us to what we believe is the central challenge for the BRI in the region: the skepticism surrounding Chinese presence in the CEE. In recent years, there has been an increased fear of what has

73 | Reuters. (2019, June 12). Hungary PM Orban's Ally to Co-build Chinese Railway for $2.1 billion. *Reuters*. Retrieved June 29, 2019, from https://www.reuters.com/article/us-hungary-china-railways-opus-global-idUSKCN1TD1JG.

74 | Budapest Business Journal. (2019, April 29). Mészáros Firm Among Winners of Massive Rail Contract. *Budapest Business Journal*. Retrieved 29 June, 2019, from https://bbj.hu/business/meszaros-firm-among-winners-of-massive-rail-contract_164893.

75 | Belt and Road Advisory. (2017, November 11). Legal Quagmire Blocks Belt and Road Initiative in CEE?. *Belt and Road Advisory*. Retrieved September 1, 2018, from https://beltandroad.ventures/beltandroadblog/2017/11/11/legal-quagmire-blocks-belt-and-road-initiative-in-cee.

76 | Cai. P. (2017, March). *Understanding China's Belt and Road Initiative*. Lowy Institute for International Policy. Sydney, Australia

77 | The Economist. (2018, September 20). Is China's Infrastructure Boom Past Its Peak?. *The Economist*. Retrieved September 21, 2018, from https://www.economist.com/china/2018/09/22/is-chinas-infrastructure-boom-past-its-peak.

78 | Kynge, J. (2018, January 24). Chinese Contractors Grab Lion's Share of Silk Road Projects. *Financial Times*. Retrieved September 1, 2018, from https://www.ft.com/content/76b1be0c-0113-11e8-9650-9c0ad2d7c5b5.

79 | Agence France-Presse. (2018, February 22). Merkel Warns Against China's Influence in Balkans. *South China Morning Post*. Retrieved September 1, 2018, from https://www.scmp.com/news/china/diplomacy-defence/article/2134196/merkel-warns-against-chinas-influence-balkans.

80 | Emmott, R. (2016, July 15). EU's Statement on South China Sea Reflects Divisions. *Reuters*. Retrieved September 1, 2018, from https://www.reuters.com/article/us-southchinasea-ruling-eu/eus-statement-on-south-china-sea-reflects-divisions-idUSKCN0ZV1TS.

been coined, quite exaggeratedly, 'Chinese-hegemony', a sentiment which is reinforced by accusations about worsening local economies and EU non-compliance.

The phenomenon of skepticism can be divided into two branches of analysis. On one hand, there is economic skepticism, the concern that BRI projects remain China-centric. The BRI as a whole is perceived to be, in part, a global economic strategy from China to manage domestic problems such as its excess capacity and regional development inequality[76]. When that is coupled with the fact that government policy is turning towards reducing and downscaling domestic infrastructure projects[77], the BRI is increasingly viewed as a means to sustain growth via infrastructure building in other countries.

Critics of the BRI point to other facts that further cast doubt on China's motivations. For example, around 89 per cent of contractors working on BRI-related projects are Chinese[78]. In addition to a lack of inclusion of the local labour force, concerns have been raised over the supposed 'win-win' principle.

On the other hand, there is political skepticism, the fear that the BRI may, intentionally or unintentionally, lead to a divide between European states. This originates from the fear that the BRI carries underlying political agendas obscured under the guise of an economic initiative. In early 2018, Germany raised such doubts with regards to the possible ulterior motives of Chinese investment in the Balkans[79]. Skeptics portray investments as a form of economic pampering, to gain political leverage via the CEE nations, which may even pave the way to their turning away from European interests and ideologies. Notably, Greece and Hungary have digressed from the majority EU members' position regarding the South China Sea[80]. These events spark worries over China's extending reach within the EU's internal politics. Furthermore, critics argue that the EU and China are incompatible rivals in a zero-sum political competition. For instance, one of the features of the Bar-Belgrade motorway investment that is being subject to critical circumspection is that it leads to a digression from Montenegro's path to EU accession. With most of its budget focused on highway construction, Montenegro's capital spending on other investments such as waste management and water treatment, which are related to EU accession, risk being crowded out[81].

Echoing these opinions, some media outlets warn of the potential dangers of increased Chinese influence, especially in strategic industries such as internet infrastructure and maritime infrastructure[82]. These concerns are among the reasons behind the frequent obstacles that Chinese investments face in Europe, such as Germany's blocking the intended acquisition of Leifeld, a machine tool manufacturer, by Yantai Taihai Group[83]. The sentiment of wariness is best illustrated by the reactions of French President Emmanuel Macron and German Chancellor Angela Merkel towards the BRI, both of whom have expressed the view that the Initiative cannot be conceived as a 'one-way street' that solely churns advantages for China[84].

The skepticism surrounding widened Chinese involvement is likely to negatively affect Belt and Road projects set to take place in Central and Eastern Europe. For one, the politicisation of the investment as a 'tug-of-war' between the EU and China may make governments and companies in the region hesitant to cooperate with the Chinese, given their conventional inclination to favour the European market. In the long run, this would attenuate the fledgling relationship between China and the CEE nations.

In view of these challenges, China, as the prime initiator of the BRI and '16+1' format, must make clear and firm responses. China must continuously and consistently maintain the position that it is not standing in the way of EU unity. A strong and amicable signal must be sent to Europe. In this regard, changes have occurred over the years. Comparing the three Guidelines, we can observe a shift towards more clarity in naming the related EU-China agreements and cooperative

81 | International Monetary Fund. (2017). *Montenegro Selected Issues*. International Monetary Fund, Washington D.C., p. 23.

82 | Asia News International. (2018, April 23). China's 'Belt Road Initiative' Exposes Its Ulterior Strategic Motives. *Financial Express*. Retrieved September 1, 2018, from https://www.financialexpress.com/defence/chinas-belt-road-initiative-exposes-its-ulterior-strategic-motives/1141964/

83 | Delfs, A. (2018, August 1). Germany Toughens Stance and Blocks China Deal. *Bloomberg*. Retrieved September 1, 2018, from https://www.bloomberg.com/news/articles/2018-08-01/germany-said-to-block-company-purchase-by-chinese-for-first-time.

84 | Elmer, K. (2018, July 16). EU Envoys Hit Out at China's 'Unfair' Belt and Road Plans. *South China Morning Post*. Retrieved September 1, 2018, from https://www.scmp.com/news/china/diplomacy-defence/article/2142698/eu-presents-nearly-united-front-against-chinas-unfair.

85 | The State Council of The People's Republic of China. (2018). *The Sofia Guidelines for Cooperation between China and Central and Eastern European Countries.* Retrieved 1 September 2018, from http://english. gov.cn/news/international_ exchanges/2018/07/16/content_ 281476224693086.htm.

86 | The State Council of The People's Republic of China. (2016). *The Riga Guidelines for Cooperation between China and Central and Eastern European Countries.* Retrieved 1 September 2018, from http://english. gov.cn/news/international_ exchanges/2016/11/06/content_ 281475484363051.htm.

87 | Scimia, E. (2018, June 01). Chinas Belt and Road A Dilemma for Germany. *Asia Times.* Retrieved September 1, 2018, from http:// www.atimes.com/chinas-belt-and-road-a-dilemma-for-germany/.

88 | *Ibid.*

89 | United Nations Framework Convention on Climate Change. (2018, July 17). *China, EU Reaffirm Strong Commitment to Paris Agreement.* Retrieved September 1, 2018, from https://unfccc.int/ news/china-eu-reaffirm-strong-commitment-to-paris-agreement.

90 | The Economist. (2018, March 15). China is Rapidly Developing Its Clean-Energy Technology. *The Economist.* Retrieved September 1, 2018, from https:// www.economist.com/special-report/2018/03/15/china-is-rapidly-developing-its-clean-energy-technology.

91 | Timperley, J. (2018, January 10). China Leading on World's Clean Energy Investment, Says Report. *Carbon Brief.* Retrieved September 1, 2018, from https:// www.carbonbrief.org/china-leading-worlds-clean-energy-investment-says-report.

frameworks. For instance, the Budapest and Sofia Guidelines both specifically mention that 'CEECs that are Member States of the EU will cooperate within the structures of the EU-China Agreement on Cooperation and Mutual Administrative Assistance in Customs Matters and of the EU-China Strategic Framework for Customs Cooperation'[85]. This contrasts with the vaguer phrasing of the earlier Riga Guidelines, which merely state their needing to conform to 'relevant EU legislation and regulations'[86].

The articulation of these structures is helpful on two grounds. First, they are a clear indication that China and the related CEE countries are willing to operate under EU regulation. Second, they acknowledge the existing cooperative mechanisms that have been developed between the EU and China over the years, thus contextualising the '16+1' format within wider EU-China relations.

Furthermore, more can be done to highlight the shared interests between Europe and China – European priorities can be Chinese priorities, and vice versa. Prominent European powers and China share a common interest in 'safeguard[ing] free trade and promoting multilateralism'[87], particularly in light of the protectionist digression the United States has taken during the Trump administration. This extends to matters of foreign policy as well, demonstrable by instances such as their joint support for retaining the Iran nuclear deal[88]. Especially on matters of combating climate change, arguably one of the gravest challenges of our century, Europe and China are deeply committed to the cause and have expressed strong support backing the Paris agreement[89]. Currently, China is leading in many areas of renewable energy development, both in terms of generating energy from renewables as well as manufacturing related components, such as wind turbines and solar panels[90]. As a case in point, Chinese factories now account for nearly 60% of global solar cell production[91]. BRI projects in Europe should align themselves with these common interests. For example, railway projects can emphasise more about their environmental impact. Railway transport is an environmentally friendlier 'middle-option', in between airfreight and maritime shipping[92], and such global public benefits should be put at the foreground of the promotion of railway projects. Given the general acceptance of the importance of sustainable development, a closer alignment of the BRI with this cause can reinforce the current, predominantly economic focused narratives, providing a more nuanced and complete picture of the BRI. Ultimately, this makes it more likely for BRI projects to be supported by people around the world.

Beyond involving European forces outside of the '16+1' in relatively passive roles,[93] they should be invited to engage in more active parts in promoting the format. In 2018, the Chinese Foreign Minister, Wang Yi, gestured to the potential of forming a tripartite cooperation between Germany, the CEE countries, and China[94]. We argue that such moves towards a more dynamic view of the '16+1' format are positive and encouraging. The '16+1' format should not be viewed as an isolated relationship between China and the CEE countries. Rather, it is one of the networks that overlaps and coexists amongst others within the BRI, and its success will be dependent upon the support of not only the 17 countries, but also from other neighbours that have close ties with these countries.

We should also recognise that a more substantial declaration of the cooperative framework is needed, one that expresses the proper intentions of Chinese investment and the rules and regulations of cooperation. Given that the chief goal of the BRI is to create long term 'win-win' relationships that benefit all participants, China needs to be sensitive about how negotiations are carried out and the terms of agreements. Where one party has far greater bargaining power, that party should be the main actor in ensuring the fair distribution of benefits.

Although the '16+1' format is promising to bring about economic development in the region, the impact will be reduced if China chiefly relies on domestic production networks and excludes local suppliers and labour[95]. To include local businesses, we may consider formally incorporating a 'joint venture' model

92 | Hillman, J. E. (2018, March 6). *The Rise of China-Europe Railways.* Center for Strategic & International Studies. Retrieved September 1, 2018, from https://www.csis.org/analysis/rise-china-europe-railways.

93 | Austria, Switzerland, Greece, Belarus, and other European institutions such as the European Bank for Reconstruction and Development, are often invited as 'observers' at high-level meetings between China and CEE countries.

94 | Xinhuanet. (2018, May 31). 王毅回應歐洲對 '16+1' 合作" 的執法 Wang Yi Huiying Ouzhou dui '16+1' hezuo de danyiu. (Wang Yi in Response to Europe's Worries Concerning the '16+1 Cooperation'). *Xinhuanet.* Retrieved on 1 September, from http://www.xinhuanet.com/world/2018-05/31/c_1122920711.htm.

95 | The Vienna Institute for International Economic Studies. (2018). *Press Release: Sofia 16+1 Summit: Chinese Investment Could Drive Major Infrastructure Improvement.* Vienna, Austria.

mandating all BRI-related projects to take the form of shared ownerships that are co-managed by foreign and local entities. The cooperative framework should resonate with EU regulation and outline a flowchart indicating how BRI-related investments are proposed, reviewed, and concluded, including, where necessary, public procurement procedures. A joint monitory body, represented by all 17 nations as well as regional powers, can be formed to check on the projects and ensure quality standards are met. This could also enable a more transparent and systematic documentation of the progress of the projects.

Finally, there needs to be more international promotion of Chinese cooperation in a fairer light that counters the negative portrayal of their being low-quality outpours of excess capacity. This requires outreach to and interaction with the citizens around the world. Online media outlets, social media, and local celebrities need to be fully utilised to this end. Cultural activities, such as student exchanges, joint exhibitions between countries, translation of different countries' history and literary texts, should also be encouraged, which would help cultivate personal bonds and resonate with the central ideals of the BRI: connectivity, cultural exchange, peace, and sustainable development.

In summary, from an economic, political, social, and cultural perspective, the Central and Eastern European Region is one of the most important bridges between Asia and Europe. For the youth around the world, this region holds immense opportunity and development potential. Striving for new cooperative relationships will inevitably give rise to challenges, yet we believe that such challenges can be surmounted by close communication and coordination. At the same time, to enhance the benefits of BRI, the youth of different countries must increase their mutual understanding of one another. Thus, we strongly urge the continuous research into Central and Eastern Europe to understand the social conditions in the region, to grasp the needs of the youth thereof, and to promote international exchanges between young people. Ultimately, this would greatly contribute to the implementation of the Belt and Road Initiative.

Skanderberg main square and mosque in central Tirana, Albania.

by **JETNOR H KASMI**
Korea Development Institute School of
Public Policy and Management

ALTIN KUKAJ
Lake Forest College

AJSELA TOCI
Epoka University

(Ordered by last name)

CHAPTER 2

ALBANIA

ESTONIA

LATVIA

LITHUANIA

POLAND

CZECH
REPUBLIC

SLOVAKIA

HUNGARY

SLOVENIA

CROATIA

BOSNIA &
HERZEGOVINA

SERBIA

MONTENEGRO

NORTH
MACEDONIA

ROMANIA

BULGARIA

ALBANIA

1. The Politics of Albania

Since the collapse of communism in Albania in 1990, the country has aimed to strengthen its relationship with the European Union and the United States of America. Today, it remains focused on delivering the necessary reforms to meet the conditions for entry into the European Union, as well as rebuilding former international relationships. The current governing party of Albania is the Socialist Party (Partia Socialiste), which has a purely capitalist foreign policy, as seen in the increase of foreign direct investment (FDI) in recent years. For the past 10 years, China has taken initiatives to bring more FDI into Albania and establish relations with the tourism, economy, agriculture, and transportation sectors. As a result, despite Albania being focused on joining the European Union, the idea of creating new relations and rebuilding the old ones with Asian countries has become increasingly significant.

1.1. Reforming the Country

Albania has been an official candidate for accession to the European Union since 2014[1], and must adopt economic, social, and political policies in order to meet the development and governance level of other EU countries. The conditions imposed by the EU to countries aspiring to gain membership will require Albania to change its approach and policies in areas such as its political system, judicial system, and economic development.

The successful election of Prime Minister Edi Rama represented a vote for the EU accession as he views 'EU membership as the key to resolving Balkan instability'[2]. In 2017, Edi Rama served another term as prime minister, which indicates a strong level of public support for the accession and the higher likelihood that Albania is headed for further integration into the EU. Indeed, a 2015 study showed that 95% of Albanians agreed the integration with the European Union is very important[3]. Rama has been described as a 'third way' social democrat, on par with Britain's Tony Blair or France's Emmanuel Macron[4], and ideologies that match those of the German Chancellor Angela Merkel. Such ideological compatibility with other EU leaders may be expected to both smoothen and drive the reform and

1 | European Commission. (n.d.). *European Neighbourhood Policy And Enlargement Negotiations: Albania.* Retrieved November 14, 2017, from https://ec.europa.eu/neighbourhood-enlargement/countries/detailed-country-information/albania_en.

2 | Kouchner, B. (2017, June 15). Albania shows EU enlargement is far from over. *Financial Times.* Retrieved November 14, 2017, from https://www.ft.com/content/a15f9b5a-51ab-11e7-bfb8-997009366969.

3 | Dobrushi, A. (2017, June 27). Albanians Voted and the Winner is ... the EU. (Politico, Ed.) *Politico.* Retrieved November 17th, 2017, from https://www.politico.eu/article/albania-vote-winner-is-the-eu-brussels/.

4 | Nence, M. (2013). *Corruption, Albania's Biggest Challenge for Integration in E.U.* PECOB.

accession process, at least within Rama's current term.

At the same time, the constitution of Albania does not limit Rama to run only for two terms, which means he could be re-elected as prime minister depending on the will of the people. While in the past there has been a consistent change of the governing political party, the pattern seems unlikely to continue in the next eight years.

If Rama is ruled out from the opportunity of running for another mandate as prime minister, the Socialist Party has a backup candidate, Erion Veliaj, who became Mayor of Albania's capital Tirana in 2015. He did a tremendous job to bring the city out of its darker ages. As Mayor of Tirana, Veliaj transformed the city from an urban jungle into a children-friendly city by reducing the number of cars, illegal constructions, reconstructing the kindergartens and installing alternative forms of transportation. 'In his first year Veliaj took 40,000 square metres of land from illegal developments, making way for 31 new playgrounds.'[5]

5 | Laker, L. (2018, February 28). What Would the Ultimate Child-friendly City Look Like? *The Guardian*. Retrieved March 20, 2018, from: https://www.theguardian.com/cities/2018/feb/28/child-friendly-city-indoors-playing-healthy-sociable-outdoors.

6 | Koleka, B. (2017, June 26). Albanian Socialists to Get Parliamentary Majority: Partial Vote Count. *Reuters*. Retrieved March 20, 2018, from https://www.reuters.com/article/us-albania-election-result/albanian-socialists-to-get-parliamentary-majority-partial-vote-count-idUSKBN19H18L.

7 | International Foundation for Electoral System. (2018). *Election Guide Democracy Assistance & Elections News*. Retrieved April 1, 2018, from http://www.electionguide.org/elections/id/2461/.

Advocacy for children means Veliaj is an advocate of the future. But the political arena can change. The Democratic Party of Albania (DPA), the main opposition, was defeated in the 2017 elections by the Socialist Party[6]. With 46.75% of eligible voters participating, the DPA received only 28.85% of votes, or 43 seats in the parliament, while the Socialist Party won 48.34% or 74 seats[7]. Out of 140 seats, the Socialist Party won the simple majority, thus being able to establish a new government without having to form a coalition. Therefore, only an impressive political performance will give the DPA a shift upwards. However, with Lulzim Basha settled as the DPA's leader, major reforms in the party took place, such as having 35% of the parliamentary candidates come from the party's youth membership and passing a national law to bar citizens with criminal convictions from holding office. However, the reforms were not enough to win the elections. Basha must make major changes in his party and for the Albanian people if he wishes to remain as a serious candidate for future general elections.

Because of the candidacy for membership of the EU, Albania must undertake economic and constitutional reforms to satisfy the requirements of entry. However, Albania is currently facing some economic challenges that could jeopardise its accession status, such as high public debt, inefficiencies in the energy sector, high structural unemployment and public land ownership which affects other areas such as agriculture and industrial development. During the communist regime, all land was state-owned. Consequently, once the regime fell, disputes among people and institutions arose which caused a massive problem for investments in land. To reach some of its political goals, such as the EU accession, establishing new relations with foreign countries, and decreasing the unemployment rate, Albania must adopt some economic reforms. It is beyond doubt that Albania lacks employment opportunities because the country lacks the necessary resources to create sufficient numbers of jobs. This usually results in more imports than exports and trade deficit. In recent years, Albania's imports have increased, 'goods imports are projected to increase relatively strongly during the final phase of the Trans Adriatic Pipeline (TAP) construction in 2017 and expanding domestic demand will be accompanied by rising imports throughout the programme period.'[8].

Other than job opportunities, the Albanian government is attempting to undertake other economic reforms (such as reducing public debt ratio) that aim to create better conditions for Albania to enter the EU. Despite the decreasing ratio, Albania's public debt profile presents risks since the informal economy still comprises a big part of the GDP, and the research and development (R&D), transport, and energy sectors require more investment.

Corruption, crime, and poor public administration and bureaucracy are also obstacles that impede Albania's process towards joining the EU. Since 2014 there have been major steps taken to eliminate these problems. Based on the 2017 Economic Reform Programme, Albania has completed some economic reforms in order to fulfil some of its major political goals. Yet there is still a lot of work to be done.

8 | European Commission. (2017). *2017 Economic Reform Programmes of Albania, the former Yugoslav Republic of North Macedonia, Montenegro, Serbia, Turkey, Bosnia & Herzegovina and Kosovo'*. Luxembourg: Publications Office of European Union. Retrieved March 20, 2018, from https://ec.europa.eu/info/sites/info/files/ip055en.pdf.

Besides economic reforms, constitutional reforms are required in order to join the EU. The Constitution of Albania, as the highest law of the nation and the most fundamental governing principles, plays an essential role in shaping the future direction of Albania's politics. Because of the weak constitution established during the 1990s, it has been very challenging for Albania to integrate its domestic laws with international institutions and EU regulations. One of the conditions set by the EU in joining its member states is that Albanian constitutional reform has to establish institutions receptive to the EU's consociational practices in order to fully participate in the EU integration process as well as to ensure democratic stability[9]. Consociational practices exist in societies that are significantly divided along ethnic, religious or linguistic lines but where there is political power-sharing among the various groups. Since Albania is a relatively homogeneous country in terms of its religious and ethnic structures, the EU imposes their consociational practices in an 'outward-looking way', which means Albania should incorporate corresponding clauses and thinking into its constitution and institutional design to make sure the country can fit in with EU practices[10].

Because of the conditions regarding incorporating the EU's consociational practices, Albania executed judicial reforms which were approved by the Albanian Parliament in 2016. Those reforms were mostly about rewriting certain laws which had increased the judges' power, and led to increased corruption. This reform was not only suggested by international bodies such as the EU, but also by many Albanians who wish to see it come to fruition – an estimated 91% of Albanians were angry with corrupt judges. While the parliament approved the judicial reform proposal to improve the situation, 'the deal remained elusive after 18 months of talks with the EU and US officials'[11]. Based on a study by Boston College School of Law, one of the most important reforms to the Court is changing Article 126, which used to read that 'a judge of the Constitutional Court cannot be criminally prosecuted without the consent of the Constitutional Court'[12]. This article of the constitution gave power to the judges to create their own immunity against the judicial system. After the reform, Article

9 | Peshkopia, R. (2014). *Conditioning Democratisation : Institutional Reforms and EU Membership Conditionality in Albania and Macedonia* (1 ed., Vol. 1). (D. I. Balazs Apor-Trinity College, Ed.) Anthem Press.

10 | *Ibid*.

11 | Koleka, B. (2016, July 21). Albania parliament passes judiciary reform key to EU accession. (G. Crosse, Ed.) *Reuters*. Retrieved November 20, 2017, from https://www.reuters.com/article/us-albania-parliament-court/albania-parliament-passes-judiciary-reform-key-to-eu-accession-idUSKCN10202F.

12 | Albert, R., Benvindo, J. Z., Rado, K., & Zhilla, F. (2017). Constitutional reform in Brazil: lessons from Albania? *Revista de Investigações Constitucionais*, 4(3), pp. 11-34.

126 states that 'the Constitutional Court judge shall enjoy immunity in connection with the opinions expressed and the decisions made in the course of assuming the functions, except where the judge acts based upon personal interests or malice'[13]. Overall, Albania faces a major problem with its rule of law and it has since become a common goal for many politicians to achieve it. Prime Minister Rama told Reuters in July 2017 that 'stopping [the judicial reform] is not an option, we have to do it if we want to be part of the EU'[14]. From the recent reforms taken by the government of Albania, it seems that the country has moved towards collaboration with many countries to strengthen its economic and constitutional reforms. By reforming the internal politics, the government is allowing more openness and conforming with the standards in the sphere of international relations.

Albania, although part of The North Atlantic Treaty Organisation (NATO) since 2009 and an official candidate country to join the European Union, does not limit itself from engaging in more international cooperation with other non-EU countries. The signing of many agreements between Albania and China serves as an example. Some of the agreements between the two nations involve establishing sister cities, boosting trade, construction of hydroelectric power facilities, and agreement on the '16+1' cooperation framework. While Albania might sign agreements and follow conditions of the EU regarding politics, there is more flexibility when it comes to Albania's domestic economic development, and economic cooperation and agreements with other countries.

For example, the EU has flexibility over whether Albania could keep its local currency, the Lek, in use. Over a century, Albania has been using the Lek as a transition to the Euro. However, the adoption of the Euro would change Albania's economy drastically since it may result in doubling commodity prices, increasing the cost of living, and limiting Albania's control over its monetary policy. While the Euro is a legal currency in Albania due to touristic demands, the current government has no plans to change from the Lek to the Euro. In fact, the central bank aims to decrease Euro usage by 10%[15]. This will not only

13 | Ibid.

14 | Benet Koleka, M. R. (2017, June 23). Albania Judges Its Judges in Pursuit of EU Milestone. (D. Stamp, Ed.). Reuters. Retrieved November 20, 2017, from https://www. reuters.com/article/us-albania-election-corruption/albania-judges-its-judges-in-pursuit-of-eu-milestone-idUSKBN19E19H.

15 | Koleka, B. (2018, February 14). Albania C.bank Says Aims to Reduce Country's Use of Euro by 10 Pct Over 3-5 Years. Reuters. Retrieved March 20, 2018, from https://www.reuters.com/article/albania-euro-cenbank/albania-c-bank-says-aims-to-reduce-countrys-use-of-euro-by-10-pct-over-3-5-years-idUSL8N1Q44DR.

establish more effective monetary policies, but also will help the country's economy. While maintaining the Lek as the country's main currency, Albania earns when the Euro fluctuates. When the Euro becomes stronger, the government has the power to apply higher interest rates, which will decrease the money supply of the Lek, and vice versa.

1.2. Rebuilding Old Relations

Foreign trade and investment were important drivers of development in Albania before it was ruled by the isolationist communist regime. Ever since the establishment of the democratic regime, the country has taken a slew of failed reforms and it took 20 years for policymakers to realise that international engagement is vital for Albania's economic development. Barriers built during communism hampered Albania's ability to establish connections with foreign countries in the post-communist democratic regime. A weak currency and lack of exports of goods and services destroyed Albania's economy and imposed a massive challenge to reaching self-sufficiency. Twenty years passed until the political leaders of Albania finally realised the need for change if they wished to remain competitive in the international arena in aspects such as trade, economy, politics, law, and education. In a fast globalising world, to remain relevant means establishing new relations and rebuilding the old ones with other countries.

16 | The idea of the 'three worlds' is part of a theory proposed by Mao Zedong to explain international relations in the 1970s. The first world included the US and USSR. The second world icluded Jaoan, Europee and Canada. The third world included Asia, except Japan, Africa, and Latin America. Ministry of Freign Affairs of the People's Republic of China. (n.d.). *Chairman Mao Zedong's Theory of the Division of the Three Worlds and the Strategy of Forming an Alliance Against an Opponent.* Retrieved from https://www.fmprc.gov. cn/mfa_eng/ziliao_665539/ 3602_665543 /3604_665547/ t18008.shtml.

By pledging to a cooperation initiative with China, Albania made a step forward to establishing collaboration in areas of infrastructure, tourism and agriculture. In the past, Albania had great relations with China. Many goods, especially military weaponry, were imported to Albania from China. During the 1970s, the idea of the 'three worlds' displeased the infamous dictator of Albania, Enver Hoxha.[16] Simultaneously, Hoxha opposed the establishment of foreign relations between China and the United States in the 1970s. This led to a split in Sino-Albanian relations. Recently, those relations are being reconstructed with more interest from the Chinese government, as illustrated by China's proactive attitude in holding regular official meetings with the political leaders of

Albania. Primary steps towards establishing this cooperation between the two nations happened in 2009, when China offered a four-point proposal to further relations. The two countries issued a joint statement on traditional friendship and signed documents regarding cooperation on cultural matters and public health[17].

The Belt and Road Initiative is a project that will not only strengthen the relations of Albania with China, but also all countries that are part of this massive Initiative. The increasing flow of the FDI in the past 13 years has allowed Albania to advance in the process of globalisation. FDI is projected to keep increasing based on a forecast conducted by Trading Economics (TE). The TE data in 2004 shows the floor of FDI per quarter was around US$80 million, whereas the lowest in 2016 was US$159 million, and the peak during the summer reached almost US$300 million. This increase of the FDI is certainly not only due to the increase in productivity, but also the number of foreign companies investing in Albanian resources. Forecasts also depict that the FDI trends will continue to grow, which implies that Albania is headed towards a future filled with continuous cooperation with other countries.

1.3. Conclusion

Albania left its dark ages behind in 1990, when communism was officially replaced by liberal democracy. Ever since, Albania has changed in many ways. Albania is an official candidate to join the European Union and is a full member of NATO. Since 2013, Albania has seen more successful reforms than in previous years. The defeat of Lulzim Basha of the Democratic Party and the election of Edi Rama for two consecutive terms as prime minister has allowed the Socialist Party to create more progressive reforms in economic and judicial aspects. Such reforms are highly recommended by the EU. The country is still in need of larger economic reforms to mitigate problems such as the lack of employment opportunities and resources. The informal economy and public debt are two main burdens that Albania has to bear on its path to joining the EU. Albania has its own currency, the Lek, that has been in use for a long

17 | Xinhua. (2009, August 24). China Offers Four-point Proposal to Further Relations with Albania. *China Daily*. Retrieved November 14, 2017, from http://www.chinadaily.com.cn/hellochina/albaniaambassador2009/2009-08/24/content_8609434.htm.

time. While the Euro is a stronger currency than the Lek and is widely used by tourists in Albania, the government does not plan to replace the Lek with the Euro. In recent years, Albania has been working to improve its relationships with other countries and establish new ones, especially with those in Asia. China and Albania have established collaborations in areas of infrastructure, tourism, and agriculture. Chinese FDI is one of the highest in Albania, and more collaboration is underway with the establishment of the '16+1 framework'.

2. An Overview of the Challenges and Opportunities for Young People in Albania

The Belt and Road Initiative (BRI) will play a major role in shaping the global economy and politics, as well as enhancing the cultural interconnectedness among youths in different parts of the world. As the youth play a crucial role in shaping the future identity of Albania, their participation in the government's decision-making process at the community level is very important. The young generation of Albania is characterised by different values, ambitions, desires, and perspectives regarding their future careers. Therefore, an effective education system together with adequate government policies can help the youth achieve their ambitions by increasing their competitiveness in the labour market. Yet, Albanian youths today face a variety of challenges that prevent them from fulfilling their potential. In this context, it is important to pay attention to the causes of such issues such as a poor education system and the repercussions of communism. The main objectives of this paper are to 1) analyse the current challenges faced by young people in Albania, 2) outline the employment opportunities for them, and 3) provide recommendations to overcome the challenges.

2.1. Main Challenges Pertaining to Young People in Albania

Youth unemployment is one of the main challenges that the youth faces in Albania. Recent statistics show that Albania is facing high unemployment among youths, especially those who are undergoing a transition from schools to the labour market. Since they face enormous difficulties in finding a job that matches their expectation, they are dissatisfied with their status quo. However, it is believed that there is a slight improvement in the situation. For example, there has been a decrease in the youth unemployment rate from 40.1% to 30% from 2015 to 2017[18]. This decrease results from the increase in the variety of opportunities in both the business sector and professional schools. At the same time, economic freedom in Albania improved by 0.1% in 2018, compared to that of 2017, and the country ranked 65th in the 2018 Index of Economic Freedom with an overall score of 64.5%[19]. Despite the increase, Albania is still lagging its regional counterparts with a score below the regional average.

This paper argues that two major factors contribute to this challenge of youth unemployment. First, the lack of an effective education system. Second, the repercussions of 'communist legacy' in Albania that still impact the development of its overall economy.

2.1.1. Lack of an effective education system

The education system in Albania is ineffective in its 1) outdated curriculum, and 2) low standard of teaching. In reality, the subjects taught at schools do not prepare students for joining the labour market. Sometimes the curriculum deliberately increases the workload of studies and reduces the time that enables students to pursue other extracurricular activities, such as doing internships to gain work experience. In this case, students cannot learn the necessary skills to earn them employment opportunities and thus it is harder for them to find a job. Ultimately, this leads to a high youth unemployment rate.

18 | International Labour Organisation. (2017). *Unemployment, Youth Total (% of Total Labor Force Ages 15-24) (Modeled ILO Estimate)*. ILOSTAT Database. Retrieved from https://data.worldbank.org/indicator/SL.UEM.1524.ZS?locations=AL.

19 | Miller, T., Kim, A. B. & Roberts, J. M. (2018). *2018 Index of Economic Freedom*. The Heritage Foundation. Retrieved from https://www.heritage.org/index/pdf/2018/book/index_2018.pdf.

Although there are vocational schools that equip students with professional skills, students tend to choose between public or private universities as they are usually more prestigious and offer degree titles. More students prefer private universities because they offer a better learning environment, curriculum, learning support, and opportunities to participate in extracurricular activities, such as mobility programmes. The imbalance between the demand for traditional universities and vocational schools has led to the closure of many vocational schools. Compared with the courses offered by vocational schools, the programmes offered by traditional universities usually fail to cater to the changing demands of the labour market and result in the low employability of their students. In response to the problem of a declining number of vocational schools, the government decided to re-open some of the schools. Yet, the immediate effect of losing young people who possess professional skills due to the previous closures has yet to be alleviated.

Besides the curriculum, the low standard of teaching is a major factor that leads to an ineffective education system. Professors, high school teachers, and elementary teachers are neither prepared for the curriculum nor equipped with public speaking skills to effectively deliver teaching substance to students. They maintain the traditional way of didactic teaching without interactive discussions and teacher-student exchanges in the classroom. In order to increase the efficiency and effectiveness of the curriculum, it is important to combine theory and practice together, which enables students to learn useful skills that can be applied in the job market.

To tackle the problem of low employability among youths, the Ministry of Education and other organisations, such as Sports and Youth, political parties, and non-governmental organisations (NGO) are designing more inclusive and communal activities to engage the youth population. These activities aim to educate young people about duties and responsibilities with the goal of integrating them in society in the long run. Other than these activities, the government has suggested improving Albania's education system by offering more training to teachers to enhance their quality and facilitating more student exchange programmes and partnerships with other universities outside Albania. Such collaborations will allow students to broaden their horizons and learn the best practices of other countries that may be applied in their home country.

2.1.2. Communist legacy in Albania

Another factor contributing to the low employability is the residual effect imposed on the economy by the communist legacy in Albania. The majority of Albanians underwent major historic changes following the fall of the Berlin Wall, the collapse of the communist bloc, and other revolutionary changes in political and economic terms. Since many of the Central and Eastern European countries were governed by the communist authority before 1990, citizens were accustomed to a centrally planned economy, resulting in a lack of incentives to express their ideas and opinions and to travel outside the country to seek opportunities. It is difficult for the now decentralised government to develop new jobs to meet the growing unemployed population in the country. Most of the people had been used to the government's discourse about centralised planning in the past.

Today the power is decentralised among the local municipalities, giving much more freedom and authority to localities. After transitioning from communism, the Albanian government started developing an open market and allowing the freedom of movement of goods and citizens. A liberal democratic regime brought new economic reforms that benefit civil society and opened new diplomatic ties with Western European countries. Despite the goodwill of promoting reforms, it is argued that the regime lacks proper rules and effective mechanisms to implement those reforms. For example, the government proposed that small businesses shall pay the same amount of taxes as that of large businesses, which raises questions of equality.

Albania is still lagging other countries in the region in terms of transitioning from a closed, less-developed economy, into a country that is receptive to the notion of liberalism and capitalism. Comparing Albania with its neighbours, its communist roots influenced and are still influencing the country's progress in political, economic and cultural terms. While neighbouring countries such as Slovenia and Croatia experienced trade and open borders when they were a part of Yugoslavia under the Tito regime, Albania attempted to transform into a democratic system under the Hoxha regime as demonstrated by an increased level of economic flexibility and openness.

2.1.3. Other challenges

Apart from a high youth unemployment rate and the lack of sufficient job opportunities in Albania, there are other challenges faced by the young people such as corruption, gender inequality, and difficulty in climbing the social ladder in public administration.

Corruption is a prevalent political problem in Albania. Even though the government is working hard to enforce constitutional and judicial reforms to punish corrupt officials and remove corrupt judges, the problem persists. Not only does widespread corruption affect Albania's process of accessing the European Union, but it also negatively affects the credibility of the government. People who have political affiliations with government officials or personal networks tend to possess unfair advantages in obtaining jobs and other positions. This phenomenon could affect many young people's morale in seeking a job as they are young and do not possess such advantages.

Gender inequality in terms of wages and opportunities also exists in Albania. At the same time, some progress has been made towards achieving gender equality. In 2008, the Albanian parliament passed a law setting a quota of 30% for female representation in political parties and parliamentary seats.[20] In 2017 the representation of women in the parliament increased to 28% compared with 18% in 2013.[21]

20 | UN Women. (2017, July 18). *In Albania, Elections Herald Historic Increase in Number of Women MPs*. Retrieved from https://www.unwomen.org/en/news/stories/2017/7/feature-in-albania-elections-herald-historic-increase-in-number-of-women-mps.

21 | *Ibid.*

Furthermore, it is difficult for young people to climb the social ladder in the public administration sector when the older generation occupies a majority of the public-sector roles and do not make room for the younger generation to advance. This depresses young people's motivation to work hard and improve their lives.

2.2. Main Opportunities for Young People in Albania

The main opportunity for young people in Albania is accession to the EU. It is believed by many Albanians that the accession would provide much more business, work, and education opportunities to the population, as well as help the people overcome the fear of migration to other European countries. Young people have a positive attitude to the government's active fulfilment of all the obligations necessary to become part of the EU. After joining the EU, Albania could engage in consolidated cooperation in youth and policy reforms, funding programmes, grants and scholarships, Erasmus+ (EuRopean Community Action Scheme for the Mobility of University Students) programmes, etc.[22] This would likely increase social inclusion and civic participation of the young generation as well as integrate them into the labour market. In fact, some students have started to migrate to other European countries to grasp more opportunities. Although there may be a brain-drain problem, studying and working in other countries should equip young Albanians with more experiences to contribute back to their homeland when they return.

2.3. Conclusion

Unemployment is one of the most challenging phenomena in Albania. The Albanian government together with the EU are working to decrease the level of youth unemployment by providing new opportunities to the Albanian youth through a number of programmes. Some of the main activities include volunteering, community work, internship opportunities, and job facilities. Even though the wage provided by these activities may not be very high, it is a good opportunity to increase the level of employment and engage the youth in various activities.

22 | Erasmus+ is a programme launched by the EU to 'support education, training, youth and sport in Europe' (reference: European Commission. (n.d.). *Erasmus+*. Retrieved from https://ec.europa.eu/programmes/erasmus-plus/about_en).

The accession of Albania to the European Union will influence the education sector a lot and provide more incentives for the youths to receive education abroad. As a result, more opportunities will arise from the increased interaction with EU institutions and may possibly bring new ideas to reform Albania's education system.

3. Evaluate the Opportunities and Challenges Offered by the Belt and Road Initiative

The impact of rapid economic growth in China, accompanied with the 'Open Door' policy and the 'Going Out' policy adopted by the regime, is felt in both Western and Asian countries, consequently bringing forth significant investment and trading opportunities. Despite the incoming economic benefits, China's territorial and population size, plus the political atmosphere combined with its present and future economic growth rate may be identified as a possible geopolitical threat in the minds of many state leaders and scholars[23].

This part addresses questions such as: where do the new foreign policies of China and ideas regarding change come from? What is China trying to achieve with the New Belt and Road Initiative in the Western Balkans with a focus on Albania? How will the Chinese FDI in Albania and the region affect trade, economic development and youth social inclusion in regard to job and opportunity creation? It will also analyse and discuss whether this initiative would be sustainable and what are some of the measures that need to be taken to assure sustainability. In sum, this part argues that China's BRI intends to share the Chinese experiences in growth, development, and connectivity and collaborate more closely on concrete projects with participating countries (mainly the European

23 | Cable, V., & Ferdinand, P. (1994). China as an Economic Giant: Threat or Opportunity?. *International Affairs*, 70(2), pp. 243-261.

Union). China's ultimate goal through the Balkan Silk Road is to penetrate European markets, given the trade instability Brexit is causing in the region, and perhaps to help develop the infrastructure of the countries along the road.

3.1. Constructing the Balkan Silk Road

The Belt and Road Initiative (BRI) takes up a significant proportion of the world's trade routes, not just the ancient silk road that extends to central Asia and the Eurasian continent, but also maritime routes that go through India, Siri Lanka, some African countries, and as far as the Mediterranean Sea that leads into Europe. The Initiative extends its interests in the 16 Central and Eastern Europe Countries (CEECs) including those in the Balkan region, forming a new trade and investment corridor called the *Balkan Silk Road*. Despite the numerous imminent opportunities brought by the Initiative, it is also important to look at the sustainability of the development projects under the BRI, especially on the grounds of accountability and transparency, as well as economic, social and environmental sustainability.

The Chinese proposal on jointly building the Belt and Road Initiative has drawn world attention since launching in 2013. China's effort to 'reincarnate' the thousand-years-old Silk Road is praiseworthy but may also cause worries about China being too ambitious. The Initiative has undoubtedly raised some eyebrows and with good reason, because the implementation of the BRI will attract the international community for many years to come. Furthermore, the sustainability and implementation of the project may depend on domestic politics and policies of the countries which are located along the One Belt One Road[24].

Besides evaluating the Initiative's sustainability, the limitations regarding its actual implementation and execution should not be ignored. Since many of the countries located along the 'Road' are developing countries, they may not have the proper institutions to follow the regional EU and international standards, to monitor the execution of the bilateral agreements

24 | Bhardwaj, A. (2016, October 21). *Belt and Road Initiative: Potential to Tame American Imperialism?* Retrieved on September 18, 2017, from http://www.epw.in/journal/2016/43/strategic-affairs/belt-and-road-initiative.html.

and to make sure that Chinese companies abide by the standards in regard to the implementation of infrastructure projects. In a speech in Beijing, the European Commission's Vice President Jyrki Katainen stressed the fact that any new scheme connecting Europe and Asia should follow a set of principles and rules including international market rules, international standards, and the already existing networks and policies[25]. Furthermore, the sustainability of the project may also be affected by political change in those countries through elections. For instance, a new government may simply revoke the plan and set back the development.

3.2. Background: Infrastructure Development in Albania

According to the 2018 Budget, Albania experienced a four-fold economic growth compared to that of 2013. In face of the growth, the budget focuses on public investment in which the government plans to devote 5.2% of the country's GDP, i.e. ALL (Albanian Lek) 86.1 billion (approximately US$860 million). The current fiscal policy will be based on honest taxation, the reduction of the informal sector and promotion of private strategic investments, such as foreign direct investments (FDI) and partnerships with foreign investors and partners[26].

The year 2018 marked a significant increase in investment in infrastructure, with a record figure of ALL 86 billion or 16 % more than the current year. There is an increase in funding in infrastructure, which means that completing the unfinished infrastructure projects will be a priority for the government. Investments will concentrate mostly on road construction, where some road segments are expected to be completed, such as the bypass between Fier and Vlora, and the Tirana-Elbasan road[27].

Building connecting roads between the regions are one of the critical areas in Albania's infrastructure development. One of the most sighificant projects which has already been completed is the highway Corridor Durres-Kukes-Morine, also known as the 'Nations Road'. This road connects the key economic sectors of Albania and the capital of Tirana

25 | European Commission. (2017, May 15). *Speech by Jyrki Katainen, Vice President of the European Commission at the Leaders' Roundtable of the Belt and Road Forum for International Cooperation.* Retrieved on March 14, 2018, from https://ec.europa.eu/commission/presscorner/detail/en/SPEECH_17_1332.

26 | Ministry of Finances and Economy. (2018). *Buxheti 2018.* Retrieved on March 14, 2018, from http://www.financa.gov.al/al/buxheti/buxheti-ne-vite/buxheti-2018.

27 | *Ibid.*

with the port of Durres and Kosovo in the north. The main economic and business sectors, such as manufacturing and shipping sectors in Albania, are concentrated in the capital of Tirana, Durres, and the other major cities. Therefore, the road will connect the smaller cities with the capital. Another road construction project underway is expected to connect the Albanian harbour city of Durres with North Macedonia and Bulgaria, joining the neighbours in the East in one road for the purposes of fuel and general trade.

The European Bank for Reconstruction and Development has already established projects with the Municipality of Tirana as a way of making the capital and the central government more accessible to the neighbouring municipalities which lack infrastructure.

The municipalities in the far north and south of Albania lack essential infrastructures such as road and political channels, impeding cooperation between the central government and the local governments. For instance, it is easier to travel from Tirana to North Macedonia than to the municipality of Diber, which shares a border with North Macedonia, due to the lack of roads connecting between the capital and the municipalities.

In order for Albania to develop its economy and make good use of the international aid and official development assistance received for poverty reduction and infrastructure projects, it would be beneficial for the country to refer to the Chinese experiences in implementing policies regarding poverty reduction by building infrastructure, which could result in sustainable growth, strong inflow of investments, and export-led industrialisation[28].

Despite strong determination to build infrastructure, Albania is facing two major hurdles in realising the goal: 1) lack of funds, and 2) lack of transparency and accountability to ensure and monitor the sustainability of the projects. The latter obstacle is reflected in the case of the 'Nations Road' construction project. Although the road was built by the same company in both Albania and Kosovo, the quality of the section in Kosovo

28 | World Bank. (2014, June 30). *Roads Create New Livelihoods on Albania's Coast*. Retrieved on March 14, 2018, from http://www.worldbank.org/en/results/2014/06/30/roads-create-new-livelihoods-on-albanian-coast.

is better than that of Albania, probably because of Kosovo's tighter regulations and road authority rules, as well as its more intensive investment in the corridor (which took up about 25% of Kosovo's GDP in 2010)[29]. The Albanian government needs to develop better accountability measures in public investments and make sure the deadlines of the project are met, especially when public funds are used. As encouraged by the Albanian government to develop infrastructure, the country now resembles a construction site where plenty of unfinished and ongoing projects are seen. For instance, the city square in Tirana changes every time a new mayor is elected. The policymakers in Albania need to look further than their party bipartisanship and work to increase the citizens' wellbeing.

3.2.1. Chinese investment in Albania

Some of the European countries in this Initiative include countries that are not members of the EU such as Albania, Montenegro, and Serbia. Albania, despite its location in the European continent, is one of the few candidate countries to the European Union that has already adopted some of the EU guidelines and standards regarding investment and trade. However, it remains to be seen whether or not Albania will adjust its practices and laws to comply with the business standards and culture of Chinese companies. Looking back through Sino-Albanian investment and trade history, Albania has been China's partner since 1949. The relations were further strengthened in 1971 when the United Nations General Assembly Resolution 2758, 'led' by Albania, resulted in the restoration of the lawful rights of the People's Republic of China in the United Nations and maintenance of the 'One China' Policy[30]. According to the country-specific data provided by the European Commission Directorate-General for Trade, the total trade volume between China and Albania reached €424 million (around US$525 million) in 2016. China holds second place in regard to trade with Albania after the EU, which comprises 27 (in light of Brexit) countries in total.

Regarding Chinese investments in Albania, the amount increased from US$4.35 million in 2009 to US$7.3 million in

29 | Capussela, A. L. (2011, December 02). *Kosovo: The unnecessary highway that could bankrupt Europe's poorest state*. Retrieved on March 23, 2018, from https://www.balcanicaucaso.org/eng/Areas/Kosovo/Kosovo-the-unnecessary-highway-that-could-bankrupt-Europe-s-poorest-state-108430.

30 | United Nations Official Document, GA. (1971). Retrieved on October 30, 2017, from http://www.un.org/ga/search/view_doc.asp?symbol=A%2FRES%2F2758%28XXVI%29.

2014[31]. Investments in the energy sector and the exploration of raw materials (e.g. petroleum, natural gas and chromium) contributed most of the increase and it is believed that Chinese investments in this key strategic sector may help Albania tackle one of the major challenges that the country has been encountering, i.e. energy production, self-sufficiency, and sustainability. China has become one of the largest FDI investors in Albania after the purchase of Banker's Petroleum, a Canada-based oil company which operates in Albania, by a Chinese oil and gas exploration company Geo-Jade Petroleum. The acquisition enables Geo-Jade Petroleum to conduct oil exploration activities and secure oil production rights in Albania, including onshore development of oil fields, gas deposits and geothermal energy in the oil-rich cities of Patos and Kucova. Apart from the energy sector, Chinese companies are interested in developing the transportation sector in Albania as illustrated by the closed deal led by China Everbright Group and Friedman Pacific in May 2016 on acquiring concessionary rights to Tirana International Airport until the year of 2025[32].

Generally, the Chinese investments in the Balkan region are welcomed by policymakers. Like every other developing country, Albania is rewriting its guidelines and offering concessions to attract more foreign investors. Albania has adopted new tax policies that aim to hinder corruption and bypass bureaucracy as, in the past, bureaucratic procedures in obtaining operating licenses have crippled foreign investment. Nevertheless, it is important to note that Albania's primary economic partner and anchor remains the EU, despite visible inroads into the Albanian market actively led by many Chinese companies[33].

3.3. Potential Impact of the Belt and Road Initiative on Albania

The Belt and Road Initiative (BRI) is expected to profoundly influence the economic development of the participating countries. Through this Initiative, China hopes to integrate the markets of the participating countries into the world economy, providing them with opportunities to make better use of their raw materials and make sure that these countries harness

31 | European Commission Directorate-General for Trade. (2016, June 29). *Statistics*. Retrieved on November 20, 2017, from http://ec.europa.eu/trade/policy/countries-and-regions/statistics/index_en.htm.

32 | Goh, B. (2016, April 27). China Everbright, HK's Friedman Buy Albania's Airport Operator. *Reuters*. Retrieved November 20, 2017, from https://www.reuters.com/article/china-everbright-albania-airport/china-everbright-hks-friedman-buy-albanias-airport-operator-idUSL3N17U2PR.

33 | Bastian, J. (2017). The Potential for Growth through Chinese Infrastructure Investments in Central and South-Eastern Europe along the 'Balkan Silk Road'. *European Bank for Reconstruction and Development*, pp. 1-62. Retrieved on November 23, 2017, from https://www.ebrd.com/documents/policy/the-balkan-silk-road.pdf.

the demographic dividend through investments in youth and ensuring youth employment. Furthermore, the scale of this global initiative would have a significant impact on climate change and the environment. The BRI will also have a major influence on whether or not construction plans would meet international standards to protect the climate and promote sustainable development. China was very instrumental in setting goals and objectives in the Paris Agreement and has infused the BRI with the UN Sustainable Development Goals (SDG) to demonstrate its commitment to environmental protection[34]. One will have to see if such goals and objectives are reachable.

In general, foreign direct investment (FDI) is thought to have a positive impact on economic growth because it can provide direct financing, technology transfer, professional training, production networks, and access to a new market. However, given Albania agreed to expand cooperation under the 'Belt and Road, 16+1 framework' in 2017, it is difficult to estimate the impact of Chinese FDI on Albania's GDP growth and employment (especially youth employment). Similarly, a 2017 report published by the European Bank for Reconstruction and Development suggests that only a few of the countries in the Balkan Silk Road 'can currently claim that the Chinese infrastructure investment and closer trade ties have created significant spillover effects in areas such as the development of small and medium-sized enterprises (SMEs), additional job creation or a reversal of existing trade imbalances'[35]. Nonetheless, in order to reduce the deficit and shortcomings in the future, Chinese companies need to engage and cooperate with 'civil society representatives, non-governmental organisations, chambers of commerce and universities on issues ranging from the transfer of professional skills, environmental impact assessment or strategic SME development'.[36]

3.4. Belt and Road Initiative and Young People in Albania

<u>3.4.1. Connecting BRI with SDG</u>

The UN Secretary-General Antonio Manuel de Oliveira Guterres compares China's Belt and Road Initiative with the

34 | Camdessus, M. (2017, May 17). China's Belt and Road Projects Must Hold Fast to Environmental Goals. *South China Morning Post.* Retrieved on November 20, 2017, from http://www.scmp.com/comment/insight-opinion/article/2094611/why-chinas-belt-and-road-must-be-pathway-sustainable.

35 | Bastian, J. (2017). The Potential for Growth through Chinese Infrastructure Investments in Central and South-Eastern Europe along the 'Balkan Silk Road'. *European Bank for Reconstruction and Development,* pp. 1-62. Retrieved on November 23, 2017, from https://www.ebrd.com/documents/policy/the-balkan-silk-road.pdf.

36 | *Ibid.*

Sustainable Development Goals (SDG) claiming that they are both 'rooted in a shared vision for global development.' According to the principles set by the UN under the SDG, sustainable development, despite at times being a broad concept, is a systematic approach to growth and development and to manage and produce natural and social capital for the welfare of humanity's future. The Belt and Road Initiative may become a leading actor in driving Albania's economy as it resembles the UN SDG in aiming to deepen connectivity across countries and regions. While China's campaign to promote its development model is expected to change the economies in the area, China must uphold its promise of maintaining sustainable growth.[37]

Rather than having a geopolitical focus, the framework of the BRI should also focus on commercial trade and investment, depending predominantly on the private sector, which enjoys comparative advantages under market rules and international laws and standards. The BRI's sustainability will be contingent on the ability of projects to bring sustainable economic growth, improvement of regional cooperation, and social and environmental aspects of human development.

In this context, it is beneficial for the BRI to link with the United Nations Development Programme (UNDP) operations in Albania. First, the UNDP has been a crucial part in the Albanian policy development. With projects aiming to tackle market inefficiency and to increase public services that municipalities offer to citizens, the UNDP has helped develop the political infrastructure of Albania. Second, the UNDP sees BRI as a premise for the implementation of the SDGs through promoting sustainable economic growth, regional cooperation, and regional coordination.[38] The UNDP also believes that Chinese investment through the BRI will promote inclusiveness and win-win cooperation while creating sustainable development gains in social and environmental domains. Thus, BRI strengthens the implementation of SDGs in developing countries such as Albania.

The alignment of the BRI with the UN's SDG can help China

37 | Camdessus, M. (2017, May 17). China's Belt and Road Projects Must Hold Fast to Environmental Goals. *South China Morning Post.* Retrieved on November 20, 2017, from http://www.scmp.com/comment/insight-opinion/article/2094611/why-chinas-belt-and-road-must-be-pathway-sustainable.

38 | Horvath, B. (2016). *Identifying Development Dividends along the Belt and Road Initiative: Complementarities and Synergies.* Retrieved November 22, 2017, from https://issuu.com/undp-china/docs/undp-ch-bri_2017_scoping_paper1____/9.

achieve the aim of promoting trade while increasing the credibility of the Initiative; it can also help countries such as Albania where the UNDP operates. The BRI could use and develop the already-established frameworks from the UNDP to positively impact the development of Albanian institutions and municipalities. For instance, through the STAR2 project (i.e. Consolidation of Territorial and Administrative Reform), the UNDP conducted workshops[39] in Albanian municipalities to help the implementation of the Standard Operating Procedure (drafted by the UNDP) in the municipalities. China's Department of Foreign Assistance (DFA) of the Ministry of Commerce, for instance, might use this established framework by offering its experience.

Such an alignment is proven to be crucial for the coordination of the BRI together with the SDGs as it can help the BRI draw up and better implement infrastructure development plans. The coordination can also help avoid misallocating and overlapping resources in infrastructure development. Therefore, connecting the BRI with the UNDP development goals could create a great opportunity for infrastructure (road and political) development in Albania. The BRI and the SDGs can provide a framework for the individual participating countries to organise trade, investment, social and environmental interactions – with the notion of creating a greater good and sustainable human development[40].

3.4.2. Challenges to the sustainable development of BRI in Albania

There are significant obstacles that need to be overcome to achieve sustainable development in Albania. To begin with, the infrastructure that connects the Balkan countries is inadequate and outdated. Connectivity of facilities among the Balkan countries should be a priority in the Initiative while being respectful of different countries' security concerns. The countries in the Balkan Silk Road should improve the connectivity and closely coordinate with each other's ministries which are responsible for infrastructure construction to form an infrastructure that would connect all the countries. Besides

39 | The author was part of the team that conducted workshops in the municipalities of Albania as a part of the UNDP STAR 2 feasibility study.

40 | Horvath, B. (2016). *Identifying Development Dividends along the Belt and Road Initiative: Complementarities and Synergies.* Retrieved November 22, 2017, from https://issuu.com/undp-china/docs/undp-ch-bri_2017_scoping_paper1____/9.

connectivity between different countries, an infrastructure gap within Albania is seen. On one hand, significant investments in infrastructure development are concentrated in the principal cities such as the capital Tirana and the harbour city Durres. The other cities that are located in the north and the east of Albania, on the other hand, do not enjoy the same infrastructural development.

Besides, there are always tensions building up at the borders of the Balkan countries especially at the Albania-Greece border (China's entry point to the Balkan region through Athens' Piraeus Harbour). In 2016 Albania and Greece agreed to end the formal state of war that has existed between the two countries since World War II[41]. Both countries' fringe politicians claim each other's territories: Greece claims the Northern Epirus region parts of Gjirokastra and other Albanian cities, whereas Albania claims Chameria, ethnic Albanian lands whose inhabitants faced forced migration during World War II. The Serbia-Kosovo border is another challenge to the smooth implementation of the BRI with long waiting lines, slow control checks, and general hostilities.[42] For this Initiative to be successful and sustainable, there needs to be an open border, firstly confined to trade, capital and labour, and perhaps gradually developing into a fully open border – similar to the EU which bypasses border bureaucracies, corruption, and red tape.

The countries in the Balkan Silk Road should improve the connectivity of their infrastructure by coordinating joint construction plans and creating an international truck corridor, which would form an infrastructure network connecting the Balkan countries with Central Europe. At the same time, these countries should take into consideration the damage this endeavour may cause to the environment and climate change and promote a 'green and low carbon infrastructure construction'[43]. Such grievances exist in all the countries in the Balkan region. Kosovo has issues over the recognition of its independence with Serbia. Geopolitical issues prevail in the region, and perhaps open borders will help resolve some of these grievances.

41 | Mejdini, F. (2016, March 22). *Albania, Greece Agree to End Forgotten 'War'*. Retrieved on March 25, 2018, from http://www.balkaninsight.com/en/article/albania-and-greece-agree-to-abolish-the-war-law-03-22-2016.

42 | Montesano, F. S., van de Ven, J., & van Ham, P. (2016). The Geopolitical Relevance of Piraeus and China's New Silk Road for Southeast Europe and Turkey, Van der Putten, F. P. (ed.), *Clingendael Report*. Retrieved from: https://www.clingendael.nl/sites/default/files/Report_the%20geopolitical_relevance_of_Piraeus_and_China%27s_New_Silk_Road.pdf.

43 | The National Development and Reform Commission, Ministry of Foreign Affairs, and Ministry of Commerce of the People's Republic of China. (2015, March). *Vision and Actions on Jointly Building Silk Road Economic Belt and 21st-Century Maritime Silk Road.*

Legally speaking, some of the infrastructure projects under the BRI that are taking place in Central and Eastern Europe have stirred controversies regarding transparency and accountability. For example, in the case of the China-invested Belgrade-Budapest rail project, it is suspected by the European Commission that the Hungarian authority failed to comply with EU tender law by awarding a public works project to the two Chinese companies (the China Railway International Corporation and the Export-Import Bank of China) without tender[44]. Therefore, Hungary is under investigation for the violation of the EU transparency act – which requires the tenders to be public and competitive, giving other enterprises a fair chance to compete for the tenders. The incident prompts doubt about China's level of compliance with the local standards of other participating countries as well as its credibility in bidding international infrastructure projects. It is crucial to abide by the EU standards and regulations in order to foster fair business competition of tenders in Albania and ensure accountability and transparency[45].

Another example of similar magnitude in railway construction that has been halted because of transparency issues is the Mexico City-Queretaro line. The Mexican government revoked the project by invalidating a US3.7 billion investment due to questions over transparency[46]. In democratic countries such as Albania, transparency in public policies and infrastructure construction is of paramount importance since the government is held accountable before its citizens. Therefore, the Albanian government should analyse the impact of previous BRI-related investment projects on other countries (particularly on the grounds of transparency) and draw adequate policy lessons to comprehend the potential legal and political challenges brought forth by Chinese investments.

Furthermore, Albania and the countries in the BRI should take political stability into account, as some of the discontinued Chinese high-speed rail projects include Venezuela's Tinaco-Anaco, which was supposed to be Latin America's first high-speed railway with an investment of US$7.5 billion, but is now abandoned as a result of regime change and political

44 | Kync, J., Beesley, A., & Byrne, A. (2017, December 20). EU sets collision course with China over 'Silk Road' rail project. *Financial Times*. Retrieved March 22, 2018, from https://www.ft.com/content/003bad14-f52f-11e6-95ee-f14e55513608; Corre, P. L. (2017, May 23). Europe's mixed views on China's One Belt, One Road initiative. *Brookings Institution*. Retrieved September 18, 2017, from https://www.brookings.edu/blog/order-from-chaos/2017/05/23/europes-mixed-views-on-chinas-one-belt-one-road-initiative/.

45 | Menon, S. (2017, April 28). *The Unprecedented Promises and Threats of the Belt and Road Initiative*. Retrieved November 25, 2017, from https://www.brookings.edu/opinions/the-unprecedented-promises-and-threats-of-the-belt-and-road-initiative/.

46 | Kynge, J., Peel, M., & Bland, B. (2017, July 17). *China's railway diplomacy hits the buffers*. Retrieved on November 25, 2017, from https://www.ft.com/content/9a4aab54-624d-11e7-8814-0ac7cb84e5f1.

instability. Another discontinued project includes Libya's Tripoli-Sirte, which was supposed to connect the capital of Libya to Sirte, Muammar Gadaffi's birth city. It was suspended due to civil war and political unrest, and China Railway Construction Corporation was forced to suspend its investment[47]. The common denominator of these projects is that they are owned by Chinese companies. While China is not responsible for the risks caused by the civil war and the political failure in Libya, or the failing government in Venezuela, they can be liable for the transparency issues in Mexico or Hungary.

As a developing country, the Albanian government is somewhat unstable and corruption is still part of daily life. After the new territorial reforms in 2015, the powers of the Albanian municipalities and institutions have been changed.[48] The municipalities have gained more autonomy and independence from the central government. For instance, the powers that once belonged to the Ministry of Education, such as those responsible for the appointment of kindergarten and primary school teachers, have been transferred to the municipalities. The transition of the powers between ministries and local governments was not adequately communicated.

As a result, many of the employees at the local government level, particularly the municipalities in the rural areas, were unaware of such changes and lacked the necessary training to comprehend the new powers and other changes in the administration. The new territorial change in 2015 and the shrinking of the Ministries in the new government that emerged after the parliamentary elections in 2017 have brought up issues on the powers and functions of each ministry. Therefore, institutional cooperation is weak, and it needs to be strengthened in order to manage foreign investment and audit properly. It has been observed that some developing countries in the Balkan Belt and Road Initiative, such as Albania, are subject to frequent political changes and instabilities. Therefore, ensuring the longevity of the project through a special governmental taskforce can help the implementation and maintain the sustainability of the initiative[49].

47 | *Ibid.*

48 | These issues were observed during the author's personal work experience as Project Officer for the Foundation for Local Autonomy and Government (subcontracted by UNDP). During the time, the author travelled to different municipalities to conduct research and workshops for the UNDP on the current situation after the territorial reforms.

49 | Bastian, J. (2017). The Potential for Growth through Chinese Infrastructure Investments in Central and South-Eastern Europe along the 'Balkan Silk Road'. *European Bank for Reconstruction and Development*, pp. 1-62. Retrieved on November 23, 2017, from https://www.ebrd.com/documents/policy/the-balkan-silk-road.pdf.

In terms of promoting human capital, Albania faces problems of high unemployment rate and unsustainable job opportunities. The current projects executed by the BRI in Greece and, to a certain extent, in Serbia have economic spillover effects that are reasonable and sustainable over time. For example, the port expansion in Greece required high-skilled workers and created sustainable employment. However, in comparison to other countries, including Albania, Chinese infrastructure projects mainly provide low-skilled jobs such as construction, trench-digging, and truck driving[50]. The jobs are usually seasonal and not sustainable over time[51]. Often Chinese-funded projects employ Chinese labour for high-skilled jobs and leave unsustainable low-skilled jobs to the local people. Such a bargain would not be ideal for Albania.[52]. Generally, the unemployment rate is underreported because most of the population claim to be self-employed, often are working in the informal sector with no insurance or social security, and are susceptible to higher risks of job loss.

3.4.3. Identifying the spillover effects

Identifying the spillover effects in Albania and the Balkan region is challenging. It is difficult to determine the mutually beneficial spillover effects as the macro-economic data is time-sensitive and currently lacking. It is also difficult to determine whether the spillover effects would bring positive or negative outcomes, as there is no simple conclusion that the spillover effects will be mutually beneficial for both Albania and China. In addition, it is difficult to discuss and analyse the positive outcomes for countries along the Balkan Silk Road, as the spillover effects in the Balkans (excluding Greece and to a certain extent Serbia) may not be feasible and sustainable over time.

The Sino-Balkan relationship will have different implications in each individual Balkan country given the differences in economies, domestic markets, political stability, levels of corruption, and the geographical features (e.g. accessibility to port and mountainous terrain). The Chinese investment in infrastructure may only be limited highways connecting Albania with the neighbouring countries, which would mean that the

50 | *Ibid.*

51 | *Ibid.*

52 | Rapoza, K. (2016, June 13). Albania Becomes Latest China Magnet. *Forbes*. Retrieved on November 25, 2017, from https://www.forbes.com/ sites/kenrapoza/2016/06/13/ albania-becomes-latest-china-magnet/#30e4d8572490.

secondary roads that connect one city with the other may not enjoy the same level of infrastructural development. It is still uncertain if the infrastructural development led by the BRI would provide the needed infrastructure to bring the capital and the central government closer to the outskirts of the country.

Nevertheless, this project will undoubtedly bring benefit to China and Chinese companies and will also bring some advantages for the Albanian government as a partner in this endeavour. The roads and the infrastructure will be there to stay if they are built under the EU standards and may benefit the locals for years to come.

The BRI is expected to bring forth development and create new job opportunities; it will however also have some spillover effects in the region and the individual countries. China's engagement in Southeastern Europe can be seen as a proving ground for further ambitions in Western Europe either through the construction of roads and railways, or investments in harbours and energy. In this early stage of BRI development, the spillover effects would be difficult to calculate, nevertheless.

Albania, China, and the other Balkan countries have begun emerging as partners with common interests. However, from the experiences in African countries where China has been heavily funding FDI, developing countries should be aware and expect great economies like China to bring workers from China to work on the local infrastructure projects instead of giving those job opportunities to the local population. Albania had similar experiences when strong ties with China were established during the Hoxha regime, hundreds of Chinese workers came to work on Albania's development projects. Similar concerns were raised among the Albanian policymakers in regard to investments coming from China. Often Chinese investment is made so as to export its own labour to the foreign market[53]. Therefore, the Albanian government needs to ensure that the BRI project provides the means of sustainable employment opportunities. Furthermore, the multiplier effect of job creation and SME (small and medium-sized

53 | Ibid.

enterprises) growth could be expected to change and increase exponentially over time. The potential for such benefits could only be possible if Chinese companies choose to involve local experts and domestic firms, instead of bringing the majority of the workforce and qualified personnel from China. Enacting a domestic approach as a policy requirement in the contract for public tenders could be considered a conditionality, which could ensure the creation of new and sustainable jobs, and the participation of local firms in the implementation of the BRI projects.

4. Suggestions on Improving the Belt and Road Initiative in Albania

As mentioned previously, there will be a number of challenges facing the Balkan Belt and Road Initiative implementation in Albania. One of the most prominent challenges may be the risk of political instability and corruption in the countries along the Belt and Road.

For Albania to manage investments of large magnitude, it needs to implement a number of institutional changes. First, to prevent any form of corruption and funds misappropriation, the State Supreme Audit Institution should conduct tighter and more independent audits on government bodies and businesses. This would create a system of checks and balances to ensure that financial statements of the project are fair and accurate. Furthermore, different governmental institutions in Albania need to improve cooperation with one another because the sustainability of infrastructure projects, such as railways, highways, and power stations, greatly depends on continuing governmental support.

Second, ensuring the longevity of the FDI projects in general through a special governmental taskforce can help the implementation and maintain the sustainability of these projects[54]. For example, Albania took the accession to the EU very seriously and changed the Ministry of Foreign Affairs to the Ministry for Europe and Foreign Affairs. Similarly, for Albania to manage and ensure the sustainability of different projects in the future, it needs to create a governmental body or a directory under the Ministry of Infrastructure and Energy to ensure clear and transparent guidelines of business objectives in the proposed infrastructure projects. Such a directory may lead the investment negotiations, serve as a monitoring and evaluation institution, and become a contact point for future investments. This approach will focus on emphasising the credibility of China's ambitions to address the potential uncertainties of the BRI, and other projects in the future.

Another significant challenge is related to the different regulatory and legal systems in the developing countries along the Balkan Belt and Road Initiative. Those countries, including Albania, must implement proper international standards and develop the necessary legal and regulatory institutions that could oversee the bilateral agreements and ensure that all parties included in the project are held accountable in accordance with such standards. Connectivity of facilities is a key pillar of China's activities in the region of Southeastern Europe. The governments of the region must work together to foster cooperation and advance infrastructure facilities that would help promote connectivity such as highways and railroads. For such an endeavour to be accomplished, however, Albania needs to liberalise its border control and perhaps adopt an EU 'no-border area' approach within the Balkans, which can be achieved with the cooperation of other Balkan countries. Therefore, as mentioned before, the countries of the BRI should improve their connectivity and invest in developing a joint infrastructure grid that would better facilitate trade. This connectivity could be done through the creation of an international trunk corridor, which at first would focus on trade, labour, and capital, connecting the Balkan countries with Central Europe.

54 | Bastian, J. (2017). The Potential for Growth through Chinese Infrastructure Investments in Central and South-Eastern Europe along the 'Balkan Silk Road'. *European Bank for Reconstruction and Development*, pp. 1-62. Retrieved on November 23, 2017, from https://www.ebrd.com/documents/policy/the-balkan-silk-road.pdf.

The BRI in Albania would offer opportunities and create options such as the maximisation of development outcomes. At the same time, it has to ensure that the impact created in the years to come is positive and foster long-term sustainable outcomes. Managing risks is a necessary exercise for the BRI to achieve sustainable development outcomes. Therefore, to maximise the positive impact, the BRI needs to align itself with the SDG implementation as the UNDP can provide experiences and data for monitoring and evaluation of the implementation of the BRI. The UNDP has already developed close cooperation with the central and local Albanian government structures through different SDG projects. Aligning itself with the UNDP mission, as UN Secretary-General Guterres has suggested, the BRI initiative could earn trust from the government[55]. Looking ahead, the BRI is keenly anticipated in the Balkan countries, especially in Albania – which is in dire need to improve its infrastructure and energy sectors. The BRI will not only help China to establish itself as a global reference in achieving sustainable and innovative human development but will also benefit the countries along the Belt and Road in terms of infrastructure and economic development.

55 | Horvath, B. (2016). *Identifying Development Dividends along the Belt and Road Initiative: Complementarities and Synergies*. Retrieved November 22, 2017, from https://issuu.com/undp-china/docs/undp-ch-bri_2017_scoping_paper1____/9.

Aerial view of Sarajevo, Bosnia & Herzegovina in 2019.

by   TONI ČERKEZ
University of Aberystwyth

ENNA ZONE ĐONLIĆ
University of Sarajevo and University of
Bologna, MA in Democracy and Human
Rights (Class of 2020)

(Ordered by last name)

CHAPTER 3

ESTONIA

LATVIA

LITHUANIA

POLAND

CZECH
REPUBLIC

SLOVAKIA

HUNGARY

SLOVENIA

ROMANIA

CROATIA

SERBIA

MONTENEGRO

BULGARIA

NORTH
MACEDONIA

ALBANIA

BOSNIA &
HERZEGOVINA

BOSNIA &
HERZEGOVINA

I. The Politics of Bosnia & Herzegovina

Understanding Bosnia & Herzegovina's (B&H) political system is a difficult task for both academics and laymen. This country's specific socio-institutional framework is a result of its complex post-socialist, post-conflict, and multi-ethno-national[1] socio-historical experience. With that in mind, deeply analysing and understanding Bosnia & Herzegovina's political system requires a holistic socio-political analysis that cannot focus only on one aspect of state-building (i.e. constitution-making) but must account for the specificity of her history and society. In that sense, parts I and II of this chapter will seek to understand B&H's political system through a prism of a unique socio-cultural and historical experience of its peoples in combination with a plethora of academic works that understand B&H quite differently, which also indicates a lack of consensus on the nature of its politics.

Part I of this chapter provides an overview of B&H's political system, offering a perspective on B&H's political system and its legal-institutional framework. This sub-chapter aims to introduce B&H as a developing multi-cleavage political system that is ascending towards EU integration. This comes notwithstanding several institutional and political difficulties that arise from the General Framework Agreement for Peace in Bosnia & Herzegovina (commonly referred to as the Dayton Peace Accord, which established the modern B&H state. Part II will offer perspectives for the future of B&H, outline its EU and NATO path, and attempt to give a clear picture on political-institutional problems in the country by overviewing two contentious constitutional-political cases – the Sejdić-Finci Decision and the Electoral Law. Given the limitations of this paper, we must concede that this sub-chapter cannot serve as a comprehensive analysis of B&H's political system, but rather a helpful introduction to the specificity of its nature.

1.1. The Bosnian & Herzegovinian Political System

Before outlining Bosnia & Herzegovina's political system more carefully, we need to ascertain that the three main ethno-national groups in the state (that is, Bosniaks, Serbs, and Croats) are in the Constitution referred to as the *constituent*[2]

1 | Be advised, in Bosnia & Herzegovina the 'multinational' is generally referred to as 'multiethnic', however, since Bosnian and Herzegovinian *constituent* (see below) peoples are ethno-national and not purely national or ethnic groups, we shall refer to it as a multi-ethno-national state.

2 | That is, the peoples (or ethno-national groups) that are specifically recognised by the Constitution of Bosnia & Herzegovina as its constitutive ethno-national groups in demographic, social, historical, and political terms.

peoples. Those groups (that is, minorities) not recognised by the Constitution as a constituent are officially titled the 'Others'. This means that the entirety of the B&H's legal-political institutional setup is based on the parity between the three constituent groups (see below). Furthermore, according to the census of 2013, B&H is a country of 3.5 million people, out of which 50.11% are Bosniaks (predominantly Muslim), 30.7% are Serbs (predominantly Orthodox Christian), and 15.4% are Croats (predominantly Catholic), thus indicating the complex composition of her society[3].

Bosnia & Herzegovina's political system envisages B&H as a consociational state with confederal elements. This hybrid system is a direct product of the war of the 1990s wherein the territorial divisions and political semi-autonomies conferred by the Constitution reflect and capture the wartime divisions within B&H. The Bosnian-Herzegovinian political system is intended to reflect ethnic divisions and appease the delicate balance of power within the three groups. With that in mind, consociational democracy means 'government by elite cartel designed to turn a democracy with a fragmented political culture into a stable democracy'.[4]

By referring to the government as an elite cartel, Professor Emeritus Rudy Andeweg speaks of the Lijphartian principle of power-sharing amongst elites in multi-cleavage societies, which is essential for establishing stable democratic constitutionalism[5]. Therefore, we come to a more concise overview of how consociationalism functions (given by political scientist Arend Lijphart) which states that:

'First, government must include a powerful 'grand coalition' of political leaders of the different segments of society, placing a strong focus on these elites to resolve potential group disputes. Second, there is a 'minority veto' mechanism, by which minorities can block decisions that they find harmful to their identity, without needing power in numbers to do so. Third, there should be proportionality among groups in representation across several areas. For elections, the recommended electoral system is proportional

3 | Al Jazeera Balkans. (2016, 6 29). Rezultati Popisa: U BiH Zivi 3.531.159 stanovnika. [Census Results: There Are 3,531,159 Inhabitants in BiH]. *Al Jazeera*. Retrieved from http://balkans. aljazeera.net/vijesti/bih-danas-rezultati-popisa-iz-2013-godine.

4 | Andeweg, B. R. (2000). Consociational Democracy. *Annual Review of Political Science*, pp. 509-536.

5 | Lijphart, A. (2004). Constitutional Design for Divided Societies. *Journal of Democracy*, pp. 96-110.

representation (PR), so that groups are represented in proportion to their demographic percentages in the population. Divided societies that are just exiting violent conflict tend to exhibit party cleavages that automatically divide along group lines'[6].

However, the Bosnian-Herzegovinian political system is not purely consociational, it is rather a hybrid system combining different elements of democracy:

'The 1995 Dayton Peace Accord envisages Bosnia & Herzegovina as a special-type consociational state. Constitutionally it is not defined either as a federation or a confederation, but an analysis of the fundamental legal documents shows that it is a sort of an 'asymmetrical confederation' made up of two entities: the unitary Republika Srpska (The Republic of Srpska) and the multiethnic Federation of Bosnia & Herzegovina'[7].

With that in mind, we can already discern two fundamental levels of governance in B&H – that is, the central government with limited powers and two entities (see above) with state-like characteristics (such as their own parliaments, presidencies, and constitutional courts). Consequently, the state government (called the Council of Ministers) has powers over foreign policy (directed by the Presidency of the country) and trade, customs, monetary and migration policies, plus regulatory powers, as well as air-traffic control[8]. The two entities are asymmetric in their organisation. In that sense, the Federation of Bosnia & Herzegovina (FBiH, the multi-ethno-national entity) consists of 10 state-like cantons (see below), whereas the largely Serb-populated Republika Srpska (RS) consists of regions which have no genuine impact on its legislative-political constitution. Cantons in FBiH possess state-like characteristics (such as their own assemblies, governments, control over education, healthcare, and internal affairs) since their purpose is to further decentralise the power from FBiH and make it impossible for one dominant group in FBiH (Bosniaks) to take all the power in relation to Croats. These cantons are essentially viewed as: 'territorial-political units of the dominant national

6 | Stroschein, S. (2014). Consociational Settlements and Reconstruction: Bosnia in Comparative Perspective (1995–Present). *The Annals of the American Academy*, pp. 97-115.

7 | Kasapovic, M. (2005). *Bosnia & Herzegovina: Consociational or Liberal Democracy?* Republic of Croatia: Zagreb, pp. 3-30.

8 | *Ibid.*

communities.'[9] Therefore, most of the cantons, albeit not ethno-nationally homogeneous, have a discernible majority ethno-national group. The Dayton Peace Accord further outlines the creation of a special district, called district Brčko, which has its own principle of self-governance and is subject to neither of the two entities, nor the central government.

The Presidency of Bosnia & Herzegovina is a council consisting of three members, each of whom represents one ethno-national group, and has a rotating mechanism wherein one member of the Presidency assumes the Chairmanship every eight months. Bosnia & Herzegovina's Presidency respects the principle of simple majority whereby acts, if not brought under full consensus, can be promulgated by two out of three members of the body. Consequently, each member of the Presidency has a right to call upon the rule of the 'vital national interest'[10] of his own ethno-national community. This means that, if one of the members of the Presidency considers the Presidency's decision detrimental for the 'vital national interest' of his/her own group, he/she has the right to veto it. The veto is then sent to 2/3 of representatives in the National Assembly of RS (if the mechanism was invoked by the Serbian member of the Presidency) or the House of Peoples in FBiH (either to Croats in this body or Bosniaks, depending on which representative in the Presidency invoked the mechanism)[11]. If these bodies uphold the claim, the Presidency's decision takes no effect[12].

In Bosnia & Herzegovina, all state, entity and cantonal institutions are structured according to the principle of proportionality and parity[13]. With that in mind, the parliaments (National Assemblies) of the central level of government and the entities are bicameral, consisting of the Chamber of the Representatives and the House of Peoples, all of which have their own rotating presidencies (that is, a president and two vice-presidents). The House of the Representatives of B&H is composed on the principle of 'entity proportionality' wherein 1/3 of delegates come from Republika Srpska (selected in the National Assembly) and 2/3 from the Federation of Bosnia & Herzegovina (elected by Croat, or Bosniak, representatives in

9 | Ibid.

10 | 'Vital national interests are: the right of the three constitutive peoples to be represented in the legislative, the executive and the judicial bodies; the right to the preservation of their identities; the right to territorial organisation; the right to the organisation of public bodies of government; the right to education; the right to the use of language and alphabet; the right to the use of national symbols and flags; the right to the protection of spiritual legacy, especially to the nurture of religious and cultural identity; the right to the preservation of the integrity of BiH; the right to a public information system; the right to submitting amendments to the Constitution of BiH; the right to veto any issue that two thirds of the members of one national family in the House of Peoples declares to be an issue of vital national interest.' See Kasapovic, M. (2005). Bosnia & Herzegovina: Consociational or Liberal Democracy? Republic of Croatia: Zagreb, pp. 25-26.

11 | BiH, P. (2002, August 12). Ustavne i zakonske nadleznosti Predsjednistva BiH. Retrieved from Predsjednistvo BiH: http://www.predsjednistvobih.ba/nadl/default.aspx?id=18267&langTag=bs-BA.

12 | Kasapovic, M. (2005). Bosnia & Herzegovina: Consociational or Liberal Democracy? Republic of Croatia: Zagreb, pp. 3-30.

13 | Ibid.

the House of Peoples)[14].

In state-level National Assembly, the decisions are promulgated according to the principle of qualified majority wherein: 'the representatives shall put in maximum effort for that majority to include at least one-third of the votes of the delegates or members from each entity's territory'[15]. Both National Assemblies of entities implement the rule of a simple majority (except in the case of the vital national interest and other specified instances where the qualified majority is necessary) when making decisions[16].

Additionally, we must understand that both the state National Assembly and the entity National Assemblies have the power to invoke the vital national interest. Representatives (that is, 2/3 of representatives in one of the clubs of House of Peoples) and members of the presidency of the National Assemblies in their respective Houses of Peoples can invoke this mechanism[17]. The mechanism is considered effective when 2/3 of a club of a respective ethno-national component proclaim the law/decision/statement to be detrimental to their vital national interest[18]. This means that, apart from the Presidency, all three legislative bodies in two different levels of government have the veto power, thus illuminating the extent to which power-sharing in B&H tries to respect ethno-national cleavages.

1.2. The Office of the High Representative

The Office of the High Representative serves as the official representative of the international community in Bosnia & Herzegovina. Since the legal basis of functioning of B&H rests on the provisions of the Dayton Peace Accord, which was brokered largely with the help of the international community (that is, the European Community, the USA, and the UN), the OHR serves as the face of the presence of international community within the state.

Although having been recognised externally (that is, B&H has external sovereignty) and internally (internal sovereignty), B&H is often viewed as an 'international protectorate'. Indeed,

14 | Kasapovic, M. (2005). *Bosnia & Herzegovina: Consociational or Liberal Democracy?* Republic of Croatia: Zagreb, pp. 3-30.

15 / Stroschein, S. (2014). Consociational Settlements and Reconstruction: Bosnia in Comparative Perspective (1995–Present). *The Annals of the American Academy*, pp. 97-115.

16 | *Ibid.*

17 | Federacija Bosne i Hercegovine. (2017, September 9). *Ustav Federacije Bosne i Hercegovine.[Constitution of the Federation of Bosnia & Herzegovina].* Retrieved from: https://predstavnickidom-pfbih.gov.ba/bs/page.php?id=1024.

18 | *Ibid.*

the creation of the Office of the High Representative did stimulate the resentment of many within B&H towards the state. Furthermore, it caused many to refer to the state as the one merely sustained by international powers due to the organisation of the Peace Implementation Council (PIC)[19], to which the High Representative for Bosnia & Herzegovina regularly reports:

> 'The final product, the Dayton Constitution, is a complex, even inconsistent document plagued by ambiguity and highly dependent on the will of the international community. Bosnia is a cautionary tale of what to avoid because of the fundamental ambivalence between partition and integration, which has created a situation that has been sustained only by the presence of an international military force and large infusions of foreign assistance'[20].

The claim that B&H is an 'international protectorate' is further exaggerated by the fact that from the Bonn Conference in 1997 until today, the High Representative has been allowed to possess near-total power over the political process in B&H:

> 'When the OHR was created, it had no institutional base to carry out its goals. As of January 2003, it had a staff of over 700 and an annual budget of about US$20 million. Its growth in numbers matches the increasing powers it has been given in recent years, as the international mission transitioned from reconstruction to economic and political reform. After two years of 'working at the margins,' members of the PIC decided in 1997 to broaden the powers of the High Representative. The 'Bonn powers' provide the OHR with nearly unchecked power, including the authority to impose legislation and dismiss from office any public official who stands in the way of the implementation of Dayton and interethnic cooperation'[21].

In addition to chairing the Steering Board of the PIC, the High Representative for Bosnia & Herzegovina reports to the UN Security Council twice a year on the situation in B&H and her compliance with the provisions of the Dayton Peace

19 | The Peace Implementation Council was founded in December of 1995 in London, following the successful negotiation of the Dayton Peace Accord. Its main function is: 'The PIC comprises 55 countries and agencies that support the peace process in many ways – by assisting financially, providing troops for EUFOR, or directly running operations in Bosnia & Herzegovina. There is also a fluctuating number of observers.' Furthermore, the PIC has a Steering Board chaired by the OHR Representative, composed of ambassadors of nine states, the Presidency of the EU, and the EU Commission, and: 'The Steering Board provides the High Representative with political guidance. In Sarajevo, the High Representative chairs biweekly meetings of the Ambassadors to BiH of the Steering Board members. In addition, the Steering Board meets at the level of political directors twice a year.' Quotes are cited from Office of The High Representative. (2017, September 24). Peace Implementation Council. Retrieved from Office of The High Representative: http://www.ohr.int/?page_id=1220.

20 | McMahon, P. C. (2004). Rebuilding Bosnia: A Model to Emulate or to Avoid? *Political Science Quarterly,* Vol. 119, No.4, pp. 569-593.

21 | *Ibid.*

Accord. The reports are seen as an important aspect of OHR's functioning in the state as they assess Bosnia & Herzegovina's progress on its path towards the EU and NATO.

Although the power of OHR in B&H seems tremendous, its primary role is not to rule over Bosnia & Herzegovina's system or to maintain the status quo, rather it aims at sustaining stability and peace within the state and cultivating a culture of inter-ethnic cooperation and agreement. However, the office itself has become an object of stark criticism from within B&H because of its involvement in the changes of certain provisions of the Dayton Accord and its involvement in filling the House of Peoples of FBiH in the aftermath of the 2010 elections. These will be covered in Part II and it is crucial to understand them if we wish to understand the political process in B&H.

Finally, this brief overview of the basic workings of Bosnia & Herzegovina's political system serves as a solid introduction to the necessity for its reform. Bosnia & Herzegovina is a state whose political existence was not only cultivated by generations of its citizens but also a state whose political existence depends, now more than ever, on the consensus between those who have helped both tear it down and rebuild it again.

2. A Brief Overview of Bosnia & Herzegovina's Political Future

This part seeks to present current existential struggles B&H is undergoing on her path to the EU. With that in mind, this part seeks to outline two main things related to her political future: the reform of the Electoral Law (the implementation of the Sejdić-Finci verdict) and the road of B&H towards

EU and NATO integrations. Special attention will be paid to the Sejdić-Finci verdict itself as well as the much-discussed 'Croatian Issue' (due to political reasons, it often appears that these two come in a package of reform) since both are fundamental preconditions for the democratisation of the Bosnian-Herzegovinian Electoral Law. Before discussing the two main issues below, we must state that the matter this article discusses could easily change at any time as B&H has an unstable political landscape and is undergoing political reforms.

Andeweg, in his article 'Consociational Democracy', states that in consociational systems majority voting is eschewed in favour of elite compromise and deal-making[22]. In that sense, many compromises can come in the form of a package wherein each 'social segment loses on some issues and wins on others'[23]. With that in mind, the aforementioned reform package referring to the Constitution of B&H deals with the verdict of the European Court of Human Rights (ECHR) in the case of Jakob Finci (a Bosnian-Herzegovinian Jewish lawyer) and Dervo Sejdić (a Bosnian-Herzegovinian Roma activist) against Bosnia & Herzegovina. These two Bosnian-Herzegovinian citizens were denied their fundamental right of pursuing political office. In 2006, Finci wanted to run for the B&H Presidency and parliament position, much like Sejdić, and was denied that possibility by the Central Elections Commission which informed him that he could not run for the office since he was not a member of one of the three constituent ethno-national communities[24]. Sejdić and Finci then decided to pursue a case against the state and they won in 2009 when the ECHR voted in their favour. The hybrid consociational-federative Constitution of B&H (see Part I) had failed in this instance since it denied basic human rights to its citizens:

'They fought for the right to be eligible as candidates for BiH's Presidency and the House of Peoples of the Parliamentary Assembly. BiH's Dayton Constitution, however, reserves this right for individuals from one of the three 'constituent peoples': self-declared Bosniacs, Croats or Serbs. Its preamble makes a distinction between them and 'Others' (Jews, Roma, other minorities, as well

22 | Andeweg, B. R. (2000). Consociational Democracy. *Annual Review of Political Science*, pp. 509-536.

23 | *Ibid.*

24 | European Parliament. (2015, June 12). *Bosnia & Herzegovina: The Sejdić-Finci Case.* Retrieved from: http://www.europarl.europa.eu/thinktank/hr/document.html?reference=EPRS_ATA(2015)559501.

as citizens who have not declared any affiliation). 'Others' can participate in power-sharing at entity level, but not at state level. They were not taken into account since the main political institutions were designed to balance power between the direct parties to the 1992-95 war'[25].

The political elites had spent more than seven years negotiating on how to implement the decision without losing the power parity in the Presidency. It also must be noted that abiding by the ECHR's decision is a condition B&H has to fulfil on its road towards the EU. The issue with implementing the decision is that the verdict indicated deeper problems in the way the Presidency of B&H is constituted within the three constituent peoples. The verdict launched a campaign of B&H Croatian parties to overhaul the Electoral Law in two ways. The first one on the level of the Presidency of B&H and the second one on the level of the House of Peoples of FBiH. This campaign, colloquially known as the 'Croatian Issue', is a type of package proposed mainly by the B&H Croatian political parties (spearheaded by the Croatian Democratic Union of B&H-HDZ B&H) that seeks to resolve the verdict in a wider package of Electoral Law reforms that would prevent discrimination at any level within B&H institutions (currently, the Croat member of the Presidency of B&H can be elected by Bosniaks, which happened twice between 2007-2014).

The package itself was opposed mainly by Bosniaks (led by the Party of Democratic Action/SDA) and some multi-ethnic parties in B&H for two reasons. Firstly, for them, it implies the creation of Herzeg-Bosnia, a third entity mainly composed of Croats that was abolished in 1994 when FBiH was formed, in the form of a special Croat electoral district. Secondly, SDA, in particular, rejected the idea that there should be an asymmetric election of the members of the Presidency[26]. In that sense, SDA argued that if we change the way members of the Presidency are elected, then we should do it for both entities, not just for FBiH. Serbs[27], although interested in changes to the Electoral Law, were stringently opposed to a symmetric solution since it would disrupt power at home in Republika Srpska. Furthermore, they have been largely absent

25 | *Ibid.*

26 | Čavčić, I. (2017, October 12). SDA bi podržala indirektan izbor članova Predsjedništva BiH, ni partnerima u vlasti to ne smeta [The SDA would support the indirect election of members of the BiH Presidency, and its partners in government do not mind]. *Klix.* Retrieved from: https://www.klix.ba/vijesti/bih/sda-bi-podrzala-indirektan-izbor-clanova-predsjednistva-bih-ni-partnerima-u-vlasti-to-ne-smeta/171009102.

27 | The Alliance of Independent Social-Democrats/SNSD (led by the President of RS Milorad Dodik) and the Alliance for Change/SzP (a coalition of parties such as the Serbian Democratic Party) – the main opposition block in RS and an alliance that is featured in the coalition of the central government and in the Presidency of B&H (Mladen Ivanić is a member of one of the parties in SzP).

from this debate as both the implementation of the verdict and the Croatian Issue seem to fall in the scope of Croat-Bosniak relations.

Most recently, however, SDA officials have stated that they would support an asymmetric indirect election of Croat and Bosniak members of the Presidency from the Parliamentary Assembly of B&H[28]. This move found agreement with both Serbs and Croats, thus indicating a potential momentum for implementing the Sejdić-Finci verdict (by stating that members of the Presidency need not be Croat, Bosniak, or Serb) and partially solving the infamous Croatian Issue (by making it impossible for one constituent nation to elect representatives to the Presidency for another)[29].

When it comes to Bosnia & Herzegovina's road towards the EU and NATO, we must again acknowledge the complex meaning these two organisations have for her constituent peoples, especially NATO. With that in mind, the road of B&H to the EU is accepted by all major parties, across all three ethno-national groups, and has a universal significance as 'the' road B&H has to take on its way of becoming a modern and stable state. When it comes to NATO, the story is a bit different. In August 2017, during an interview with the bestselling Bosnian-Herzegovinian newspaper *Dnevni Avaz'*, Mladen Ivanić, the Serbian representative to the B&H Presidency from November 2014 to November 2018, best elucidated this complexity by stating that it makes no sense for B&H to enter NATO if the Republic of Serbia does not want to associate with it[30]. The complication arises because the Serbian official policy is to attain EU membership without NATO integration, due to the 1999 NATO bombing of Belgrade (Serbia's capital).

In that sense, regardless of the Croat-Bosniak pro-NATO axis, Serbs are fiercely against NATO integration without Serbian support for the organisation. Bosnian Serb leadership (both SNSD and SzP) is not against the activation of the Membership Action Plan (MAP) agreement and is in support of Bosnian-Herzegovinian troops participating in NATO missions, but are

28 | Čavčić, I. (2017, October 12). SDA bi podržala indirektan izbor članova Predsjedništva BiH, ni partnerima u vlasti to ne smeta [The SDA would support the indirect election of members of the BiH Presidency, and its partners in government do not mind]. *Klix*. Retrieved from: https://www.klix.ba/vijesti/bih/sda-bi-podrzala-indirektan-izbor-clanova-predsjednistva-bih-ni-partnerima-u-vlasti-to-ne-smeta/171009102.

29 | *Ibid.*

30 | Avaz. (2017, August 5). Ivanić: BiH neće uskoro u NATO [Ivanic: BiH will not join NATO soon]. *Dnevni Avaz*. Retrieved from: http://avaz.ba/vijesti/bih/288366/ivanic-bih-nece-uskoro-u-nato.

against full-fledged membership in the alliance. The NATO civilian exercise which took place around Tuzla in 2017 had the wholehearted support of the Bosnian-Herzegovinian leadership and was organised by the Ministry of Defence and the Euro-Atlantic Disaster Response Coordination Center (EARDCC)[31]. The final obstacle for Bosnian-Herzegovinian activation of MAP is the registration of all the immovable defence property in the state as state property (some of it is under the jurisdiction of entities, contrary to NATO's prescriptions), a process which was strongly opposed by politicians from RS who viewed the registration as an ingression by the state against entities[32]. However, since a landmark Constitutional Court verdict cleared the path for the registration (by stating that the immovable defence property falls under the jurisdiction of the state, not entities), the process was reinvigorated and is expected to be completed by the end of 2017[33].

Finally, when it comes to B&H-EU relations, it can be stated that they have been stalling since 2008 with intermittent periods of activity and progress. Bosnia & Herzegovina signed a Stabilisation and Association Agreement (SAA) with the EU in 2008, however, this agreement only entered into force in 2015, after a lengthy period of stalled relations due to instabilities in B&H's political system[34]. The entry into force of the SAA in 2015 was preceded by months of intense diplomatic activity under the joint British-German Initiative (that was initiated and catalysed by the Republic of Croatia under the Foreign Minister Vesna Pusić) which re-started the Bosnian-Herzegovinian EU road. While the Initiative was presented in late 2014, B&H signed the SAA in 2015 and submitted its application for EU membership on February 15th, 2016[35]. Although the two years in between were marked by increased political turbulence (not always connected to the EU process), B&H managed to comply with the EU's pre-accession demands.

In December 2016, Bosnia & Herzegovina received the Opinion Questionnaire from the European Commission, a document which serves as the foundation for the EU Council to assess B&H's membership application.[36] The Questionnaire is a standard element in EU accession procedures and acts as

31 | FENA. (2017, September 25). Počela civilna NATO vježba u BiH [A civilian NATO exercise has begun in BiH]. *N1*. Retrieved from: http://ba.n1info.com/a216655/Vijesti/Vijesti/FOTO-Pocela-NATO-vjezba-u-BiH.html.

32 | NATO. (2017, June 23). *Relations With Bosnia & Herzegovina*. Retrieved from: http://www.nato.int/cps/en/natolive/topics_49127.htm.

33 | Maksimović, D. (2017, August 17). Nakon odluke Ustavnog suda BiH - Republika Srpska (ne)knjiži vojnu imovinu [After the decision of the Constitutional Court of BiH - Republika Srpska (does not) register military property]. *DW*. Retrieved from: http://www.dw.com/bs/nakon-odluke-ustavnog-suda-bih-republika-srpska-neknji%C5%BEi-vojnu-imovinu/a-40130651.

34 | European Commission. (2016, December 6). *European Neighbourhood Policy and Enlargement Negotiations*. Retrieved from europa.eu: https://ec.europa.eu/neighbourhood-enlargement/countries/detailed-country-information/bosnia-herzegovina_en.

35 | *Ibid.*

36 | European Commission. (2018, April 17). Key findings of the 2018 Report on Bosnia and Herzegovina. Retrieved from https://ec.europa.eu/commission/presscorner/detail/en/MEMO_18_3408.

an assessment of states' institutional capacity to deal with the EU reforms. In February 2018, B&H delivered the answered Questionnaire to the EU.

In conclusion, this brief overview of Bosnia-Herzegovina's political future emphasises the dyad of challenges it faces: internally (the Sejdić-Finci and the Croatian Issue), and externally on its EU and NATO integration road. This process will be complex and, judging by recent experiences, quite turbulent, but it appears that Bosnia & Herzegovina can withstand its challenges, albeit less efficiently than many other states. All that is left is to hope that she will be able to reform in order to progress.

3. An Overview of the Challenges and Opportunities for the Youth in Bosnia & Herzegovina

3.1. Economic, Political and Social Challenges for the Youth in Bosnia & Herzegovina

In economic terms, the youth in Bosnia & Herzegovina (B&H) are the most vulnerable group. The future of the country's youth is tied in with the stagnating socio-political and economic construct that is the Dayton Peace Accord. The future of the youth in B&H seems grim indeed, especially given the youth unemployment data (see below). However, their future is not lost and there are still ways to extricate them from the socio-economic abyss.

This part will examine three main issues. First, the problematic economic prospects for youth in B&H, with many young people leaving the country given the high unemployment rate and non-diversified economic sectors, in combination with a traditionally socialist economic worldview (categorical, stringent economic development) and the legacy of the 90s war. Second, the Bosnian-Herzegovinian path to the EU, which could introduce important reforms that could animate its youth not only economically but also socio-politically (thereby making them a stronger part of the civic society). Finally, the social alienation of youth from political participation and an increasing emigration rate are making it harder for youth to be present in the public sphere.

As mentioned above, the economic prospect of youth in Bosnia-Herzegovina is deeply problematic. Examining statistics from 2005-2014 shows that the youth unemployment rate in B&H (age range: 15-24) grew from around 53% to 58%[37]. However, according to the World Bank, this number went as high as 64.3%[38]. Compared to the global average of 13.8%, we can see that B&H's youth unemployment rate is very high[39].

The economic issues are primarily related to the poor state of B&H's economy, the social challenges youth face in B&H in connection with the social alienation produced by the political system and economy, as well as the prevalence of emigration in the society. According to the Ministry of Civil Affairs of B&H, political participation of young Bosnia-Herzegovinians in the country is strikingly low. The alienation from the system due to the lack of strong social programmes and laws instituted by the various levels of government in the country, in combination with poor economic prospects, is taking its toll on B&H's youth. For instance, only 1% of youth in B&H are engaged in different youth representative bodies (such as youth councils), 5% are members of youth organisations, and only 6% are members of various political parties[40]. Combining this data with the recently published data stating that in the period from 2013 to 2017, around 150,000 people left B&H and moved elsewhere, the state of youth in the country is strikingly problematic[41]. It is not surprising then that almost 14% of Bosnian-Herzegovinian GDP comes from outside the country by way of remittances[42].

Bosnia & Herzegovina is in a dire economic situation and its youth face great challenges. However, with some economic

37 | Ali, A., Berahab, R., Boot, N., Wolff, G. B., et al. (2016). *Seven Years after the Crisis: Intersecting Perspectives.* OCP Policy Centre. Retrieved from https://www.bruegel.org/wp-content/uploads/2016/08/Livre-Web-Seven-Years-After-the-Crisis.pdf.

38 | World Bank. (2017). *Unemployment, youth total (% of total labour force ages 15-24) (modelled ILO estimate).* Accessed October 30, 2017. Retrieved from: https://data.worldbank.org/indicator/SL.UEM.1524.ZS.

39 | *Ibid.*

40 | Ministry of Civil Affairs of Bosnia & Herzegovina. (2017). *Podaci o mladima u BiH [Data on youth in BiH].* Accessed October 30, 2017. Retrieved from: http://www.mladi.gov.ba/index.php?option=com_content&task=view&id=46&lang=hr.

41 | Klix.ba. (2017). *Od 2013. godine do danas iz BiH iselilo 150 hiljada ljudi [From 2013 until today, 150 thousand people have emigrated from BiH].* Retrieved from: https://www.klix.ba/vijesti/bih/od-2013-godine-do-danas-iz-bih-iselilo-150-hiljada-ljudi/171030011.

42 | UNDP Bosnia and Herzegovina. (n.d.) *Diaspora for Development (D4D) Project.* Retrieved from: https://www.ba.undp.org/content/bosnia_and_herzegovina/en/home/development-impact/Diaspora4Development.html.

recovery since 2013 and the Bosnian-Herzegovinian road towards Euro-Atlantic integrations, there is a renewed sense of optimism, albeit a weak one, in the improvement of the situation. Notwithstanding all that, the first step in creating a better environment for youth in Bosnia-Herzegovina is tripartite, that is: fighting the endemic, systemic corruption; improving the country's social policies (including education); and reforming its outdated political system.

The first aspect is especially hard, given that corruption in B&H is a problem endemic to the system of organised political elites. According to Transparency International's Corruption Perceptions Index for 2016, B&H is in the group of highly corrupt states. On a scale from 0-100 (0 indicating most corrupt and 100 least corrupt states), B&H is assigned the number 39[43]. Political elites and their cronies are engaged in largely small-scale corruption, but many scandals related to high-scale corruption have hit the country. For example, the former Minister for Communications and Traffic at the state level of B&H (Social-Democratic Party of B&H) was convicted for abuse of official position and high-level corruption after damaging the budget of the municipality of Novi Grad (Sarajevo) for 1 million BAM (Bosnian Convertible Mark – national currency)[44]. Although everyone is aware of corruption in B&H, people generally do not act against it. Reporting corruption is, unfortunately, rare due to fear of repercussions.

Reforming the outdated political system is also a deeply problematic issue. The Elections Law, as discussed in previous parts, is an issue of great contention and the efforts to change it seem to be at a standstill and will most likely have to be pushed for by the international community (the EU in particular). While the system itself currently works – laws are passed, budgets are enacted – parts of the Electoral Law pertaining to elections of the members of the Presidency and assembling the House of Peoples in FBiH are deemed unconstitutional and will cause serious disturbance to her political-economic development if not altered. Reforming the outdated political system in B&H will not be possible without first reforming her Electoral Law and then focusing

43 | Transparency International. (2017). *Corruption Perceptions Index 2016*. Retrieved from: https://www.transparency. org/news/feature/corruption_ perceptions_index_2016.

44 | N1 Televizija. (2017, January 19) *Damir Hadzic pusten iz pritvora [Damir Hadzic released from custody]*. Retrieved from: http://ba.n1info.com/a133353/ Vijesti/Vijesti/Damir-Hadzic- pusten-iz-pritvora.html.

on implementing EU-delegated reforms as a part of the EU integrations.

Finally, when it comes to education and social policies in the country, B&H faces a very complex situation. On one hand, education in FBiH is divided between cantons with a Federal Ministry having only nominal power. On the other hand, education is often used to entrench ethno-political divisions instead of building a multi-layered national identity. With that in mind, the fact that all cantons use their own high-school curricula, that are adjusted to ethnic-dominated areas without inter-canton-coordination, presents a big problem for wholescale reform. While cantons can be efficient when dealing with local-level politics and can be breeding grounds for B&H's democratic development, the disparities between cantons in their legislation and curricula are striking. For example, the West-Herzegovina Canton (predominantly Croat) uses an adjusted curriculum from the Republic of Croatia and respects Croatian school holidays. This is but one example of systemic divergences regarding education. Another example is from the RS, where Bosniak students are denied their right to call their language 'Bosnian'. Instead, they have to use 'the language of the Bosniak people' on official school transcripts.

The next two sections will outline opportunities for youth in B&H, but it is of utmost importance to understand these opportunities in the context of problems the country and the society have. While the youth have the ability to be connected to European-scale economic developments and opportunities, they often have constrained access to such opportunities at home. Hopefully, the economic growth B&H has been experiencing will be more visible in the short-term and will open more opportunities to her youth.

3.2. The Reform Agenda and Opportunities for the Youth in Bosnia & Herzegovina

Referring to the previous chapters regarding the politics of Bosnia & Herzegovina, one may notice a trend that has developed in recent years. This trend is connected to the

45 | International Monetary Fund. (2017). *Bosnia & Herzegovina.* Retrieved from: www.imf.org/en/Countries/BIH.

46 | Čolaković, V. (2017). *Siva ckonomija čini 25 posto BDP-a BiH.* October 30. Accessed October 30, 2017. http://ba.nlinfo.com/a223736/Vijesti/Vijesti/Siva-ekonomija-cini-25-posto-BDP-a-BiH.html.

47 | Ministry of Civil Affairs of Bosnia & Herzegovina. (2017). *Podaci o mladima u BiH* (The grey economy accounts for 25% of BiH's GDP). Retrieved from: http://www.mladi.gov.ba/index.php?option=com_content&task=view&id=46&lang=hr.

48 | The Reform Agenda was signed in 2015 as a part of the reinvigorated EU approach to B&H. It calls for wholesale economic and social reform of the country, thus putting focus on economic growth and reform instead of wholesale political reform in order to make B&H better prepared for upcoming EU integration process. This has had positive impacts on B&H's economy and society, although the undertaking itself is stalling due to the increasing politicisation of her public sphere and the focus of political elites on ethno-national issues, instead on economic and social policies.

49 | Ministry of Civil Affairs of Bosnia & Herzegovina. (2017). *Podaci o mladima u BiH* (The gray economy accounts for 25% of BiH's GDP). Retrieved from: http://www.mladi.gov.ba/index.php?option=com_content&task=view&id=46&lang=hr.

modernisation of the country and its society on the road to Euro-Atlantic integrations. The result of the reinvigoration of B&H's road to the EU is the Reform Agenda (see below) and the growing interest of the international community in the development of her economy has brought increasing optimism to the country. Positive economic growth further increases this optimism[45]. However, given the sluggish economic-political development of B&H and the fact that there is a considerable 'grey market' (comprising 25% of the GDP), many of the positive economic trends developed cannot be seen[46]. Furthermore, political corruption is identified as one of the biggest reasons for poor youth development in the country[47]. Poor connection between higher education and industry is also identified as problematic. This is especially important given that in B&H, education is divided into multiple levels of governance that also entrenches ethnic division from elementary to high school.

Changes that are encouraging, however, are the implementation of the Reform Agenda[48], which is modernised and harmonised with the EU standards, as well as the greater involvement of the EU in pushing B&H elites to prioritise more efficient governance and reform in light of future EU integrations. The latter development has had a great political impact on the country. For example, the Ministry of Civil Affairs identifies Europeanisation of Bosnia-Herzegovina's system of governance as one of the main conditions for further development of the youth sector, thus confirming the need for further reformation of the state apparatus[49]. The former (i.e. the Reform Agenda) has yielded results as well, albeit being a politically contentious issue at times. B&H's economy is growing, and this is largely due to the progress of the Reform Agenda and the fact that B&H now shares a border with the EU following Croatia's accession to the EU in 2013. However, the inefficiency of her political system is seriously hindering any chance of more visible economic improvement, which directly impacts youth opportunities in the country.

If the reforms are implemented, and there is an indication to believe this would happen, then we can see a more prosperous future for B&H's youth. However, if the inefficiency of the system prevails and if the reforms are not conducted in a consistent and continual manner, then the future of the country and its youth remains bleak. Therefore, B&H's participation in international initiatives such as Belt and Road Initiative can prove to be fundamental for the improvement of the economic situation. In order to reap the benefits of BRI, her political establishment must produce necessary reforms in order to make the country more visible to international investors and not wait for international economic initiatives to anticipate B&H's economic development. In the next part of this research paper, the opportunities for B&H's youth are investigated more closely and further contextualised with the developments within and without the state, and in relation to the BRI.

4. The Belt and Road Initiative and its Opportunities for Young People in Bosnia & Herzegovina

Opportunities are what Bosnia & Herzegovina and its citizens were looking for from the moment the country gained independence. B&H needs new possibilities and the Belt and Road Initiative seems to be offering exactly that. In order to better understand and elaborate on the opportunities and challenges of the BRI within B&H, certain basic provisions of the BRI must be understood.

The target of the BRI is a multipolar world characterised by wide regional cooperation that embraces cultural diversity, economic globalization, open economy, and the regime of free trade. The Initiative not only offers to lend provisions for cooperation along the so-called 'Silk Road', but it also promises to help strengthen the cooperation between neighbouring states by connecting them with an open, inclusive, and well-balanced economic partnership that will suit each and every country[50]. The Initiative is all-inclusive to both countries and organisations and it seeks to provide mutual benefits.

The BRI is meant to focus on five key areas: policy coordination; infrastructure development; investments and trade facilitation; financial integration; and cultural and social exchange. Accordingly, its successful implementation will lead to an increase in foreign direct investments (FDI), economic growth, increase in trade and public infrastructure, development of markets, and eventually the provision of grounds for prosperity and new jobs[51]. It is obvious that these areas and their outcomes represent direct opportunities. For Balkan countries in general, and B&H in particular, it is quite important to take part in the BRI as it will help embed China in their economies, provide needed additional economic partnerships that will directly influence economic growth, and boost the connectivity between neighbouring countries, a necessary development given the historical context and current relations between said countries[52]. The BRI will help Bosnia & Herzegovina to strengthen relations with China politically, economically, and culturally. The embedding of China into B&H's economy and eventual economic cooperation will significantly boost its economy[53]. Moreover, the BRI does not only raise the possibility of cooperation with China but also with all other states that are part of the Initiative.

Creation of new jobs has proven to be a problem for the government of B&H due to the lack of economic strength and resources at their disposal[54]. With the implementations of the BRI and the opportunities it offers, the government would be able to provide new job opportunities within a stronger market, boosted by FDI. Stemming from this, other opportunities may arise for B&H, namely infrastructure innovation and investment in bridge-building or highway construction, while at the same time, the country's transport and logistics networks will improve significantly. Policy coordination as one of the key factors should also be considered as an opportunity, at least for the countries in the Balkans, since it will bring them much-needed modernisation and harmonisation with international

50 | Klein, B. A., & Kwok, S. (19. September 2017). *China's 'One Belt, One Road' Initiative Creates Opportunities and Regulatory Challenges*. Retrieved from: https://www.lexology.com/library/detail.aspx?g=a6826af9-a2ea-42c0-bef6-37cf88a97879.

51 | Wolff, P. (2016). *China's 'Belt and Road' Initiative – Challenges and Opportunities*. German Development Institute. Retrieved from: https://www.die-gdi.de/uploads/media/Belt_and_Road_V1.pdf.

52 | Chan, L. (2017, 15 August). *Bosnia & Herzegovina: Market Profile*. Hong Kong Trade Development Council. Retrieved from: http://china-trade-research.hktdc.com/business-news/article/The-Belt-and-Road-Initiative/Bosnia-and-Herzegovina-Market-Profile/obor/en/1/1X000000/1X0A3IWM.htm.

53 | For more information about economy of Bosnia & Herzegovina visit Central Bank of Bosnia & Herzegovina, see http://www.cbbh.ba/?lang=en.

54 | Ministry of Civil Affairs of Bosnia & Herzegovina. (2017). *Podaci o mladima u BiH* (The gray economy accounts for 25% of BiH's GDP). Retrieved from: http://www.mladi.gov.ba/index.php?option=com_content&task=view&id=46&lang=hr.

trends. It will also lead to better and easier cooperation with China directly, as well as with other members of the Belt and Road Initiative, and beyond. Economic growth, investment and increase in trade are factors every state in the world wants to see developing positively, and it is no different for B&H, which has struggled for years to maintain sustainable economic growth and achieve incremental growth in trade, or at least sustain it at a positive level.

Cultural and social exchange is extremely paramount for societies and citizens of the countries that are involved in the Initiative. These exchanges may range from academic visits to tourism. Academic visits would be quite easy to achieve through the coordination and cooperation of universities. These visits will greatly benefit the perspectives and views of academics and students, with special emphasis on the possibility to experience the culture of other countries directly. Tourist visits will, to put it simply, help citizens of one country to experience the culture and traditions of another country directly through personal exploration. Tourist visits will be easier to organise once the no-visa policies are regulated for the benefit of all sides – China and Bosnia signed the no-visa policy agreement on November 27, 2017, in Budapest[55].

Transparency, rules-based public tenders, and reciprocal market access are some of the challenges that are being addressed[56], especially with regards to the Initiative implementation in the Western Balkans. Some of these challenges will be applicable to Bosnia & Herzegovina (B&H), and there are other hurdles arising due to a variety of reasons such as political incompetency or unwillingness to commit to necessary reforms and policy implementations, as well as the lack of understanding and necessary knowledge to name a few.

Opportunities set out by the BRI are of significant importance for B&H as a country. However, given its history of lengthy policy implementation, it is reasonable to expect that certain policies may take several years to come into effect, which would indirectly influence the eventual outcomes and success of the implementation[57]. Eventually, continuation and the

55 | Fokus. (2017, November 27). BiH i Kina potpisale sporazum o ukidanju viza [BiH and China sign visa waiver agreement]. *Fokus*. Retrieved from: https://www.fokus.ba/vijesti/bih/potpisan-sporazum-gradjani-bih-mogu-u-jos-jednu-zemlju-bez-viza/931914/.

56 | Bennett, V. (2017, September 11). *What China's 'Belt and Road Initiative' means for the Western Balkans*. Retrieved from: http://www.ebrd.com/news/2017/what-chinas-belt-and-road-initiative-means-for-the-western-balkans.html.

57 | *Ibid.*

amount of the time spent on the implementation of the Reform Agenda in B&H will have a direct influence on the challenges faced and possible outcomes of the BRI.

The implementation of the BRI in B&H carries with it numerous opportunities for young people. These opportunities are directly drawn and derived from the five major opportunities the Initiative offers to the country. For example, modernisation and adaptation of the public institutions will most probably lead to the development of stronger markets and economic growth which will eventually result in the creation of new jobs.

There is a place for young people in each and every part of the implementation of the BRI in Bosnia, as they might be part of the solution for a better and timely implementation of the Initiative. Young, bright, well-educated and motivated people will be offered an opportunity to exercise their expertise and share their opinions on certain policies – they will have a channel to submit their proposals and ideas – as more and more young entrepreneurs will be able to contribute to the market and find themselves and their businesses in a better position within the market. Young entrepreneurs and businessmen will benefit largely from the modernisation of public institutions, the adaptation of new policies and the eventual opening-up of the market which will help them construct and widen their network of possible partners.

Most importantly, as mentioned before, successful implementation of the Belt and Road Initiative will lead to the creation of new employment opportunities, as it will widen the market and open possibilities for new actors, particularly from China, to be present and to invest in B&H. The formation of new jobs is of utmost importance for the young population in B&H. In the past few years, Bosnia has been struggling to cope with the scarcity of talent, as young and well-educated people are leaving the country on a quest for better opportunities and job prospects[58]. The presence of new actors in the market and increased investment would lead to a better perception of opportunities in B&H and will give the young population there hope that they can find the job they seek within their country.

58 | Latal, S., Ahmetasevic, N., Jegdic, Z., & Panic, K. (2015, May 18). *Mass Depopulation Threatens Bosnia's Future*. Retrieved from: http://www. balkaninsight.com/en/article/ mass-depopulation-threatens-bosnia-s-future.

Cultural and social exchanges are one of the greatest opportunities for the young people of B&H. These exchanges can include previously mentioned academic visits, cultural events, social activities, and tourist visits. They would be to the benefit of youth from all the countries involved in the BRI, including Bosnia and China. Students and young academics who take part in the academic visits will be able to learn from their peers from other countries through the experience of education, research and academic opportunities in completely new settings. They will be able to benefit exponentially from experts who are knowledgeable and well-known in their fields.

These academic exchange programs will help young academics and students to become better researchers and global citizens. They will be able to gain knowledge from their colleagues, as they will possibly trigger different understandings of certain events through their unique perspectives and evaluations. Youth will be able to build fresh perspectives and contribute to the whole picture, with some integral segments being completed and added to by colleagues from different countries. Cultural exchanges will help young people connect and understand each other better. It will also allow them to hone their skills and attitudes, which would aid them in understanding the culture of another country better and eventually find success in their endeavours.

Youth education, skills training and research, as well as development opportunities, will be held on a much bigger scale due to the BRI. Education and research in different countries as well as in various fields will be much more accessible to the youth of B&H. As this Initiative has a long-term goal, it also requires long-term commitment, buttressed with valuable ideas and support from youth. In fact, youth from each and every country in the BRI will help to connect countries and foster deeper communication on all levels.

Young people who have spent time abroad have experienced living in well-developed countries and have been witness to how they implement policy changes. They will be of great help to Bosnia & Herzegovina. Direct experience of good practices of policy implementation will help the youth of B&H to connect their knowledge with experience and use it in practice to help Bosnia achieve its goals. Young people with their knowledge and experience of different cultures and countries in combination with the understanding of how those countries operate and implement policies will be key players in the policy harmonisation process. Hence, Bosnian and Herzegovinan youth should be directly included in policy creation, harmonisation and implementation.

Challenges are an integral part of Bosnia's day-to-day politics. It must be acknowledged that the challenges that Balkan countries, in general, and B&H, in particular, face have become transnational in character. B&H's neighbouring countries face similar problems – such as the lack of democratic capacity and marginalisation of civil society[59]. A majority of the challenges B&H faces are directly related to the political system in place in the country, as well as the competency knowledge of politicians in leading positions[60].

Challenges outlined in achieving the BRI eventually apply to B&H directly, in addition to other pre-existing ones. B&H will need to hire new experts in the field to help design new policies and plan their implementation within the country. Mutual agreement on the employment of said expert(s) will be difficult to achieve due to the structure of the political system in B&H, as well as its prescribed functioning by the Constitution[61]. Similarly, reaching a consensus on the creation and implementation of the needed policies will be quite challenging, as there will again be the need for agreement among the three constitutive peoples of B&H and its respective entities. Accordingly, the slow implementation of policies and the Initiative's key points is inevitable due to the complex structure of the bureaucracy in Bosnia & Herzegovina[62].

Last but not least, we need to bear in mind that in order to implement the policies and enact the provisions of the Initiative, the government of B&H will need young, well-educated, skilled and motivated people. The role of young people is crucial to the development and prosperity of B&H in general, and for

59 | Đonlić, E. Z. (2015, February 7). *My Contribution to Regional Reconciliation –'Heal the Region.'* Retrieved from: http://iea. rs/2015/02/07/my-contribution-to-regional-reconciliation-heal-the-region/.

60 | Đonlić, E. Z. (2016, October 6). *Bosnia & Herzegovina on the path towards the European Union.* Retrieved from: https:// europeanstudentthinktank. com/2016/10/06/bosnia-and-herzegovina-on-the-path-towards-the-european-union/.

61 | For more information about the political system in Bosnia and & Herzegovina, see Karić, M. (2016). *Consociational democracy in theory and practice: the case of post-war Bosnia and & Herzegovina.* Sarajevo: International University of Sarajevo.

62 | Manning, C. (2004). Elections and political change in post-war Bosnia & Herzegovina. *Democratization,* 11 (2), pp. 60-86.

the BRI in particular. Younger generations possess skills and knowledge, but most often, they lack experience. Accordingly, governments should provide their younger generations with the possibility of practising their knowledge and expertise under the supervision of persons who have been working in the same field for a long time. They should cultivate channels for the young population to give their opinions and express their ideas, which will possibly help Bosnia & Herzegovina grow in the right direction.

5. The Importance of the Belt and Road Initiative to Bosnia & Herzegovina

This final part of this chapter might sound both pessimistic and optimistic at the same time, as it will elaborate on why the Belt and Road Initiative (BRI) is important for Bosnia & Herzegovina (B&H) and why it might be Bosnia's only choice for achieving development, stability and prosperity.

For quite a while, B&H was in search of initiatives and provisions that would help the country embark on the path towards stability and prosperity. Since its political system made reaching any agreements an arduous and a complex endeavour[63], B&H needs policy changes and new initiatives that will change its economic system and lead to much-needed economic growth. For B&H, there is no alternative for either the European Union or the BRI. From a political point of view, it is undesirable to have no alternatives as the sole option available might turn out to be the wrong one. But in this case, the situation is quite different[64]. For a while, B&H politicians were not able to initiate any kind of reform on their own, as they got used to relying on help from the international community and their preparation of initiatives and mechanisms for its implementation. The BRI may

63 | For more information consult Karić, M. (2016). *Consociational democracy in theory and practice: the case of post-war Bosnia and & Herzegovina.* Sarajevo: International University of Sarajevo.

64 | Đonlić, E. Z. (2015, February 7). *My Contribution to Regional Reconciliation – 'Heal the Region.'* Retrieved from: http://iea.rs/2015/02/07/my-contribution-to-regional-reconciliation-heal-the-region/.

be an external enterprise, but at the same time, it requires full commitment from all 'participant states'. The political leadership in B&H has seen the importance of the implementation of the Initiative and committed itself to the timely implementation of necessary policies. A great example of the commitment and dedication is the signing in 2017 of the agreement on Agricultural cooperation between the Ministry of Foreign Trade and Economic Relations of Bosnia & Herzegovina and the Ministry of Agriculture of the People´s Republic of China. This agreement has started the process of intensive communication between institutions and business communities in the field of agriculture and rural development in both countries[65].

As discussed in this chapter, the BRI brings a lot of opportunities for the country in general and its economy in particular. Bosnian politicians, as well as the international community, are aware that B&H needs changes immediately. In that regard, it is significantly important for B&H that it becomes a part of the BRI.

Policy harmonisation and incorporation that are part of the BRI are necessary for changes to occur in B&H, particularly in regard to the constitutional reform. B&H can move closer to policy harmonisation through the careful examination of current policies and their comparison with policies in countries which proved to be excellent in that sector and policy implementation. In the case of missing policies, the government should examine successful countries and employ experts who will help create policies in a way that they will fit the Bosnian political system yet be successful and harmonise with pre-existing policies in the region. Reform of policies regarding the openness and equality of the market, trade and the economy in general, will largely depend on and be positively influenced by the BRI. So far, economic growth has been neither easily achievable nor always positive for B&H. Therefore, the best solution for B&H is to implement policies and build up infrastructure, both essential parts of the BRI.

Accordingly, for former Yugoslavian countries (including Bosnia & Herzegovina), there is a need for policies and mechanisms

65 | Fokus. (2017, November 27). BiH i Kina potpisale sporazum o ukidanju viza [BiH and China sign visa waiver agreement]. *Fokus*. Retrieved from: https://www.fokus.ba/vijesti/bih/potpisan-sporazum-gradjani-bih-mogu-u-jos-jednu-zemlju-bez-viza/931914/.

that will help them achieve a higher level of cooperation and allow them to share good political practices. Ideally, they could create some kind of regional cooperation mechanism through which they will be able to share ideas and expertise that will help the development of each country individually and the region as a whole. The BRI will help each country individually but will also help nurture stronger relationships between former Yugoslavian countries and other countries that are a part of the Initiative. Policy incorporation and regional cooperation through the Belt and Road Initiative will prove to the Western Balkan electorates that their countries are connected and that only by cooperation will development, stability and prosperity be possibly achieved. The BRI will be the first step towards higher cooperation between former Yugoslavia countries outside the European Union. Cooperation through the Initiative might open new opportunities for young experts and researchers to use their skills and knowledge to help countries of former Yugoslavia to prosper. It might also be the first step towards broader cooperation between the Western Balkan countries.

Bosnia needs this process and its good practices in order to be able to provide a better future for its citizens and finally bring stability and prosperity that will last. Bosnia should strive to be a well-developed, structured and organised country with various opportunities for its citizens, and it may be able to achieve that through the BRI.

The above analysis has highlighted the many challenges and opportunities the BRI brings to Bosnia & Herzegovina. Opportunities show us direct reasons why B&H should fully commit to the implementation of the Initiative. Challenges do not represent reasons why the Initiative should not be implemented. Rather, they represent an outline of the problems the country may face in its path towards the successful implementation of the BRI. For the sake of B&H, it is important to find experts and knowledgeable individuals to help resolve the issues and overcome challenges in the future. Preparations for all scenarios must be made and challenges have to be dealt with in a timely manner. Early identification of said challenges and adequate addressing will significantly increase the success of the implementation of the Initiative. Bosnia & Herzegovina should form a panel or a commission that will be in charge of identifying all the challenges and problems the country might face in its bid to implement the Belt and Road Initiative. That commission would also be in charge of finding intuitive solutions, involving knowledgeable and skilful youth who will be able to create long-term plans and strategies which will allow for economic growth and prosperity.

Policymakers of countries in the Balkan region should not only focus on the implementation of policies that will better their country individually, but also take into consideration the greater picture and commit themselves to work for the good of the whole region and its development through the BRI.

Alexander Nevsky Cathedral in Sofia, Bulgaria in 2019.

by   ALEXANDER GEORGIEV
University of Amsterdam

MARK GEORGIEV
Erasmus University Rotterdam

NIKOLAY NEDKOV
Helsinki University of Technology

MIHAIL TSVETOGOROV
University of Glasgow

(Ordered by last name)

CHAPTER 4

BULGARIA

ESTONIA

LATVIA

LITHUANIA

POLAND

CZECH
REPUBLIC

SLOVAKIA

HUNGARY

SLOVENIA

CROATIA

ROMANIA

BOSNIA &
HERZEGOVINA

SERBIA

MONTENEGRO

NORTH
MACEDONIA

ALBANIA

BULGARIA

The research for the case study of Bulgaria within the framework of 'Young Belt and Road' will be separated into four different parts – firstly, the political overview which was conducted by Alexander Georgiev following secondary research in order to present contextual information which will be important for subsequent analyses. Secondly, the expert opinion of Nikolai Nedkov was utilised to make predictions on Bulgaria's future. Thirdly, critical research was done by Mark Georgiev regarding the opportunities and challenges facing Bulgarian youth. On a final note, interviews involving different age groups and occupations were carried out by Mihail Tsvetogorov in order to analyse the potential involvement of Bulgaria in China's Belt and Road Initiative (BRI).

This chapter will first present the relevant historical implications of Bulgaria's political past to make valuable predictions regarding a potential change in the political status quo and the economic opportunities for the Bulgarian youth and for the national state of affairs in general.

1. Historical Overview and Implications

In order to fully understand the contemporary potential opportunities available to youths in Bulgaria with regard to the BRI, it is important to understand and acknowledge the historical development of the current Bulgarian political system and its implications. The history of Bulgaria consists of significant developments following its liberation from Ottoman rule to the modern-day. A historical overview of essential events – the country's involvement in the wars in the first half of the 20th century, communist rule, communism's fall, the transition to a constitutional liberal-democracy and their implications will be discussed. Then, inferences regarding the contemporary government system and constitution will be presented.

Following the five-century-long Ottoman yoke, the Russo-Turkish Wars and the Treaty of San Stefano (1878), Bulgaria was freed. The treaty was made favourable for Bulgaria in terms of giving it borders that accessed two seashores – the Black and the Aegean Seas. However, it was limited just a few months later by the Bismarck-led Congress of Berlin due to a fear of the emergence of a strong nation within the Balkan Peninsula[1]. This significantly cut the Bulgarian territories and was immensely consequential for the subsequent foreign policy decisions of the country. The newly independent Bulgaria engaged in the First Balkan War in 1912, hoping to benefit from the weakened Ottoman Empire. However, the territorial success was offset by the atrocious effects of the Second Balkan War (1913) against the former allies from the previous war. As such, Bulgaria was left with a weakened military, a frail internal political state, and unfortified borders[2].

These events had seminal influence over Bulgaria's involvement in the following two World Wars. During World War I, an inevitable partnership was established with Germany in order to reacquire the territories lost to Serbia during the Second Balkan War[3]. Following the defeat of the Central Powers, the Neuilly-Sur-Seine Treaty during the Paris Peace Conference in 1919 crippled Bulgaria economically and territorially, significantly affecting national self-esteem[4]. During the Interwar Period, there was political turmoil due to the power vacuum created by the disappointing defeat in World War I. The turmoil escalated when Bulgaria was threatened by both the USSR and Nazi Germany in 1940 before the signing of the Tripartite Pact in 1941, making Bulgaria an ally of the Axis Powers[5]. The subsequent defeat in World War II and the political state of affairs in that period has influenced contemporary political condition in Bulgaria to a great extent.

In 1946, a referendum was conducted to reject the monarchy and to establish a people's republic[6]. This was the result of continuous efforts of the Bulgarian Communist Party to

1 | Kosev, K. (2016). Bismarck Forces Russia to a War with Turkey. *Epicenter*. Retrieved from: http://epicenter.bg/article/Bismark-tlaska-Rusiya-kam-voyna-s-Turtsiya/94289/11/33.

2 | Hall, R. C. (2000). *Balkan Wars, 1912-1913: Prelude To The First World War (Warfare And History)*. Routledge.

3 | Tsanev, S. (2008). *Bulgarian Chronicles Vol. III*. Sofia: Trud, p. 427.

4 | Tsanev, S. (2008). *Bulgarian Chronicles Vol. III*. Sofia: Trud, p. 489.

5 | Filov, B & Dimitrov, I.I.. (1990). *Dnevnik* (Diary). Sofia: Izd-vo na Otechestveniia front, p. 221.

6 | Curtis, G. (2013). *Bulgaria: A Country Study*. CreateSpace Independent Publishing Platform, p. 176.

overrule the monarchy and attempt to establish a Soviet-like communist state[7]. The latter was executed by the adoption of a new constitution in 1947, which vastly resembled the Soviet one from 1936[8]. The following period was a 42-year long era of communism with the shifting powers of Georgi Dimitrov, Valko Chervenkov, and Todor Zhivkov (the three general secretaries of the Bulgarian Communist Party in that period); with Zhivkov being the last head of state before the fall of the regime in November 1989. One important aspect of the regime was the abolishment of governmental institutions, giving rise to the 'unity of state power'[9]. The possibility for private property was present if it did not serve as a 'detriment to the public good'[10].

Purges against political rivals occurred and churches were placed under the control of the state. However, in 1971, a new constitution was drafted under Zhivkov, which established a governmental body with the utmost power within the state – the State Council, with Zhivkov as Chairman and Head of State. The council could create and approve legislation and could focus on the ultimate power of the Bulgarian Communist Party, as opposed to 'the state' in the previous constitution[11].

Towards the end of the 1980s, liberal movements propelled by Mikhail Gorbachev in the USSR were mimicked by some Bulgarian officials, in the same vein of Bulgarian officials copying the Soviet-style governance model[12]. Short appointments of new prime ministers in a pseudo-democracy followed in the next two years until the first democratically elected non-communist President in 1990 – Zhelyu Zhelev. With Zhelev's election, a new liberal-democratic constitution was drafted and adopted in 1991, which represents the modernistic and liberal values upheld in Bulgaria.

The new constitution centred around Western liberal values, meaning the core aspects of communism had to be abolished. Private property rights are inviolable and protected by the state. The economy has to be shifted from state-governance to free economic initiatives, incentivised by competition and having people's economic rights protected by the state. The collectivist focus was changed to an individualist one. The unrestricted

7 | *Ibid.*

8 | *Ibid.*, p. 177.

9 | Curtis, G. (2013). *Bulgaria: A Country Study.* CreateSpace Independent Publishing Platform, p. 177.

10 | *Ibid.*, p. 178.

11 | *Ibid.*, p. 181.

12 | *Ibid.*, p. 184.

expression of religion, foreign investment following the entry to the European Union, and the separation of powers within the three independent branches – judicial, executive and legislative – became key features of the state[13]. Furthermore, one major implication of the new constitution was political and economic pluralism. As such, the single-party state concept was cautiously outlawed and the Western model of polyarchy was adopted. A more prosperous liberal democracy is possible because of the legislation that allows freedom to protest and constitutional protection of fundamental individual rights and freedoms. The drastic change and complete adoption of constitutional liberal values have created a favourable foundation for the difficult political and economic transition to follow.

The rule of law was reinforced, which would replace all previous arbitrary purges with fair trials in an independent court. Essentially, the wrongs of the single-party state were hoped to be corrected by the establishment of the new constitution and the drastic change in values and legislation. This shift allowed considerable augmentation of political parties, some of which will be discussed further. From an economic perspective, this allows Bulgaria to become a more accessible stakeholder for foreign investments and form partnerships and gain memberships in different organisations, which will be discussed in detail in this paper later.

From a political perspective, Bulgaria has a multi-party system for its National Assembly, or parliament, for a four-year term. To enter the parliament, political parties must gather a minimum of 4% of the national vote. Two-thirds of the seats in the parliament are chosen proportionately from the lists that each party puts up for a popular vote[14]. In proportion to the votes obtained in each electoral district, the various parties recommend members of their party list to the parliament. The remaining one-third of the seats are chosen based on the simple majority principle, meaning one wins all in the respective electoral regions[15]. Such a structure is based on the liberal pluralist foundation that allows political deliberation and access for all members of society when tackling economic, social, and cultural issues among other things.

13 | Parliament of Bulgaria. (2011). *The Bulgarian Constitution*. Retrieved from http://www.parliament.bg/en/const.

14 | Parliament of Bulgaria. (2011). *The Bulgarian Constitution*. Retrieved from http://www.parliament.bg/en/const.

15 | *Ibid.*

Despite decades of political turbulence in the 20th century, the adoption of liberal democratic values at the beginning of the 1990s has allowed Bulgaria to conduct progressive reforms to transit from communism to capitalism. Although Bulgaria has encountered challenges that prevented the country from a fast growth rate as compared to other post-communist countries, its positive state of affairs and strong foundation in political and economic values favour more progressive policies and partnerships in the future.

2. Political and Economic Trends of Bulgaria

As one of the most recent and least economically developed members of the European Union (EU), there is little doubt about the present and future Bulgarian government's strong determination to maintain a course of further integration with the Union, which it joined in 2007. In 2016, 57% of Bulgarians supported the country's membership of the EU[16]. This seems to be a genuine sentiment of the majority of the population, while the government would expect to enjoy substantial electoral support on the EU issue. Thus far, the benefits of Bulgaria's membership in the EU have been obvious. Bulgaria has enjoyed access to €20 billion of funds for integration[17] that has significantly contributed towards the industrial, agricultural, cultural, and regional development of the country[18].

However, Bulgaria faces a number of challenges with its EU membership as well. Bulgaria remains near the bottom of the league table concerning its economic growth [19]. The majority of Bulgarians harboured widespread hope that when the country entered the EU, they would enjoy swift improvements but those improvements did not materialise overnight[20].

16 | National Statistical Institute. (2012, April 17). *Households Income, Expenditure and Consumption in 2011*. Retrieved from: http://www.nsi.bg/sites/default/files/files/pressreleases/HBS2011p_en.pdf.

17 | International Monetary Fund. (2016). *Bulgaria*. Retrieved from: http://www.imf.org/external/pubs/ft/weo/2016/02/weodata/weorept.aspx?pr.x=53&pr.y=5&sy=2011&ey=2016&scsm=1&ssd=1&sort=country&ds=.&br=1&c=918&s=NGDPD%2CNGDPDPC%2CPPPGDP%2CPPPPC%2CLP&grp=0&a=.

Nevertheless, the number of relatively well-paid middle-class professionals have been increasing. In comparison to the other member states, Bulgaria remains at the bottom of the league table concerning its economic growth[21] and this has become a constant reason for complaints on part of the population and an embarrassment to the government in power.

Membership of the EU overlapped with Bulgaria's active membership in the North Atlantic Treaty Organisation (NATO), which it joined in 2004. Bulgaria played a relatively minor role prior to the occupation of the Crimean Peninsula by Russia, the war in Syria, and the political destabilisation in Turkey. However, all three instances increased the importance of Bulgaria in the alliance and placed higher demands on subsequent Bulgarian governments, causing the country to take a more active stance on issues such as increasing NATO bases and personnel in the country, as well as exhibiting a more determined stance on issues such as the confrontation with Russia[22].

Bulgaria is a parliamentary republic with a one-chamber parliament. The system is based on three subsystems: the representative power rests with the president, the executive power with the parliament and the government it appoints, and the judiciary remains an independent third branch for checks and balances[23].

Recently, there have been some public voices urging for a change in the character of the Constitution and turning the Republic from a parliamentary to a presidential one[24]. However, they remain isolated and are not supported by either the majority of the political elites or the general public. Circles of society, which tend to blame the system more than the people responsible for the state of it, have been unceasingly pressing for reforms of basic elements of the executive constitutional arrangement. Their main demands are twofold: 1) reform of the electoral system and 2) of the parliamentary structure[25].

The first demand for change is to transform the existing electoral system into a majority arrangement. There seems to be a prominent group, albeit a minority, behind this project

18 | Central Intelligence Agency. (n.d.). *The World Factbook. Europe: Bulgaria.* Retrieved from: https://www.cia.gov/library/publications/the-world-factbook/geos/bu.html.

19 | Eurostat. (2017). *Regional gross domestic product (PPS per inhabitant), by NUTS 2 Regions.* Retrieved from: https://ec.europa.eu/eurostat/databrowser/view/tgs00005/default/table?lang=en

20 | Popkostadinova, N. (2014). *Angry Bulgarians Feel EU Membership Has Brought Few Benefits.* EU Observer. Retrieved from: https://euobserver.com/eu-elections/123199.

21 | Eurostat. (2017). *GDP per capita in PPS.* Retrieved from: https://ec.europa.eu/eurostat/databrowser/view/tec00114/default/table?lang=en.

22 | NATO. (2004). *Seven New Members Join NATO. NATO Update.* Retrieved from: https://www.nato.int/docu/update/2004/03-march/e0329a.htm

23 | Parliament of Bulgaria. (2011). *The Bulgarian Constitution.* Retrieved from http://www.parliament.bg/en/const.

24 | Seiler, B. & Lilov, E. (2013, June 26). Bulgarians Protest Government of 'Oligarchs'. *Deutsche Welle.* Retrieved from: https://www.dw.com/en/bulgarians-protest-government-of-oligarchs/a-16909751.

25 | *Ibid.*

who believe that such a system would increase the chances of having strong, determined, and highly capable representatives enter the parliament, who will focus on realising the Bulgarian people's desires and less on their own political interests. According to the supporters of this idea for change, this development would be a way to eliminate the powerful party elites of faceless political spin doctors who, under the current system, are permanently assured of a seat in the parliament[26].

The second demand for parliamentary change involves decreasing the number of members of parliament. It is suggested to decrease the number from the current 240 to between 120 and 150 members – the key intention is that such a decrease would reduce the expenses for the maintenance of members of parliament[27]. This of course strikes a chord with the less well-off part of the population, who seem to believe that the political elite should undergo some sort of suffering and punishment for their lavish and corrupt practices.

Third, there has been popular support for reforms of the legislative power. Here, as in all other parts, the existing Constitution mirrors the best practices of its analogues in the more developed democracies. However, the problem lies within the politicians who work in the system and their general lack of ethics.

While the president and members of the parliament can be changed every four years, judges and prosecutors are professionals employed in their positions without any specified time constraint. Any electoral bodies in that system are a product of internal elections[28]. Magistrates are specially educated professionals, who usually have long years of practice and experience in their expertise. All these make changes in the system very difficult. Hence, when magistrates are inclined to corrupt practices, reforms are difficult to implement and are easily jeopardised[29].

In Bulgaria, there is a widespread belief that changes in the legislative system are long overdue, as many people have encountered lengthy and ineffective legislative practices, weak

26 | Popkostadinova, N. (2014). *Angry Bulgarians Feel EU Membership Has Brought Few Benefits*. EU Observer. Retrieved from: https://euobserver.com/eu-elections/123199.

27 | Hauser Global Law School Program. (2006). *The Bulgarian Legal System and Legal Research*. Retrieved from: http://www.nyulawglobal.org/globalex/Bulgaria.html.

28 | Parliament of Bulgaria. (2011). *The Bulgarian Constitution*. Retrieved from http://www.parliament.bg/en/const.

29 | Bulgarian National Radio. (2012). *Transparency International Report: Bulgaria Perceived As EU's Most Corrupt Country*. Retrieved from https://web.archive.org/web/20121101112317/http://bnr.bg/sites/en/News_eng/Pages/en0112_B2.aspx.

investigative process, and incompetent prosecution verdicts from judges, all of which have fuelled popular discontent[30]. Probably the best example to illustrate the weak legislative system in the country is the fact that there has not been a single verdict against a corrupt politician[31].

In the first decade after the end of communism, Bulgaria transformed from a single-party state into a multi-party system. Some old pre-communist parties were revived, and new parties were formed. The Turkish minority formed its own ethnic-centric party. These developments marked a period of sharp controversies and political conflicts largely along the lines of the traditional divide between the left and right[32]. However, when Bulgaria entered the new millennium, it began to exhibit tendencies closer to the current trends in the rest of the developed western democracies – characterised by diminishing general interest in political affairs that led to sharp declines in voter turnout, low party membership, and higher mobility of votes away from traditional parties[33]. The decreased interest in traditional politics among the electorate opened a space for populist and nationalist movements which exploited the general disinterest and discontent of the population by offering captivating rhetorical promises and quick remedies.

This rise of political movements with multiple internal factions, rather than traditional parties or coalitions of parties, began to set a model for domestic political development. The past decade was filled with several attempts to establish such amorphous formations in the political landscape and almost all of them were either of a populist or nationalistic nature. Most came and went depending on the financial resources they had received from sources both in and outside of the country. They operated according to a well-known trajectory: when funds were present, they gained popularity, but when those funds dried up, the movements lost momentum[34].

Here, it is worth mentioning a few political formations that are related to the movement: 'Bulgaria Without Censorship', 'The Bulgarian Business Bloc', 'Bulgaria for Citizens', 'Order, Law and Justice', 'Lider', 'Movements for Rights and Freedom', 'National

30 | Seiler, B. & Lilov, E. (2013, June 26). Bulgarians Protest Government of 'Oligarchs'. *Deutsche Welle*. Retrieved from: https://www.dw.com/en/ bulgarians-protest-government- of-oligarchs/a-16909751.

31 | *Ibid.*

32 | Ghodsee, K. R. (2010). *Muslim Lives in Eastern Europe: Gender, Ethnicity and the Transformation of Islam in Postsocialist Bulgaria*. Princeton University Press.

33 | *Ibid.*

34 | *Ibid.*

Front for Salvation of Bulgaria', 'VMRO', and 'Ataka' – with the last three forming the coalition of 'United Patriots'. With their founders viewed as heavily compromised characters, some of those movements are now history or have simply faded out. Nevertheless, the trend is still very much in vogue. The 'United Patriots' entered parliament and have even become junior partners in the government. The 'Movements for Rights and Freedom', formed by a wealthy businessman also entered the parliament. This trend of populist movements is likely to continue, the latest attempt being a TV show celebrity Slavi Trifonov trying to ride the back of popular discontent to enter politics[35].

The disillusionment of the electorate stems from the way traditional politics has involved alleged 'shady deals' between the political parties as well as rampant corruption scandals[36]. The fact that Bulgaria continues to lag behind other European Union members in terms of personal wealth and social benefits also provides fertile ground for populism to grow, leading the electorate to blame the system or the current political elite for slow or no progress in economic development[37]. Moreover, unstable relations with minorities, namely members of the Roma Minority (who are accused of working less and not abiding by the laws yet receive better social benefits,) adds fuel to nationalist movements[38].

The nationalist organisations are the quickest to capitalise on ethnic differences and weaknesses. They use propagandistic slogans like 'Bulgaria for the Bulgarians'. Nationalist movements aim to galvanise audience support by proposing initiatives such as: restricting the social benefits of members of the Roma Minority; limiting the voting rights of Bulgarian citizens of Turkish origin who live in Turkey but hold Bulgarian passports; and proposing outright refusal to participate in the European Union refugee policy.

By default, nationalist movements in Bulgaria are pro-Russian in orientation because they often accept financing from Moscow. These nationalist movements also take on an anti-Turkish sentiment and exhibit great scepticism regarding multilateral

35 | Peycheva, V. (2018). *Кастингът на Слави: Иновативен проскт или пак разочарование? [Slavi's Casting: Innovative Project or a New Disappointment?].* Retrieved from: https://www.dnes.bg/politika/2018/02/05/kastingyt-na-slavi-inovativen-prockt-ili-pak-razocharovanie.367159.

36 | Seiler, B. & Lilov, E. (2013, June 26). Bulgarians Protest Government of 'Oligarchs'. *Deutsche Welle.* Retrieved from: https://www.dw.com/en/bulgarians-protest-government-of-oligarchs/a-16909751.

37 | Popkostadinova, N. (2014). *Angry Bulgarians Feel EU Membership Has Brought Few Benefits.* EU Observer. Retrieved from: https://euobserver.com/eu-elections/123199.

38 | Bogdanov, G. & Zahariev, B. (2011). *Bulgaria Promoting Social Inclusion of Roma A Study of National Policies.* European Commission DG Employment, Social Affairs and Inclusion. Retrieved from: http://ec.europa.eu/social/BlobServlet?docId=8963&langId=e.

organisations like the EU and NATO.

Apart from the populist movements, there are actors who are more firmly oriented within the framework of civil society. They can generally be subdivided into three groups: environmentalists, minority-based parties, and civil reform parties, who all carry different potential in influencing the political life of Bulgaria. Although the environmentalists are quite well-established in society and receive steady financing from different organisations abroad in addition to the EU, they have been unable to form a political organisation that can enter the parliament to officially participate in domestic politics and legislation.

The minority-based parties present another potential actor of change in the future. Although the constitution forbids ethnic-based parties, all of them gain political recognition and continue to operate by registering under some ideological cover and by allowing Bulgarians to assume them as figureheads. The Turkish minority in Bulgaria – who make up almost 0.6 million out of the total number of 7.2 million Bulgarian citizens[39], is represented by the DPS (Movements for Rights and Freedom). The DPS claims to be a liberal party and is even a member of the liberal faction of the European Parliament[40]. However, it is highly doubtful that the majority of its members and even the party's leading politicians acknowledge or abide by key liberal values in practice. The DPS excels at gaining votes *en masse* for their party. It has by far the most stable electoral base in the country ever since the downfall of the communist regime and since its registration as a political party. With approximately the same number of seats in every parliament, the DPS tends to always be a potential candidate for a minority coalition partner with any of the other big players when those big players lack sufficient seats to form a government on their own.

In this way, the DPS' influence on Bulgarian politics is likely to remain significant. As the party largely represents the interests of the Turkish minority, there are also suspicions that it is affiliated closely with the government of Turkey.[41]

39 | National Statistical Institute. (2011). *2011 Population Census – Main Results*. Retrieved from: https://www.nsi.bg/census2011/PDOCS2/Census2011final_en.pdf.

40 | Ghodsee, K. R. (2010). *Muslim Lives in Eastern Europe: Gender, Ethnicity and the Transformation of Islam in Postsocialist Bulgaria*. Princeton University Press.

41 | Communist Consolidation. (2011). *Domestic Policy and Its Results*. Retrieved from http://lcweb2.loc.gov/cgi-bin/query/r?frd/cstdy:@field(DOCID+bg0062).

Another minority that has been trying for years to establish a political organisation is the Roma, who officially amount to 325,343 citizens[42]. However, the members of the Roma Minority are divided into different clans and differ by region, which undermines any attempts at political unity. Thus, the members of the Roma Minority are easy prey to other political formations who often buy the members of the Roma Minority's votes during elections. DPS has benefitted chiefly from such vote-buying, and the DPS has extended its electoral hold into areas densely populated by gipsy communities. While the chances for the gipsy minority to form a stable political entity are very slim, they will still play an important role in the political plans and calculations of other parties because of the rapidly increasing Roma population due to their consistent high birth rates and perceptions about their way of life, which often allegedly borders on the edge of legality, if not beyond it, and causes discomfort to other non-Roma members of the communities they inhabit[43].

Lastly, civil movements are important regarding the future development of Bulgarian politics. Civil movements differ from populist ones because they have a genuine and clear political orientation, either towards the socialist left or towards the liberal and conservative right[44]. While they share similar political values with the traditional parties and ideologies, they differ because of their emphasis on priority lists of reforms. The leftist organisations are few, but all of them share the similarity of opposing the corruption and conformism of the Bulgarian Socialist Party (BSP). They stand for a clear-cut social policy where governmental institutions would have a stronger say in the distribution of wealth and social benefits and where the influence of wealthy individuals over the policy of the Socialist party will be significantly reduced. So far, the political chances of the leftist organisations have been limited as shown in their election results[45]. However, if the political right commits a mistake, the leftists could be the major winners of swing votes and that might give them a significant say in the political future of the country.

On the right, there is 'Yes Bulgaria', which was created just

42 | National Statistical Institute. (2011). *2011 Population Census – Main Results*. Retrieved from: https://www.nsi.bg/census2011/PDOCS2/Census2011final_en.pdf.

43 | The Economist. (2015, June 4th). *The Roma -- Left Behind*. Retrieved from: https://www.economist.com/europe/2015/06/04/left-behind.

44 | Judice, J. (2016). *The Populist Explosion: How the Great Recession Transformed American and European Politics*. Columbia Global Reports.

45 | National Assembly of the Republic of Bulgaria. (2018). *43rd National Assembly (27/10/2014 - 26/01/2017) Parliamentary Groups*. National Assembly Archives. Retrieved from https://www.parliament.bg/en/archive/51.

before the 2017 elections by the former Minister of Justice, Hristo Ivanov. This movement is based on the support of a lot of young professionals who do not believe that the other traditional rightist parties represent their interests or understand the need for reform[46] Their main priority is to begin the reform from the legislative system, which, they believe, if functioning properly, would push all other branches of government to reform. The movement has come up short in elections so far.[47,48]. However, the chances for 'Yes Bulgaria' to enter the parliament remain considerable, provided it stays united and its political development continues to progress.

The current system can be greatly criticised, but neither democracy nor capitalism is a perfect system. Capitalism is based on greed of the elites, but also on trust in order to reinvest the capital that is an integral part of the current state of affairs. Democracy is not an ideology, but rather a system where there is no guarantee that people would use their right to vote for the cleverest, most intelligent, and best-suited person to lead them. It is most likely that people would rather vote for a candidate who best represents their interests. Hence, the free market can be a curse if not checked by the state and if allowed by the corrupt judicial system, which allows the interests of politics and business to proliferate. This means that there is a lot to be done by the Bulgarian people in order to upgrade their level of mental preparedness and maturity to deal with the challenges of the modern world. Can they do it? We believe they can. Their 13 centuries of history clearly demonstrated that they have overcome bigger challenges in the past on their own.

46 | Da Bulgaria!. (2018). *Manifesto of Political Party 'Da Bulgaria' Movement.* Retrieved from: https://dabulgaria.bg/en/manifesto-of-political-party-movement-da-bulgaria/.

47 | *Ibid.*

48 | National Assembly of the Republic of Bulgaria. (2018). *43rd National Assembly (27/10/2014 - 26/01/2017) Parliamentary Groups.* National Assembly Archives. Retrieved from https://www.parliament.bg/en/archive/51.

3. An Overview of the Existing Opportunities for Bulgarian Youth

In order to accurately assess the current possibilities for young people in Bulgaria, one has to look at a number of different factors including the state's proactivity, the opportunities brought by EU-funded programmes and initiatives, the overall level of engagement of the youth, as well as young people's capacity to utilise potential opportunities accordingly. The list of these factors is not exhaustive but will provide a good overview analysis of the opportunities currently available for Bulgarian youth. What can also be added to this combination of factors is the socio-economic status of Bulgaria, 25 years after the inception of democracy and nine years after joining the EU.

Before we start to analyse the existence (or the lack) of opportunities for young people, it would be valuable to focus on the social and economic problems young people face. The main issues faced by youth in Bulgaria are employment exploitation, emigration at an early age, increased criminal activity, and challenges in educational institutions. Due to the high poverty rate in Bulgaria[49], many young people are forced to join the labour force at an early age. Many 'early age workers' are illegally employed and do not enjoy social benefits. Bulgarians who are unable to provide satisfactory living conditions for their families choose to emigrate abroad in search of economic opportunities. It is indisputable that this influences the life of young Bulgarians who have been forced to adapt to new environments at a very young age[50].

Another significant factor is the increasing amount of criminal activity amongst young people, which is a result of poor state policies that address youths' welfare. However, the most detrimental problem is the outdated educational system, which largely fails to provide relevant teaching methodologies to students. Also, there is a lack of resources devoted to motivating young students' involvement and attentiveness in class[51].

Despite the problematic and unstable socio-economic situation in Bulgaria, there are opportunities for the young population. These opportunities are created through numerous governmental and commercial programmes, which mainly aim

49 | Kanev, P. (2012). *Information Template on Social Inclusion of Young People in Bulgaria*. Youth Partnership, Council of Europe.

50 | Iliev, V. (n.d.). *YSPDB Report about Situation of Young People in Bulgaria, Youth NGOS and Youth Policy and Relations with Government*. Youth Society for Peace and Development of the Balkans.

51 | *Ibid.*, p.2.

to offer youth more prospects for self-improvement, better career paths, and an overall increase in the quality of life.

Since Bulgaria's entry to the EU, it has started to initiate a number of governmental programmes establishing opportunities for young people both within the country and through European cooperation. An example of this approach is the 'First Job' initiative, which represents a national agreement signed in 2012[52]. The focus of this scheme was to decrease the number of unemployed young Bulgarians by providing free vocational training and aiding them in adopting high-value skills and competencies. The initiative presented the possibility for subsidised employment for a total of 12 months to a number of young individuals who previously lacked the capacity or the desire to seek better career paths. In fact, it has been proven that the 'First Job' programme helped prepare a significant number of young Bulgarians for specialised placements in the job market.

Other examples of such 'possibility streams' include the 'Support for Employment' scheme which also focused on better employability with the aid of training courses. The list also includes 'A New Beginning' programme, which offered various internships to provide young people with the opportunities to gain more experience in the early stages of their careers.

Alternatively, there are a lot of major corporations and big entities that also target the unemployed and disengaged young individuals in Bulgaria in order to encourage them to become more competitive and skilled. Those companies frequently target job seekers aged up to 29 years. For instance, Samsung Bulgaria established 'Trends of Tomorrow', where young students had the chance to learn more about opportunities in the professions and the different ways to choose a suitable career depending on their skills and interests[53]. Moreover, Samsung Bulgaria created a small competition related to online platforms where the winners received scholarships[54].

To fully analyse the youth situation in Bulgaria, the right approach will be to investigate the root of the high

52 | Ministry of Labour and Social Policy (2012). *First Job Initiative*. Retrieved from: https://www.az.government.bg/bg/news/view/pyrva-rabota-za-mladeji-do-29-godishna-vyzrast-143/.

53 | European Centre for the Development of Vocational Training, Bulgaria. (2013). *Opportunities for youth Training in Bulgaria*. Retrieved from: https://www.cedefop.europa.eu/en/news-and-press/news/bulgaria-opportunities-youth-training-bulgaria.

54 | National Network for Children. (2012, July 12). *What Are the Opportunities for the Young People in Bulgaria?*. Retrieved from: http://nmd.bg/en/what-are-the-opportunities-for-the-young-people-in-bulgaria/.

unemployment rate and the high number of uneducated young Bulgarians, which is undoubtedly the poor education system. In general, if Bulgarian youths had easier access to significantly better schooling, they would be more engaged in their future career choices. Unfortunately, the education system in Bulgaria is outdated. From an early age, children are bound to face the reality of studying more than 10 subjects in school without having the chance to choose what to focus on and what to drop out of. From the author's own experience and observations, the latter plays a major role in the loss of motivation and lack of engagement at school.

In the past 20 years, many scholars have proven that obligatory classes should be reduced to the fundamental sciences and arts. Students should be allowed to choose their own subjects based on their interests and personal beliefs. This not only helps students understand how choices can affect their development, but may also improve their general satisfaction with education as they are not being pressured into studying subjects that they have no particular interest in.

It is worth mentioning that the more developed EU member states, like Finland and the Netherlands, have already adopted systems that are based on freedom of choice at a very young age starting from around 13-14 years old. Throughout the years, these innovative methods have produced lower unemployment rates in the aforementioned countries, and these successes are often combined with higher overall satisfaction amongst young people[55]. Issues in education are connected with the large number of Bulgarian students who choose to attend higher education abroad. In fact, after joining the EU, the number of Bulgarian students who prefer to study outside their home country has increased dramatically. It is fair to say that the lacking quality of the local education system is one of the main causes for the occurrence of brain drain.

At the same time, the more significant and large-scale issues such as low satisfaction with education, need to be addressed accordingly as well. In order to examine general levels of satisfaction, one must turn to the different elements

55 | Eurostat. (2016). Unemployment Rate – Annual Data. *Eurostat.* Retrieved from https://ec.europa.eu/eurostat/ databrowser/view/tipsun20/ default/table?lang=en.

contributing to higher overall satisfaction. The two most dominant factors are the quality of education and the existence of diverse and fulfilling career options. Less than half of Bulgarian students are satisfied with the quality of education and some of the students believe that this is due to the lack of uniformity amongst the different pillars of the education system itself[56]. Very little has been done to create a more holistic education environment to increase career opportunities for youths. Another important aspect of this issue is that there is a large gap between the opinions of the high achievers and those with average or lower grades. Recent studies have suggested that the latter group has lower expectations in general, thus they have expressed higher levels of overall satisfaction in relation to education and career opportunities[57]. Therefore, when it comes to diagnosing problems affecting the maximisation of the country's potential – the corrupt practices in the labour market, as well as the lack of quality in the education system, can be identified as the most detrimental causes.

Furthermore, 'early leavers' from school also represent an obstacle to the realisation of the full potential amongst young individuals in Bulgaria[58]. In 2013, at least 12% of children dropped out of school too early and this number has continued to grow over the years as more education institutions close down, especially in the rural areas[59]. As the early leavers are the ones least likely to succeed in the job market, even if more opportunities are created, the chances that the benefits will reach the ones who need them most remains low.

Therefore, it can be concluded that there is a large gap between the availability of self-improvement possibilities and the accessibility of young individuals to such opportunities. This stems from the fact that most of the valuable opportunities suffer from unequal distribution. It is not the low number of chances which prevents the uneducated and unemployed young people from expanding and fulfilling their potential. It is rather the phenomenon where more favourable circumstances are offered to those who are already in possession of some

56 | Mitev, P. & Kovacheva, S. (2014). Young People in European Bulgaria, A Sociological Portrait. *Friedrich-Ebert-Stiftung*, p. 118. Retrieved from: https://library.fes.de/pdf-files/bueros/sofia/12569.pdf.

57 | *Ibid.*

58 | *Ibid.*, pp. 118-120.

59 | *Ibid.*, p. 12.

basic fundamental skills required to improve quality of life. This leads to an absence of exposure to meaningful projects by those with less favourable circumstances, especially young Bulgarians who simply cannot obtain the adequate amount of skills in order to adapt to the more dynamic work markets.

Having identified the main issues regarding Bulgaria's poor education system, there are a number of things that can be done to mitigate existing problems. Initiating a better focus on developing the necessary skills and dropping the burdensome 'study-all' approach will improve the number of opportunities for the younger generation. Bulgarian students nowadays rarely have the chance to become specialists in given fields, because most schools have the tendency to overlook the power of well-qualified young individuals. In other words, there are some schools that offer training in development paths like economics or engineering, but even those can hardly deliver opportunities, which might be available to them. Sometimes this is a result of problematic management of the institutions and sometimes obsolete ideas and methods which need innovative change.

In the author's opinion, there are two main channels Bulgaria can explore within the BRI. The first step is to eradicate the problematic education system on a wider scale and try to adopt successful teaching and schooling techniques from other countries that have fixed similar issues. Better connectivity across all countries participating in the BRI will definitely lead to valuable conversations and fruitful exploration of ideas that will help the education steps taken by Bulgarian youths. We strongly believe that Bulgaria can actively learn from other European countries. Successful schooling methods, referenced from other countries, can be adapted efficiently into Bulgaria's education system.

There are also external sources like non-profit organisations and projects enabling the acquisition of new skills and to learn about fields they are interested in. An example for that is the organisation 'Together in Class' whose main mission is to provide quality education to all of those early leavers, the ones who do not fit in the 'study-all' paradigm, and the ones which simply want to learn more, but their school is not offering it. Such organisations are strong advocates of the idea that every child has the potential to become great, do better, and learn fast about things they like. Therefore, by tackling the heart of the problem, organisations like 'Together in Class' are not only offering lessons to the ones who cannot attend them through traditional channels, but also creating new horizons for people who never believed in the existence of these new disciplines.

The second main channel is the establishment of specialised schools and colleges. Not very long ago, Bulgaria had a strong tradition in specialised professional colleges. Those institutions resemble normal high schools, to which students can apply after they turn 13. These schools prioritise subjects around a particular industry or field. This means that 20 years ago Bulgaria was producing a very large number of all kinds of engineers, maritime professionals, textile designers, hospitality specialists, and other specialists. By letting young individuals choose a particular field they want to develop new and better skills in, they can start improving their decision-making process. Hence, they will always be looking for new opportunities within that business sector. Transport Colleges can be a great example of that if the right steps are taken in that direction. In my own experience, we have seen cities with traditions in the transport industry like Rotterdam and Copenhagen create a unique education ecosystem following their successful developments in the sector. Thus, given the fact that the success of the BRI partly depends on large infrastructure projects and smoother trade systems, we are certain that reviving Bulgaria's traditions in the transport industry is a huge potential, which will create a variety of jobs, opportunities, and local trends for specialised colleges that equip people with required skills.

In conclusion, the poor local education system has made the Bulgarian youths' transition from education to the labour market difficult. The decision-making processes amongst policymakers should consider a wide array of factors for improvements beyond job-creation; such as efforts in constructing a friendly ecosystem where younger individuals can develop skills and knowledge. The BRI may provide such opportunities to young people in all countries because the possibilities of different projects can bring internships and employment. The BRI may offer knowledge in different fields to Bulgaria. For example, it could offer advice on how Bulgaria can improve its infrastructure, economy, education system, laws, and governance apparatus amongst others. These ideas have surfaced in the interviews where the main problems in Bulgaria are discussed in the next part.

4. Advantages and Challenges of Bulgaria in the Belt and Road Initiative: A Discussion Based on Interviews with Bulgarians

During this research, we interviewed four people of different ages and backgrounds. All interviewees provided their personal opinions on the current state of Bulgaria and were highly critical of Bulgaria's current political state. The first (and youngest) participant, Kaloyan Kutiysky, is a 17-year-old high school student. The second participant, Radost Stefanova-Encheva, is a 45-year-old woman with a family of four and is an executive employee in a small family firm. Moreover, she has been an active contributor in her town, such as helping the development of the Sport Orienteering Club 'Sini Kamani' in Sliven and helping her alma mater by sourcing financial opportunities and sponsors for talented students and for the high school gymnasium. During the 2016-2017 school year, she managed to retain the same number of classes despite the threat of a possible decrease in pupils due to the passing of new laws governing education institutions. She was also the Chairman of the Public Council of that same school. The third participant, Valentina Vasileva-Filadelfefs, is a Bulgarian politician. Lastly, the fourth participant, Nadya Nencheva, is an English teacher.

Despite the attempt to interview a sample from different occupations and from different social backgrounds, all of the interviewees held similar views on the present government. Their views echo the general sentiment of many Bulgarians towards the government, as Bulgarians believe the country has been going through a rough patch since 1945 and they demand change. This survey aims to provide an understanding of why Bulgarians feel dissatisfied with the current authorities and the election process. This will most definitely be an important issue for the BRI if a large volume of citizens is becoming disillusioned with the electoral and political process. The prevailing situation now is that the same politicians rule the county, they 'migrate' from party to party and change their political views easily, leading to a decrease in trust levels and a low voter turnout during elections. In two parliamentary elections in the 2010s, the turnout rate was below 50%.

As Radost Stefanova-Encheva, one of our interviewees stated: 'The situations with our social, health, pension, and legal systems are almost as chaotic as the situation with our Ministry of Foreign Affairs' (See Appendix III). Regarding the Ministry of Foreign Affairs, Stefano-Encheva was referring to its inadequate solutions concerning its administration work.

During our interviews, nearly all of the interviewees raised issues regarding corruption. Vassileva-Filadelfefs accused the government of corruption and stated the opinion that the government is run by the mafia. According to her, bribery is rampant in the upper levels of the government (See Appendix II). Due to the fact that she is the leader of a party, she is somewhat 'leading a war' against the 'GERB' [60]. We view this as normal rhetoric among all parties in Bulgaria that compete for government power.

Nadya Nencheva said: 'The whole model of electing members of parliament is corrupt – the people who are eligible professionals do not participate in politics or are not allowed to do so since parties themselves choose their candidates. This leads to incompetent people managing the country, which then leads to the lack of managerial skills and complete absence of a

60 | The majority party of the Bulgarian government as of 2017.

visionary approach to running the country. Things are done for the day and are not part of a strategy.'

Nencheva was criticising the election candidates selection process in which the same people circulate around to manage different facets of the country, even if they do not have the right competency for the particular Ministry. For example, a Minister of Health was an economist, therefore lacking the competence to fill the role, and the Bulgarian Minister of Defence was not connected to the military at all. In fact, he was a leader of one of the parties in the parliament.

In Kaloyan Kutiysky's view, the state apparatus is corrupt and that means youngsters would have a hard time to develop both professionally and individually because resources are not allocated to address pressing needs (See Appendix V). These results suggest that people from all age groups observe that there is a corruption problem in the Bulgarian government. Yet, the administration has not changed for good in 30 years, so people continue to mistrust politicians and what they say.

If a full change of the state apparatus does occur, Bulgaria will develop positively and the education system will undergo massive change and the young members of the community would be able to develop in almost all professional aspects. There is also a higher likelihood of attracting Bulgarians who have emigrated to return to Bulgaria to live and work here again.

'The only solutions to these problems that could be effective are to replace the current political figures with intelligent, educated people who want society to flourish. A way has to be figured to separate the independent Supreme Court from the government. A change is needed in the country's system because as of now it works for the people who rule, not for the people in the country,' Kutiysky stated (See Appendix II), 'Major political reforms are required to solve these problems, and the progress seems to be slower than that of most other Eastern European EU member states.' (See Appendix V)

This is an ambitious and wide-scale reform, but if it does happen, the country can maximise its current assets and resources to become a real economic giant. These 'resources' are that '...Bulgaria has the needed intellectual resources...' as expressed by Stefanova-Encheva, 'Bulgaria is a tiny country as compared to most others; However, it has the potential to be a self-sustainable one, bearing in mind its geographical location, the combination of sea and mountains, rich soil suitable to grow every kind of cereals, fruit and vegetables, the mild climate clearly divided into four seasons... Natural resources of every kind here are a prerequisite for the production of renewable energy. Furthermore, the country is sparsely populated, which gives scope for development and construction of many factories, farms, and project implementation' (See Appendix IV). Again, these present opportunities for collaboration in the BRI.

Nikolai Nedkov (one of this chapter's authors) expands on the latter point by identifying that Bulgaria has the advantage of offering a unique experience as a country that stands at the confluence of three civilisations. For six centuries Bulgaria has been at the edge of the European region, while at the same time it is at the gates of the Asian region and has been an active player in the Russo-Soviet world. The experience in the dealings between those three has helped Bulgaria to develop a unique understanding of the specifics of the three different civilisations.

The geographic situation of the country ensures its central position as both a communication and transport hub between Europe and Asia, hence making it attractive for the construction of optic cable networks, pipelines for the transportation of oil and gas and logistic centres for rail and road transportation.

The problem Bulgaria faces is that its private sector is composed of companies with modest sizes that do not possess sufficient financial strength on their own to either develop its own resources or ensure penetration to foreign markets. These limitations are largely a consequence of the rather small domestic market, which in the sector of fast-moving consumer goods (FMCG) cannot provide enough returns to Bulgarian companies to fund adequate marketing campaigns to promote their products even in relatively small markets like neighbouring Romania and Serbia, never mind Turkey, Russia, or any of the big Western European markets.

This lack of international clout in trade prevents the country from developing its natural resources on its own, so it has to invite foreign investors to do the exploration instead as is the case with 'Dundee Precious Metals' for its gold resources or 'Umicore' for its ore resources.

To conclude, it may be worth observing that even today Bulgaria is a bridge between those three worlds. If properly developed and supported in an international context, Bulgaria can turn from a bridge to a highway of cooperation between different civilisations. That is one of the main reasons for the country to look forward to the development of initiatives such as the BRI. It offers more and bigger opportunities for Bulgaria as it breaks the boundary of international bilateral relations and provides a more inclusive and eclectic cooperation framework for those countries who participate in the initiative.

However, any collaboration might be hindered by the problems faced by the education system. There are four main causes of the crisis with our education system. First, the birth rate has decreased to 9.1%[61]. Second, the mortality rate of 15.1% is high[62]. Third, the disproportionately large number of schools and universities, which leads to credential inflation. For instance, in the school year 2016-2017, there were 2,505 primary schools and gymnasiums and 54 universities for a small country with a population of around seven million.

Fourth, despite a large number of schools and universities, only 1 in 11 people in the country are students, and the ones receiving higher education were only 243,199[63]. In Bulgaria, primary and secondary education are not what they should be – there is no connection between education and the job market when the market would determine what specialisations are demanded. The education system also does not encourage understanding and focuses only on memorisation.

61 | National Statistical Institute. (2012, April 17). *Household Income, Expenditure and Consumption in 2011*. Retrieved from http://www.nsi.bg/sites/default/files/files/pressreleases/HBS2011p_en.pdf.

62 | *Ibid.*

63 | *Ibid.*

Thus, students are not taught how to think, they are merely trained to memorise – which poses a potential for disaster later on, since there is no encouragement for teamwork or producing innovative ideas. It can be said that most activities are individualistic, and the grading system is more subjective than objective as students receive grades that do not properly reflect the knowledge they have (See Appendix IV). The system created by the Bulgarian Ministry of Education is based on facts and knowledge rather than creativity. For example, students in high school have to analyse a literary text and most teachers do not accept students' personal opinions, because every opinion is benchmarked against a professionally written marking scheme. Education should adapt constantly to changing realities around us and reforms will be needed. Since the BRI is economically oriented, there is a high demand for qualified human resource, thus there is a need for a good education system for nurturing those professionals.

The situation of higher education is even worse. Due to the huge number of vacancies in universities, higher education has turned into a business. For every person who studies in university, the Ministry offers grants. This means more learners means a bigger budget. Yet, no scheme determines the necessary skills required for grants, and everyone can graduate from any kind of university. Moreover, when students graduate, they have no practical experience because only very few universities offer internship programmes. Internship opportunities are rare due to the lack of connection between business and education. As most of the graduates believe that a higher education certificate translates into a high pay without requisite experience, graduates have high expectations of their future careers and demand high salaries. When their expectations do not match reality, they end up being unemployed or dissatisfied with their jobs. Even if they manage to overcome these challenges and start their own businesses, they face anti-competitive and inefficient laws, which create high barriers to entry for start-ups and corrupt procedures discouraging new businesses from developing any further (See Appendix III). As mentioned earlier, Bulgarian students lack active participation in their chosen profession, for which they receive no practical training, and that will be a drawback for them if they decide to study or work abroad, because most companies and employers seek workers with experience, not fresh graduates. In a lot of European universities, it is common practice to have internships for students to get in touch with experts and benefit from real work experience.

With the trade potential offered by BRI, Bulgaria will have the chance to develop and stabilise its economy. With new workplaces and opportunities provided by the BRI, expertise will be in demand, and the educational facilities within Bulgaria will need to create the much-needed connections to businesses.

'The good news is that we still have a somewhat good education system. A lot of younger members of the community manage to develop nicely both outside and inside the country. Yearly, we see youngsters who leave the country and do a great job of proving themselves in all manner of spheres. Unfortunately, only a small number of young people accomplish the same in Bulgaria. Ostensibly, the government does make an effort to change things but practically there are words and no actions...', said Stefanova-Encheva.

Young people are essential to a country's development. So it is necessary to make Bulgaria more attractive to younger members of the community by the creation of a stable economy. Through the BRI, Bulgaria can receive financial aid and move in the direction of a more balanced economy.

Through the interviews conducted, we can observe that the BRI can provide opportunities to young people. However, due to previous experiences and their influence on Bulgarian politics and the economy, people are cautious about the possible implications of the BRI. Now, the level of foreign influence is also observed in Bulgaria's membership in the EU and some Bulgarians are concerned with the foreign influence the BRI might bring. 'Belt and Road' has a great concept and mission. The only thing that should not happen is any kind of intervention with the member state's internal affairs. 'Belt and

Road' could be a chance for Bulgaria to stabilise its economy without political interference.

Sefanova-Encheva felt it would lead to corruption when politics are connected to business. Some interviewees voiced concerns about the fact that the agreements may not be fruitful for all and even these plans can fail due to political causes. Valentina Vassileva-Filadelfefs said there could be an enforced political influence in some regions. Although she added that only time would tell. She was concerned about the fact that Bulgaria is strategically located, and foreign influence could propel actors to manipulate and benefit from Bulgaria's valuable geography. The government needs to assess whether there is a risk of foreign influence in its calculations and the possible impacts the BRI could bring to the Bulgarian people and economy.

Bulgaria already has basic agreements giving it a strong and motivated position to fend off any interference from external power. Those basic agreements are the membership agreements with the EU and NATO. In those Bulgaria has undertaken certain obligations, while in return she has received guarantees from the other state members of these alliances ensuring Bulgaria's relative protection from interference in its domestic policy, which could come from countries outside those alliances.

In addition to the above, Bulgaria has a number of bilateral agreements with all great powers and regional players such as Russia, China, India, Turkey, Japan, Iran, Saudi Arabia, and even distant Brazil. Those treaties ensure mutual respect and define what would be considered the interference in the domestic affairs and what are the mechanisms to avoid such incidents.

However, this is the official picture. In reality, the domestic political life of the country is very diverse and there are a number of volatile political players with influence who pursue short-term goals and who seem to be open to all types of influence from foreign parties. Sometimes these foreign parties are virtually considered to be interfering in the domestic affairs of the country.

A good example is the behaviour of the Turkish government when there was a split in the DPS, known as the Movements for Rights and Freedom, in December 2015 and the then party leader Ljutvi Mestan was discharged from the party. His first reaction was to go to the Turkish embassy in Sofia and, just a few days after that visit, he established a new party known as the Democrats for Responsibility, Freedom and Tolerance (DOST). This party took an active part in the next elections and was financially supported by Turkish foundations related to the government. It was also helped by public appeals from ministers in the Turkish government to Turks who held Bulgarian passports and who resided in Turkey asking them to vote in support of the Bulgarian party. At the same time the Turkish ambassador in Bulgaria engaged in the electoral campaign by publicly agitating in favour of DOST, action which was deemed unacceptable to many Bulgarians. The situation was quite complicated because Turkey is a member of NATO and a supposed ally of Bulgaria. Also, since Turkey hosts a lot of refugees from Syria, Turkey could easily encourage the refugees to migrate to Bulgaria. So, despite the public outrage, the Bulgarian government limited its actions to only an official protest presented to the Turkish government. There was no apology nor the replacement of the ambassador in Sofia as would be expected in such a case.

The core of the matter is that the main interferences in Bulgarian domestic affairs come through political players who are part of the everyday political life of the country rather than direct pressure from abroad. There are no international treaties protecting a country from such incursions, so the only way for Bulgarians to withstand such interferences would be to show national responsibility and not to empower such political formations that are financed from abroad or are promoting foreign interests at the expense of national interests.

In conclusion, we return to the question: why are young people vital to BRI's success? Youths are the leaders of tomorrow, but more importantly, they are the partners of today. Youths are the ones playing the leading role in change and progress. Therefore, the youth's contribution is key to any successful collaboration between China and Bulgaria in the BRI. Some of the interviewees have the same ideas – creating different projects where young people can gain experience, network, and be exposed to the different cultures of countries participating in the BRI. Also, they felt that a project which encompasses international teamwork by the global community of youngsters would be a good idea as it may offer different opportunities to young people (both inside and outside the country) to broaden their work opportunities.

Appendixes

Appendix I: Content of the Questionnaire

- In your opinion, what are the main obstacles that prevent Bulgaria from fulfilling its full potential, and does the government do anything to eliminate them?

- Does a young person in Bulgaria encounter any obstacles on his way to development? If yes, why and what measures have to be taken/are being taken to eliminate them?

- In your opinion, can 'the Belt and Road Initiative' help youngsters in Bulgaria develop? Why? How?

- What does Bulgaria offer the younger members of the community in terms of personal development, business, career, education etc. and how can they be improved?

Appendix II: Transcript of the interview with Valentina Vassileva-Filadelfefs

The biggest obstacles for the younger members of the community in Bulgaria to be successful are the adhesion of the mafia and government which leads to corruption on all levels of authority in the country's rule and the non-existent independent, unbiased and effective Supreme Court. As a result, the country cannot fulfil its potential. The government of PM Boyko Borisov (assumed office in 2017) does not have the means to get over these obstacles since it serves the mafia to rule the country (it can be said that it is part of the mafia).

It is very difficult for a young person to develop due to these main reasons:

- The colossal levels of bribery and the existence of a biased and government-ruled Supreme Court break the dreams of opening their own business and of advancement. And even further — it causes them to flee the country.

- Cartels, racketeering and, yet again, corruption induce bigger investors to be driven away from Bulgaria and young people have a choice either to leave Bulgaria and pursue their careers or to stay here, struggling to get by

- Pay is extremely low, due to the reasons mentioned above

The only solutions to these problems that could be effective are to replace the current political figures with intelligent, educated people who want society to flourish. A way has to be figured out to separate the Supreme Court from the government. A change is needed in the country's system because for now it works only for those who rule, not for the people in the country. A normally functioning market economy will be needed, too, instead of one that works not by the rules of fair trade, but by the 'rules' of the monopoly. Only by making these schemes possible will Bulgaria be made desirable for younger members of the community.

As for BRI, it is hard to say anything since there is nothing concrete in the project, which could show if this kind of plan could be genuinely achievable and most importantly if it will it be fruitful for all sides. Only time could show us if there is a real political impact. One concern is that there may be a problem in the customs agreements with the EU. The changes after BREXIT will make an impact and there could be many more unknown variables.

As for Bulgaria, the country will be a strategic point of logistics as the country has a prime geographical location. If BRI turns out to be cost-effective, it will help the development of youngsters in our country.

Appendix III: Transcript of the interview with Radost Stefanova-Encheva

In my opinion, Bulgaria has the needed intellectual resources but weak economic development and that is the main obstacle to fulfilling the country's full potential. One of the serious issues is corruption – almost every level of government requires bribery for reaching certain goals. Another serious obstacle is the choice of the staff in important administrative positions, picked in a not-so-random way and, of course, there is no need to mention their incompetence. The situation without social, health, pension, and legal systems is almost as chaotic as the current situation in our Ministry of Foreign Affairs.

The good news is that we still manage to keep a somewhat good level of education. A lot of younger members of the community manage to develop nicely both outside and inside the country. Yearly, we give the world youngsters who have left the country's bounds, and they do a great job in proving themselves in all manner of spheres. Unfortunately, only a small number of young people accomplish the same in Bulgaria. The mentioned issues in the previous paragraph have an enormous effect on young people, as well. Ostensibly, the government does make an effort to change things but practically there is empty talk – the peak of the lies are told when elections are near. Sadly, the birth rate is extremely low, recent surveys found out that the mortality rate in Bulgaria is one of the highest in the EU, therefore the number of children in the state is decreasing, there are too many schools and universities and most of them turn out to be almost unattended, so our education quality is becoming poorer.

BRI is a great mission and concept. By the looks of it, the world is on the path of globalisation (which in my opinion is the correct path, since man is a social animal). There definitely should be international cooperation networks like BRI, so that the countries who are part of them can become partners in different aspects (social, economic, education and cultural). But there definitely should not be ANY kind of intervention in the internal affairs of the countries who are part of this kind of society. Bulgaria has had a very rough experience with the so-called 'intervention' for decades before 1989 (the fall of the Bulgarian Communist Regime). But can BRI be an opportunity for Bulgaria to finally get out of the abyss in which it has been for decades? Probably yes, but only on the condition that Bulgaria is free from any intervention. Without any doubt, a framework like BRI can aid the advancement of youngsters. The smartest ways to do that, in my opinion, are through projects, which include exchange programmes and acknowledging the culture of the members of the community. Internships for young citizens from different countries will also keep youngsters who are well-educated in Bulgaria motivated.

At the moment, we were speaking from personal experience, all businesses are lacking both highly and lowly qualified personnel. The cause for this is the way our education system is designed since the qualified people go abroad to work for foreign companies for better payment and better workspaces. Small business is slowly but surely destroyed and this causes the advent of monopoly and also ruins the chances of good economic development.

Appendix IV: Transcript of the interview with Nadya Nencheva

Bulgaria is a tiny country as compared to most others. However, it has the potential to be a self-sustainable one, bearing in mind its geographical location, the combination of sea and mountains, rich soil suitable to grow every kind of cereals, fruit and vegetables, the mild climate clearly divided into four seasons – you name it, we've got it. Natural resources of every kind here are a prerequisite for the production of renewable energy. Furthermore, the country is sparsely populated, which allows scope for development and construction of many factories, farms, and project implementation. So, how come a potentially rich country like this is so poor?

It's a tough mission being a young entrepreneur or career-seeker in Bulgaria. The explanation lies in a few detrimental aspects:

First of all, it's the no-hands-on experience approach when it comes to primary and secondary education. Students are taught to cram and learn facts by heart – not to think, evaluate and find solutions, which has a disastrous impact on a young mind. Bright ideas and teamwork are not encouraged in the slightest. Discipline is out of any control – democracy and human rights are misinterpreted to the level of chaos.

The situation becomes even worse at the next level. When it comes to university, the vacancies are more than the number of applicants and higher education has turned into business because of poor legislation. Nobody has insight into what kind of specialisms they want to get into.

As we are not well-informed about the projects in detail, we would say that any kind of initiative leading to international teamwork among young people would have a beneficial effect.

In my opinion, the only two fields worth developing and studying at universities in our country are Medicine/Dentistry and Pharmacy.

The whole model of electing members of parliament is corrupt - the people who are eligible professionals do not participate in politics or are not allowed to do so since parties themselves choose their candidates. Which leads to incompetent people managing the country, which leads to a lack of managerial skills and complete absence of visionary approach. Things are done for the day and are not part of a strategy.

Appendix V: Transcript of the interview with Kaloyan Kutyisky

Bulgaria faces a set of barriers to reaching its full potential, mostly stemming from the widespread corruption in the state apparatus and an oversized-yet-inefficient administrative sector. Major political reforms are required to solve these problems, and the progress seems to be slower than in most other Eastern-European EU member states.

Bulgaria offers a lot of possibilities to ambitious young people to receive education, work and start businesses, but some of the most talented people leave for abroad, where possibilities are even greater.

This initiative could be a chance for Bulgaria to get its economy going and raise the living standards of the country.

As the project would employ a significant number of managers, specialists, and help the economies of the Mediterranean countries, 'Belt and Road' may offer career opportunities for educated Bulgarians to make a better living.

Plitvice Lakes National Park in Croatia (photo provided by the author).

by ANASTASYA RADITYA LEŽAIĆ
Research Assistant at the Institute for Development and International Relations

CHAPTER 5

ESTONIA

LATVIA

LITHUANIA

POLAND

CZECH
REPUBLIC

SLOVAKIA

HUNGARY

SLOVENIA

ROMANIA

BOSNIA &
HERZEGOVINA

SERBIA

MONTENEGRO

BULGARIA

NORTH
MACEDONIA

ALBANIA

CROATIA

1. Overview of the Political Structure

The first part of this paper presents an overview of the Croatian political system. It first provides a brief description of important historical events, such as the break-up with Yugoslavia and the independence war, to lay out the necessary context in order to understand current Croatian politics, then a summary of the country's current national ideology and political structure, including an explanation of the executive, legislative and judicial branches.

1.1. Separation from Yugoslavia and the Independence War

The nation of Croatia dates back to the 7th century. However, the most definitive period of modern Croatia is when it was a part of the Socialist Federal Republic of Yugoslavia (SFRY). Since the end of World War II in 1945, as a part of Yugoslavia, Croatia underwent a rebuilding process of industrialisation and development of the tourism industry. Under Yugoslavia's communist rules, privately-owned factories and estates were nationalised while free healthcare and secured pensions were provided.

A turn of events came in 1980 when the death of SFRY leader Josip Broz Tito marked the beginning of the slow disintegration of Yugoslavia. Individual republics, including the Socialist Republic of Croatia, asserted their desire for independence. In 1990, when the communist party lost to the conservative nationalist Croatian Democratic Union (*Hrvatska demokratska zajednica* or HDZ) led by Franjo Tuđman, Croatia held its first free election in more than 50 years. On December 22nd, 1990, the Croatian Parliament ratified a new constitution. On May 19th, 1991, an independence referendum was held and passed with 94% in favour of independence. Soon, on June 25th, the parliament declared the independence of Croatia and dissolved its association with Yugoslavia. To respond to Croatia's declaration, Yugoslavia immediately declared the result of the referendum illegal and in violation of the Yugoslavian Constitution[1].

In an attempt to salvage the legacy of Yugoslavia, the National Army of Yugoslavia started military operations in predominantly

1 | Cohen, L. J. & Dragović-Soso, J. (2008). *State Collapse in South-Eastern Europe: New Perspectives on Yugoslavia's Disintegration*. West Lafayette: Purdue University Press.

Croat areas. Just a month after Croatia's declaration of independence, the Yugoslav army and other Serb forces took over around a third of Croatian territories[2]. As the war continued, cities of Dubrovnik, Gospić, Šibenik, Zadar, Karlovac, Sisak, Slavonski Brod, Osijek, Vinkovci, and Vukovar have all came under the attack of Yugoslav forces[3]. As a result, around 220,000 Croats and 300,000 Serbs were estimated to be internally displaced during the war[4].

The war peaked in November after the Vukovar Massacre, in which a massive number of Croatian prisoners of war and civilians were killed by Serb paramilitaries and Yugoslav arms, despite the European Economic Community's (EEC) simultaneous efforts to set up an Arbitration Commission of the Peace Conference on Yugoslavia on August 27th, 1991, to resolve the issue of Croatia's self-determination and declaration of independence. It was the largest massacre during the war and the worst war crime in Europe since World War II up to that point[5]. A lasting ceasefire agreement was finally reached on January 2nd, 1992.

Soon after the ceasefire, The EEC recognised Croatian independence on January 15th, 1992. However, Croatia gained first diplomatic recognition long before that when Iceland recognised its independence on December 19th, 1991 and Germany following four days later. Croatia became a member of the UN on May 22, 1992.

1.2. Croatia's Constitution and National Ideology

The Constitution of the Republic of Croatia, adopted on December 22nd, 1990, is the highest law in Croatia. It has been amended several times in 1997, 2000, 2001, 2010 and 2013. With the latest amendment in 2010, a new article was added to its preamble about a 'righteous, liberating, and defensive Homeland War' (The Constitution of the Republic of Croatia, I. Historical Foundations). This latest amendment also completed the European Union (EU) accession negotiations by fully harmonising Croatian legislation with the EU community *acquis,* which is the body of common rights and obligations

2 | Powers, C. T. (1991, August 1). Serbian Forces Press Fight for Major Chunk of Croatia. *Los Angeles Times*. Retrieved from http://articles.latimes.com/1991-08-01/news/mn-177_1_defense-force (Last Accessed: 4 October 2017).

3 | Williams, C. J. (1991, November 4). Belgrade Gets a Final Warning from EC. *Los Angeles Times*. Retrieved from http://articles.latimes.com/1991-11-04/news/mn-719_1_final-warning (Last Accessed: 5 October 2017).

4 | Zanotti, L. (2011). *Governing Disorder: UN Peace Operations, International Security, and Democratisation in the Post-Cold War Era*. Pennsylvania: Penn State Press.

5 | Brcic, E. (1998, June 28). Croats Bury Victims of Vukovar Massacre. *The Independent*. Retrieved from https://www.independent.co.uk/news/croats-bury-victims-of-vukovar-massacre-1168387.html (Last Accessed: 11 October 2017).

that is binding on all the EU member states[6]. With that, Croatia became a member state of the EU on July 1st, 2013.

The Constitution defines Croatia as a sovereign, unique, democratic, and social state[7]. Its power derives from the people and belongs to the people as a community of free and equal citizens. The highest values of the constitutional order of Croatia are: freedom, equality, national equality, peace, social justice, respect for human rights, inviolability of property, preservation of nature and human environment, rule of law, and democratic multiparty system[8].

1.3. The Political System in Croatia

Croatia is a unitary democratic parliamentary republic. The president of the republic is the head of state, while the prime minister is the head of government in the multi-party system. The two biggest parties are the right-wing Croatian Democratic Union (*Hrvatska demokratska zajednica* or HDZ) and the left-wing Social Democratic Party (*Socijaldemokratska partija* or SDP). The Government exercises executive power together with the president. Legislative power is vested in the Croatian Parliament with democratically elected members as the highest representatives of the people. The Judiciary is independent of the executive and the legislature.

The president of the republic is directly elected and serves a five-year term, with a maximum of two terms. The president is also the commander-in-chief of the armed forces and has an influence on foreign policy. As the constitution forbids the president to be a member of any political party, the president-elect normally withdraws their party membership before the inauguration.

The Government of Croatia, as the main executive power in the country, is subordinate to the prime minister. Under the prime minister, there are 16 ministers in charge of different sectors, which are all elected by the prime minister with the consent of the parliament. The executive branch is responsible for proposing legislation, budgeting, executing laws, and guiding

6 | Eur-Lex. (n.d.). *Glossary of Summaries: Acquis.* Retrieved from: https://eur-lex.europa.eu/summary/glossary/acquis.html.

7 | Croatian Parliament. (2010). *The Constitution of the Republic of Croatia.* Retrieved from https://www.usud.hr/sites/default/files/dokumenti/The_consolidated_text_of_the_Constitution_of_the_Republic_of_Croatia_as_of_15_January_2014.pdf (Last Accessed: 27 October 2017).

8 | *Ibid.*

domestic and foreign policies.

The Parliament of Croatia is the only legislative body in Croatia, with a minimum of 100 and a maximum of 160 members stipulated by the constitution. The current parliament is composed of 151 members, with the seats allocated according to the electoral districts: 140 members elected in multi-seat constituencies, 8 from the minorities and 3 from the Croatian diaspora. Members are elected in the parliamentary election and serve a four-year term. The legislative power includes deciding on the enactment and amendment of the Constitution, passing laws, adopting the state budget, declaring war, and calling referenda, etc.[9] In most cases, parliamentary decisions are based on majority vote where more than half of the members of parliament are present. In order to form a Government, the winning party and its coalition partners need to secure 76 seats (out of the total 151) in the parliament.

The independent judicial system is governed by the Constitution, with national legislation enacted by the Parliament. The system comprises three tiers, with the highest court being the Supreme Court, to which judges are appointed by the National Judicial Council. The president of the Supreme Court is elected by the Parliament upon the proposal of the president of Croatia. The two lower tiers of the judicial system are the 15 County Courts and 67 Municipal Courts[10]. There are also other specialised courts, such as: commercial courts and the superior commercial court; misdemeanour courts that try trivial offences and the superior misdemeanour court; administrative courts and the superior administrative court; and the Croatian Constitutional Court that rules on constitutional compliance of legislations.

9 | WIPO. (n.d.). *The Constitution of the Republic of Croatia.* Retrieved from: https://www.wipo.int/edocs/lexdocs/laws/en/hr/hr049en.pdf.

10 | Croatian Parliament. (2010). *Odluka o proglašenju zakona o područjima i sjedištima sudova [Decision on promulgation of the law on areas and seats of courts].* NN 144/2010. Retrieved from https://narodne-novine.nn.hr/clanci/sluzbeni/2010_12_144_3625.html (Last Accessed: 1 November 2017).

2. Croatia's Political Direction

After understanding the political structure of Croatia, this part describes the current situation in Croatia regarding its constitutional, political, economic, and legal affairs. Recent movements that are largely supported by the people will also be reviewed, followed by a prediction of the country's political direction in the future.

2.1. The Rise of the 'Third Option'

For more than 20 years after independence, two political parties were dominant in Croatia: the right-wing nationalist-democratic HDZ and the left-wing social-democratic SDP. Both parties took turns in winning majority seats in the parliament. However, there has been a new trend arising, which we may note as the 'third' option.

The bipartisan dominance persisted until the parliamentary election on November 8th, 2015, when, unlike previous elections, neither party won the simple majority number of seats in the parliament to form a Government. HDZ and its coalition partners were the 'relative winner' of the election with 59 parliament seats, and SDP and its coalition partners won 56 seats. For the first time, a 'third option' other than the two dominant parties won 19 seats and became essential in the formation of the coalition government. This third option was an independent list of candidates called 'Most', which also means 'bridge' in the Croatian language.

Both HDZ and SDP negotiated with 'Most' for a post-election coalition in order to guarantee a sufficient number of seats to form a Government. The negotiations lasted for almost three months and a Government was finally formed at the end of January 2016. 'Most' decided to make a post-election coalition with the right-wing HDZ, but they appointed an independent person (neither a member of the party nor an elected Member of Parliament) as the prime minister to head the Government. This Government did not last long. It fell apart in September of 2016 and was officially the shortest-lasting Government in the history of modern Croatia[11].

11 | Šurina, M., Žapčić, A. and Sučec, N. (2016, June 17). Oreškoviću izglasano nepovjerenje. Vlada je pala! [Orešković was given a vote of no confidence. The government has fallen!]. *TPortal*. Retrieved from https://www.tportal. hr/vijesti/clanak/oreskovicu-izglasano-nepovjerenje-vlada-je-pala-20160616 (Last Accessed: 19 November 2017).

An early election was held immediately, and a similar scenario repeated. HDZ won relatively, with 61 seats, insufficient to form a Government. SDP and its coalition partners won 54 seats. While 'Most' won 13 seats, less than the previous election, a new 'third option' secured eight seats by the Živi zid party. 'Most' again agreed to form a post-election coalition with HDZ, and they formed a Government in October 2016. Instead of appointing an independent person, the President of HDZ, Andrej Plenković became the prime minister to head the Government. The other third option, Živi zid, chose to stand on the opposition side in the parliament.

These events showed a new trend that had not been seen in modern Croatian history: the rise of a third option. HDZ and SDP were no longer the only two strong political parties.

2.2. The Rise of Populism

The rise of populism in Croatia has been a popular topic among political analysts in recent years. Although the original etymologies of the word populism (*populus*) and the word democracy (*demos*) are similar, in the modern context the two terms are very different. While democracy has a positive connotation, populism has the contrary.

British political theorist Margaret Canovan used populism to classify new parties, in which ideological classifications were loose, inconsistent, and would self-change to what appeals to the people[12]. More modern academics define populism as an ideology that 'pits a group of virtuous and homogenous people against a set of elites who are depicted as depriving the sovereign people of their rights, values, prosperity, identity and voice'[13]. This deprivation comes from fake and/or unrealistic promises that populists give in order to mobilise the support of the people by stimulating their nationalist feelings and passions[14].

The trend of populism in Croatia has been observed by political analysts and become a repeated topic reviewed by the media. Prior to the parliamentary election in 2015, one of the

12 | Canovan, M. (1981). *Populism*. Houghton Mifflin Harcourt.

13 | Albertazzi, D., & McDonnell, D. (2008). *Twenty-first Century Populism: The Spectre of Western European Democracy*. Palgrave MacMillan.

14 | Šalaj, B. (2013, April 18). Demokracija znači vladavinu naroda, a što je populizam? [Democracy means the rule of the people, and what is populism?]. *Večernji list*. Retrieved from https://www.vecernji.hr/premium/demokracija-znaci-vladavinu-naroda-a-sto-je-populizam-540781 (Last Accessed: 27 November 2017).

main selling points during the political campaign held by HDZ was €1,000 in financial aid promised to be given to each new-born in Croatia. The campaign was packaged as a social policy to raise the low birth rate in Croatia. However, when HDZ and its coalition formed a new Government, the policy was never implemented, and the Government soon fell apart. The next HDZ Government decided not to continue with the idea. Many accused the infamous '€1,000 for new-born' promise as a populists' move to gain votes in the election[15].

Accusations of populism have also been levelled against the above-mentioned third option Zivi zid. A member of parliament, Ivan Pernar from that party created media publicity when he decided to go on public speaking tours to high schools to meet and greet with high school students, as well as openly invite students to leave school and visit his office in the parliament. Despite gaining extreme popularity due to his concerns for the needs and opinions of the youth, Pernar was investigated by the Croatian State Attorney Office (*Državna odvjestništvo Republike Hrvatske* or DORH) for entering a high school without prior notice and inviting students to leave classes. Croatia's ombudsman for children condemned his actions and accused him of manipulating children for achieving political goals[16].

Populism has spread on a global level. Ample examples of politicians inciting populist sentiments were seen in the United States, where Donald Trump used populism to take over the presidential office; and in the UK, where Nigel Farage gained majority voter support for Brexit. In Croatia, people are increasingly dissatisfied with the social and economic state of the country. So, when alternative politicians came up and offered 'something new, something else, something that the people want and something that is in the interest of the people'[17], they became popular. Their popularity comes from the promise of hope, even though they do not have a clear plan and agenda on how to reach the promised goals[18].

Looking at the trend in Europe and on the global level, populism in Croatia will not disappear any time soon. On the

15 | Martinović, B. (2016, July 13). HDZ-ov predizborni adut: Novorodene bebe primat će €1000 samo jedan mjesec [HDZ's pre-election trump card: Newborn babies will receive €1,000 for just one month] *Novilist*. Retrieved from http://www.novilist.hr/Vijesti/Hrvatska/HDZ-ov-predizborni-adut-Novorodene-bebe-primat-ce-1000-eura-samo-jedan-mjesec?meta_refresh=true (Last Accessed: 17 October 2017).

16 | Telegram redakcija. (2017, January 31). Populizam se u Hrvatskoj, izgleda, oteo kontroli. Pernar ide po školama, a svi se pitaju kako ga zaustaviti [Populism in Croatia seems to have spiraled out of control. Pernar goes to schools, and everyone wonders how to stop him]. *Telegram*. Retrieved from http://www.telegram.hr/politika-kriminal/populizam-se-u-hrvatskoj-izgleda-oteo-kontroli-nitko-ne-moze-zaustaviti-pernarov-pohod-po-skolama/ (Last Accessed: 30 November 2017).

17 | Leinart Novosel, S. (2017, February 11). Interview in Populizam u Hrvatskoj: Zašto se širi i tko ga može zaustaviti? [Interview in Populism in Croatia: Why is it spreading and who can stop it?]. *Jutarnji list*. Retrieved from http://republika.eu/novost/61905/populizam-u-hrvatskoj-zasto-se-siri-i-tko-ga-moze-zaustaviti (Last Accessed: 30 November 2017).

18 | *Ibid*.

contrary, it might be present even more prominently as the popularity of the big two political parties' diminishes. In future parliamentary elections, the third options might get even more votes and become a significant decision-maker about the type of government that will be formed.

2.3. Towards the Schengen Area and the Euro Zone

The Schengen area is named after the Schengen Agreement. It is an area comprising 26 European states that have abolished passport and all types of border control at their mutual borders for its member states[19]. It works as a single country for international travel purposes and has a common visa policy. After accession to the EU, Croatia's focus was to join the Schengen Area and the Euro Zone. Although accession to the EU already ended the need for Croatians to travel with a passport to the Schengen area and for customs checks and duties when importing goods, border control had still not been abolished in late 2020 as Croatia was yet to become a member of the Schengen area.

Croatia borders Slovenia and Hungary, so these borders are gateways to the Schengen area. Border control with Slovenia could get very crowded and lengthy in the summer season when many European tourists travel south to Croatian coasts. Becoming a part of the Schengen area will bring convenience not only to Croatians travelling to the Schengen Area, but also to northern and western Europeans travelling to Croatia. However, with over 500 kilometres of coastline along the Adriatic and Mediterranean seas, and a land border with non-EU member states such as Serbia, Bosnia & Herzegovina, and Montenegro, Croatia will have a huge responsibility of guarding the border of the Schengen area.

Starting from June 2017, Croatia has become part of the Schengen Information System (SIS), a widely used information sharing system for security and border management in Europe[20]. By being a part of this system, Croatia is now able to 'exchange information with other Member States... on persons wanted in relation to terrorism and other serious

19 | European Commission. (n.d.). *Schengen Area*. Retrieved from https://ec.europa.eu/home-affairs/what-we-do/policies/borders-and-visas/schengen_en.

20 | European Commission DG Migration and Home Affairs. (2017, July 27). *Croatia Becomes Part of the Schengen Information System (SIS)*, Memo/16/4427.

crimes, missing persons and certain objects such as stolen vehicles, firearms and identity documents'[21].

A total of €120 million has been spent so far in regard to the implementation of the recommendations, mostly for the procurement of land vehicles, vessels, and helicopters for borders supervision; construction or renovation of border police offices and buildings; and procurement of thermal vision cameras, explosive detectors, and counterfeit document detectors.

Other than the accession to the Schengen area, Croatia is also in the process of accessing the Euro Zone. This process is predicted to take longer than that of the Schengen area accession. The recently introduced Strategy of Introducing the Euro in Croatia concluded that it is in Croatian interests to introduce the Euro as the new currency to replace the Kuna (HRK) as soon as possible based on cost and benefit analysis[22]. Introducing the Euro would decrease the macroeconomic risk and increase financial stability, as well as leading to better financing conditions and lower transaction costs[23]. The Strategy also claimed that using the Euro would positively influence the foreign investments that come into Croatia. According to this Strategy, Croatia should replace the Kuna with the Euro by 2022.

Compared to the Schengen accession process, public support is low for the Euro accession process. Media outlets in Croatia tend to analyse the negative side effects if the Euro replaces the Kuna, such as high inflation and increase in commodities prices. However, the benefit that comes with the introduction of the Euro outweighs the side effects. Based on the Italian experience of introducing the Euro, research conducted by Manasse, P., Nannicini, T. and Saia A. provided sufficient evidence that, although prices of primary and secondary goods doubled overnight for citizens, the long-term benefits of introducing the Euro were substantially higher than its cost given that it would also lead to a decrease in yields of government bonds resulting in low interest rates, reduction in loans, and an increase of GDP per capita[24].

21 | European Commission. (n.d.) *Schengen Information System (SIS)*. Retrieved from: https://ec.europa.eu/home-affairs/tags/schengen-information-system-sis_en.

22 | Hrvatska Narodna Banka [Croatia National Bank]. (2018, April). *Strategija za uvođenje Eura kao službene valute u Hrvatskoj [Strategy for the introduction of the Euro as the official currency in Croatia]*. Retrieved from: https://euro.hnb.hr/documents/2070751/2104255/h-strategija-za-uvodenje-eura-kao-sluzbene-valute-u-HR.pdf/69a1c208-c601-4df3-95f6-d336f665b5f9.

23 | *Ibid.*

24 | Manasse, P., Nannicini, T., & Saia, A. (2014). *Italy and the Euro: Myths and Realities*. CEPR Policy Portal. Retrieved from http://voxeu.org/article/italy-and-euro-myths-and-realities (Last Accessed: 28 November 2017).

3. Opportunities and Challenges for Young People in Croatia

After providing a brief account of Croatia's political direction, this section provides an evaluation of the current opportunities and challenges that are faced by the young people of Croatia. It will review the current trend of mobility in the post-EU-accession, education, youth unemployment, and the massive emigration of young people. Moreover, it will explain the political, social, and economic policies that were implemented by the Government to tackle the above-mentioned problems.

Questionnaires (see Appendix 1) were designed to gather first-hand information regarding the satisfaction levels and opinions of young Croatian nationals aged between 25 and 35. The sample was 30 randomly chosen respondents within the age framework. 6.7% of the respondents' highest level of education attained is a bachelor's degree, 66.7% in a master's degree, 13.3% in a non-doctoral post-graduate degree, and 13.3% in a doctoral degree. The gathering of research findings is a combination of 1) distributing online questionnaires to respondents who independently answered the questionnaire through a computer programme called LimeSurvey, 2) distributing paper-based questionnaires which respondents independently answered, and 3) face-to-face interviews (F2F) in which respondents were interviewed in person. Although all respondents were given a chance to attend short interviews, only less than 20% did the interview.

3.1. High Mobility in the European Higher Education System

Mobility was one of the many benefits campaigned on during the EU accession process. It gives Croatian students and young people access to the opportunities of studying and working abroad with almost no barrier. Despite the benefits brought by high mobility, it is also important to discuss the potential side effects of an increasing trend of youth emigration.

In 2007, the Croatian Government established the Agency for Mobility and Programmes of the European Union under the supervision of the Ministry of Sciences and Education to implement the then two most important EU programmes for education and training: 1) The Lifelong Learning Programme and 2) the Youth in Action Programme. In the 2014-2020 programming period, the Agency implemented the Erasmus+ (EuRopean Community Action Scheme for the Mobility of University Students) Programme in the field of education, training, youth, and sport; and the Horizon 2020 Programme in the field of research and innovations.

After the adoption of the European Credit Transfer and Accumulation System (ECTS), studying abroad for young Croatians has become even easier. With the available above-mentioned EU programmes, student and employee mobility are highly encouraged. The most popular destination for Croatian students to study abroad is its neighbouring country Slovenia. In 2015, Croatian students made up 30% of the total foreign students studying in Slovenia[25]. Similarly, Slovenes made up the largest share of international students studying in Croatia (14%), followed by Germans (9%) and Swedes (6%).

Croatian students are well-known for their active participation in mobility programmes. In 2017, the students and volunteers of the Croatian Erasmus Student Network won the Best Country's Student Network Award for their work in assisting Erasmus students who are currently in exchange programmes, and their continuous work in developing the future of the Erasmus+ Programme.

100% of the respondents of this research believe that higher mobility for Croatian students and workers brings positive impact (see Appendix 1 – Part 2 Question no. 4). This is because the programmes bring young Croatians opportunities to learn and train in other European countries, and thus create a more competitive workforce in Croatia. 80% of the respondents also believe that the accession of Croatia to the EU had a positive impact on improving the number of students and worker mobility in Croatia.

3.2. High Youth Unemployment Rate

Despite the high mobility in the EU, it is important not to neglect one of the most notable challenges encountered by

25 | Eurostat. (2017). *Share of Tertiary Education Students from Abroad by Country of Origin for the Three Largest Partner Countries, 2015*. Retrieved from http://ec.europa.eu/eurostat/statistics-explained/index.php/File:Share_of_tertiary_education_students_from_abroad_by_country_of_origin_for_the_three_largest_partner_countries,_2015_(%25_of_all_tertiary_education_students_from_abroad)_ET17.png (Last Accessed: 7 December 2017).

Croatian youths – the high youth unemployment rate. The general unemployment rate in Croatia is one of the highest among EU member states. In 2016, 13.4% of the active population was unemployed. Only Greece (23.6%) and Spain (19.6%) had higher unemployment rates than Croatia in the EU[26]. However, the unemployment rate in Croatia has been decreasing, compared to 16.1% in 2015 and 17.2% in 2014[27].

On the other hand, the youth unemployment rate is much higher. In 2016, the number of unemployed persons aged 25 or less in Croatia was 31.8% of the active population, much higher than the EU average of 18.7%. In the EU, Greece has the highest level of youth unemployment (47.3%), followed by Spain (44.4%), Italy (37.8%), and then Croatia.

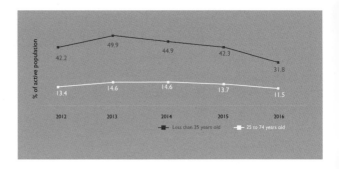

Figure 1: Unemployment rate by age in Croatia as the percentage of the active population[28]

As expected, the lowest unemployment rate in Croatia lies within the population that attained tertiary education (college or university education), with the unemployment rate at only 7.8% in 2016. The percentage is higher for those with upper secondary and post-secondary non-tertiary education (e.g. high school, gymnasium or equivalent) at 14.6%, and highest for those with less than primary and lower secondary education (e.g. elementary or junior high school) at 17.4%.

At the end of 2016, the Croatian Ministry of Labour and Pension System, together with the Croatian Employment Service (HZZ), launched the initiative called 'from measures

26 | Eurostat. (2017). *Unemployment by Sex and Age – Annual Average.* Retrieved from http://ec.europa.eu/eurostat/statistics-explained/index.php/Unemployment_statistics.

27 | Eurostat. (2017). *Unemployment by Sex and Age – Annual Average.* Retrieved from http://ec.europa.eu/eurostat/statistics-explained/index.php/Unemployment_statistics.

28 | *Ibid.*

to careers'. It contains nine measures of active employment policies with the main goals of increasing the employment level, providing education for workers, and protecting workplaces. The nine measures are as follows[29]:

1. **Support for employing** is offered to employers. This support co-finances 50% of the total cost of a worker's gross salary for 12 months if the employer employs people with no work experience who are registered in the database of the unemployed; or people with work experience and special circumstances (children of homeland war veterans, long-term unemployed youth up to 29 years-old, unemployed with no high school degree, etc.)

2. **Support for training** is offered to employers who own profitable businesses. This support co-finances up to 70%, maximum at HRK15,000.00 (US$2,474), to train their workers in applying new technologies or programmes at work to raise productivity and competitiveness.

3. **Support for self-employment** is offered to registered unemployed people who decide to start a business. The support subsidises up to HRK140,000 (US$23,088) to start a company, funding that can be used for procuring related equipment, leasing venues, and paying obligatory contributions for the first 11 months.

4. **Education for the unemployed** is offered to the unemployed registered in the Bureau. It finances 100% of the education cost in educational institutions, and 50% of the legal minimum wage to people who received education for 6 months.

5. **Training for workplaces** is offered to the unemployed who want to receive training for specific skills needed for a position in a company that can offer mentoring for 6 months. The unemployed person is given a legal minimum wage (HRK 2,620.80 or US$432) monthly for 6 months, and the mentoring company receives HRK

29 | Croatian Ministry of Labour and Pension System. (2016). *Od mjere do karijere [From measure to career]*. Retrieved from http:// mjere.hr/ (Last Accessed: 5 April 2018).

700.00 (US$115) monthly for every person mentored.

6. **Professional training without establishing a working relationship** is offered to unemployed people up to 30 years old with less than 12 months of working experience. For 12 or 24 months, the unemployed in training receive minimum wage and stipends for transportation cost, and the state, instead of the employer, pays for his/her monthly obligatory retirement insurance and health insurance.

7. **Public work** is offered to the long-term unemployed, handicapped people, and the non-competitive workforce. It finances up to 100% of the cost of the person's gross salary if he/she is employed by local authorities or civil society organisations to provide on public works such as care for the elderly in households or institutions, care for disabled persons, assistance for children and youth with disabilities, revitalisation of public space, waste management, water management, pollution control, care for cultural heritage, communal works, etc.

8. **Support for keeping a working place** is offered to the employees of a company that is suffering from problems or facing losses that are threatening the business. It co-finances up to 40% of the employees' total gross salary for 6 months.

9. **Permanent seasonal work** is offered to people who have worked seasonally for at least six months and plan to work for another season for the same employer. For six months of the gap period of the seasonal work, the support co-finances up to 100% of retirement and health insurance of the worker that the employer normally has to pay, and the worker receives up to 70% of the national average salary for the first three months and 35% of the average salary for the second three months.

Although some of the above-mentioned measures are highly popular among young people, 46.7% of the respondents to the questionnaire are not sure if they are effective at reducing youth unemployment (see Appendix 1 – Part 5 Question no. 1). 33.3% believe that the measures are not effective, while only 20% are convinced that the measures are fit for their purposes. Most of the measures are also seen to be beneficial only to the employer by providing them with a cheap workforce, rather than for the personal and professional development of young people.

In the short follow-up interviews, respondents suggested other kinds of measures and/or policies to reduce youth unemployment such as: providing timely information, increasing transparency of public procurements and competitions, creating high quality of vocational training and life-long learning education opportunities, providing free workshops with a professional orientation, providing free workshops to stimulate creative skills and foreign language skills, and creation of a more flexible labour market.

Although the HZZ cooperates with the Ministry of Labour and Pension System in implementing the measures, its performance is questionable as 46.7% of the respondents had never used the services of the HZZ to find a job, nor their other services such as counselling. Nevertheless, it is important to note that the results might be caused by the fact that the respondents are highly educated and did not need the service, or it could also be caused by common scepticism of the youth towards the institutions. Unfortunately, there is no available data that measures the percentage of unemployed youth who use the service provided by the HZZ.

3.3. Massive Emigration to Other Countries

Emigration of Croatians to other countries is an emerging trend. Many Croatian citizens, especially young people, have decided to move to other countries for better work opportunities. The issue of emigration is further worsened by the constant decrease of the Croatian population. In 2011, when the last national census was conducted, the total population of Croatia was 4.28 million people, 0.15 million less than that of 2001 when the population was 4.43 million. With the decreasing population, the trend of increasing emigration is not favourable to Croatia[30].

The migration balance of Croatia has been in deficit since 2009. The last time the balance was positive was in 2008, when there were 7,053 more immigrants moving into Croatia than the Croatians moving abroad. Since then, the balance has been negative. In 2016, 36,436 Croatians moved abroad to live and work, while only 13,985 foreigners immigrated to Croatia. The net number of Croatians leaving the country has been increasing since 2009.

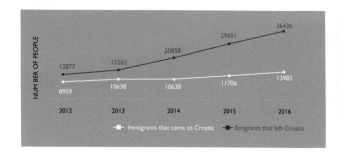

Figure 2: International migration of the population of Croatia[31]

Together with a low birth rate, net emigration would speed up the decrease of population, and in the long term, create a huge burden to the state's pension system, when the number of retired people outnumbers the number of the active working population in Croatia.

Similar to the issue of unemployment, the majority of the emigrated population is young Croatians. Out of the 36,436 Croatians who left the country in 2016, 14% were aged between 25-29, 13% were aged between 30-34, 11% were aged between 35-39, and 8% were aged between 20-24[32]. In other words, almost a quarter of the emigrated population comprised young people aged between 20 and 29.

79% of the emigrants in 2016 moved to other EU member

30 | Croatian Bureau of Statistics. (2013). *Census of Population, Households and Dwellings 2011, Population by Sex and Age.* Statistical Reports, ISSN 1333-1876.

31 | Croatian Bureau of Statistics. (2017). *Migration of Population of Republic of Croatia, 2016.* Retrieved from: https://www.dzs.hr/Hrv_Eng/publicati on/2017/07-01-02_01_2017. htm.

32 | *Ibid.*

states, and the rest moved to non-EU European countries or other countries in other continents. The most popular destination for Croatian emigrants was Germany with 20,432 people (56% of the total emigrants), Austria with 2,164 people (5.9%), and Ireland with 1,917 people (5.3%)[33]. These countries are known as the most promising lands for Croatian workers due to their rapid economic growth.

When asked about the link between youth unemployment and massive emigration (see Appendix 1 – Part 4 Question no. 3), 80% of the respondents expressed the view that massive emigration is directly linked with the high level of unemployment. Although only 33.3% of the respondents had actually left Croatia, 66.7% sometimes think about and/or consider moving abroad.

In order to keep young people in Croatia, the respondents suggest the following measures: reducing taxes and burdens for entrepreneurs; reforming the entire education system; punishing nepotism and corruption; creating better work conditions; increasing competitiveness and excellence; introducing material and non-material rewarding systems; increasing opportunities to get promoted based on work performance; and creating a more flexible labour market. These measures could create positive impacts such as the creation of more SMEs (small and medium-sized enterprises) and start-ups and a possible overall industrial growth that could enable young people to have a fulfilling job and career in the country. Punishing nepotism and corruption is also important to gain back people's trust and confidence towards the institutions that they see failing in doing their tasks.

33 | *Ibid.*

4. Croatia and the Belt and Road Initiative[34]

34 | This section heavily draws from the author's work in Raditya-Lezaic, A. & Boromisa, A. (2018). China-Croatia Cooperation: Past, Present, and Future. *How the 16+1 Cooperation Promotes the Belt and Road Initiative China-CEEC Think Tank Book Series.* Huang, P. & Liu, Z. (Eds.). Paths International Limited.

35 | This paragraph is a direct quote from the author's work in Raditya-Lezaic, A. & Boromisa, A. (2018). China-Croatia Cooperation: Past, Present, and Future. *How the 16+1 Cooperation Promotes the Belt and Road Initiative China-CEEC Think Tank Book Series.* Huang, P. & Liu, Z. (Eds.). Paths International Limited.

36 | Croatian Government. (2017). *Memorandum o suglasnosti izmedu Vlade Narodne Kine I Vlade Republike Hrvatske o suradnji u okviru gospodarskog pojasa put svile I inicijative Jedan pojas, jedan put' [Memorandum of Understanding between the Government of the Republic of Croatia and the Government of the People's Republic of China on cooperation within the framework of the Silk Road Economic Belt and the 21st Century Maritime Silk Road Initiative].* Retrieved from https://vlada.gov.hr/UserDocsImages//2016/Sjednice/2017/05%20svibanj/36%20sjednica%20VRH//36%20-%2012.pdf.

37 | This paragraph is a direct quote from the author's work in Raditya-Lezaic, A. & Boromisa, A. (2018). China-Croatia Cooperation: Past, Present, and Future. *How the 16+1 Cooperation Promotes the Belt and Road Initiative China-CEEC Think Tank Book Series.* Huang, P. & Liu, Z. (Eds.). Paths International Limited.

After discussing the opportunities and challenges faced by Croatian youths, this part of the paper briefly describes the Belt and Road Initiative (BRI) in Croatia, including the investments made. It then analyses the opportunities and challenges of the BRI in Croatia, and whether the initiative offers solutions to the problems faced by the young people identified in the previous part.

4.1. The Belt and Road Initiative in Croatia

A Memorandum of Understanding (MOU) between the Government of the Republic of Croatia and the Government of the People's Republic of China on cooperation in the framework of the Belt and Road Initiative (BRI) was signed and later approved by the Croatian Government in June 2017.[35] The MOU paves the ground for five areas of cooperation between Croatia and China[36]:

1. Transport, logistics and infrastructure focusing on cargo transportation, seaports, railways and logistic centres, and additionally establishing direct flights between the two countries;

2. Trade and investments including cooperation between companies, with a focus on wood products, machines, pharmaceutical products, cosmetics, foods, and clothing;

3. Financial cooperation for supporting trade and investments;

4. Science and technology to stimulate companies from both countries to cooperate in the sector of high and new technology, and innovations; and

5. Cultural and people exchange in the sectors of education, arts, and culture.

Even before the BRI MOU was signed, China had shown great interest in transportation infrastructure projects in Croatia.[37] In February 2016, in a meeting between the Chinese Ambassador

38 | Croatian Ministry of Maritime Affairs, Transport and Infrastructure. (2016). *Minister Butković Met with a Chinese National Development and Reform Commission.* Retrieved from: https://mmpi.gov.hr/ infrastructure/news/minister-butkovic-met-with-a-chinese-national-development-and-reform-commission-ndrc/18098.

39 | Croatian Government. (2016). *Memorandum o razumijevanju o suradnji na području luka I lučkih industrijskih parkova izmedu Nacionalne komisije za razvoj I reforme Narodne Republike Kine I Ministarstva gospodarstva, maloga I srednjega poduzetništva I obrta Republike Hrvatske [Memorandum of Understanding (MOU) On Port and Harbour Industrial Park Cooperation Between The National Development and Reform Commission of the People's Republic of China And The Ministry of Economy, Small and Medium Entrepreneurship and Crafts of the Republic of Croatia].* Retrieved from https://vlada. gov.hr/UserDocsImages//2016/ Sjednice/2016/2%20sjednica%20 14%20Vlade//2%20-%209.pdf.

40 | This paragraph is a direct quote from the author's work in Raditya-Lezaic, A. & Boromisa, A. (2018). China-Croatia Cooperation: Past, Present, and Future. *How the 16+1 Cooperation Promotes the Belt and Road Initiative China-CEEC Think Tank Book Series.* Huang, P. & Liu, Z. (Eds.). Paths International Limited.

41 | Milekić, S. (2016, May 30). *Major Chinese Business Summit Opens in Croatia.* Balkan Insight. Retrieved from http://www. balkaninsight.com/en/article/ biggest-chinese-business-summit-kicks-off-in-zagreb-05-27-2016 (Last Accessed: 8 December 2017).

to Croatia and the Croatian Minister of Maritime Affairs, Transport and Infrastructure, it was expressed that China was interested in participating in the construction project of a railway connection between Rijeka via Zagreb to the border of Hungary as well as the upgrade of the Port of Rijeka[38]. Later in November 2016, an MOU was signed between the two countries on the Port and Harbour Industrial Park, a cooperation between the Chinese National Commission for Development and Reforms and the Croatian Ministry of Economy, Entrepreneurship and Crafts. There were three areas in which cooperation was encouraged: (1) investment, construction, modernisation, and management of seaports; (2) development, construction, and management of industrial parks including investments in projects; and (3) construction of railways, highways, logistic centres, and storage in seaports and industrial parks[39].

A summit between Chinese and Croatian businessmen aimed at boosting economic relations between the two countries was held in 2016 by the Chinese embassy in Zagreb.[40] A hundred businessmen, representatives of over 50 companies from the coastal region of Zhejiang in China came to Zagreb, Croatia, to meet over 50 companies from Croatia, to discuss possibilities of investment and trade in a spectrum of sectors like construction, communications, chemicals, energy, car-making, electronics, and food. In this summit, the companies from China showed interest in investing in the ports of Rijeka in the north and in Ploče in the south for unloading containers with Chinese goods for Europe[41]. This interest is in sync with the previous Chinese interest of building a railway from Rijeka to Budapest via Zagreb, which is essential to the success of BRI.

For the BRI, Croatia offers one of the shortest connections from Asia to Western Europe given the fact that three Pan-European Corridors (V, VII and X) pass through Croatia, as well as the three Trans-European Transport Network corridors (Mediterranean, Baltic-Adriatic, and Rhine-Danube).[42] However, although the port infrastructure is available, cargo and turnover are limited. The needs are therefore mutual because China could provide cargo and turnover. Shipments from the Far

42 | This paragraph is a direct quote from the author's work in Raditya-Lezaic, A. & Boromisa, A. (2018). China-Croatia Cooperation: Past, Present, and Future. *How the 16+1 Cooperation Promotes the Belt and Road Initiative China-CEEC Think Tank Book Series.* Huang, P. & Liu, Z. (Eds.). Paths International Limited.

43 | Croatian Ministry of Maritime Affairs, Transport and Infrastructure. (2017). *Transport Infrastructure Projects.* PowerPoint presentation at a China-EU Connectivity Platform Meeting.

44 | World Travel and Tourism Council. (2017). *Travel & Tourism Economic Impact 2017 Croatia.* Retrieved from https://www.wttc.org/-/media/files/reports/economic-impact-research/countries-2017/croatia2017.pdf.

45 | This paragraph is a direct quote from the author's work in Raditya-Lezaic, A. & Boromisa, A. (2018). China-Croatia Cooperation: Past, Present, and Future. *How the 16+1 Cooperation Promotes the Belt and Road Initiative China-CEEC Think Tank Book Series.* Huang, P. & Liu, Z. (Eds.). Paths International Limited.

46 | Chinese Embassy. (2016). *Ambassador Deng Ying Gives A Written Interview in Croatian Well-known Magazine 'National'.* Retrieved from http://hr.chineseembassy.org/eng/dssghd/t1420121.htm (Last Accessed: 4 December 2017).

47 | This paragraph is a direct quote from the author's work in Raditya-Lezaic, A. & Boromisa, A. (2018). China-Croatia Cooperation: Past, Present, and Future. *How the 16+1 Cooperation Promotes the Belt and Road Initiative China-CEEC Think Tank Book Series.* Huang, P. & Liu, Z. (Eds.). Paths International Limited.

East to Europe could be eight days shorter in transit times if the Port of Rijeka is used instead of the Port of Hamburg or Rotterdam[43].

4.2. Opportunities for Young Croatians in the Belt and Road Initiative

Chinese investments in Croatia will, directly and indirectly, create jobs. For the young population, this could be an alternative to emigration. In the long term, this will prevent Croatia from losing its young and educated generation.

In 2016, the direct contribution of tourism to GDP was US$5.5 billion or 10.7% of the total GDP, and it created 138,000 jobs, which is 10% of the total employment in the country[44,45]. Investments made in travel and tourism in 2016 amounted to 11% of total investments in Croatia and should rise by 2.5% in 2017.

In an interview in 2016, the Chinese ambassador to Croatia highlighted the need for the Croatian government to promote its tourism to the Chinese market because the rich tourism resources of Croatia are not well known to Chinese tourists and it is a hidden gem for them[46,47]. The number of Chinese tourists visiting Croatia has been rapidly increasing, with 88,000 Chinese tourists visiting in 2015. Although this number is relatively small compared to the total overseas visits of Chinese citizens that amounted to 120 million in 2015, there is huge potential in the sector.

Investments in the framework of the BRI will create jobs for young Croatians, whether it is in the tourism or other sectors. When asked whether Chinese investments within the BRI framework would create jobs for Croatians (see Appendix I – part 5 Question no. 4), 53.3% of the respondents in the questionnaire answered maybe, and 46.7% answered yes. None of the respondents answered no, although the option was offered. This optimism shows that the young people of Croatia think BRI is a relevant alternative for solving the common problems faced by young people, such as high unemployment.

4.3. Main Challenges of the Belt and Road Initiative in Croatia

48 | This paragraph is a direct quote from the author's own work in Raditya-Lezaic, A. & Boromisa, A. (2018). China-Croatia Cooperation: Past, Present, and Future. *How the 16+1 Cooperation Promotes the Belt and Road Initiative China-CEEC Think Tank Book Series.* Huang, P. & Liu, Z. (Eds.). Paths International Limited.

49 | The Economist. (2017, May 18). *Progress and Next Steps for China's Belt and Road Initiative.* Retrieved from http://country.eiu.com/article.aspx?articleId=19 05436574&Country=Croatia&topic=Politics (Last Accessed: 8 December 2017).

50 | Podgornik, B. (2016, November 6). Kinezi nude, Hrvati odbijaju: Europa I svijet zapljusnuti ulaganjima, a mi smo dosad bili – neodlučni *Novilist.* Retrieved from http://www.novilist.hr/Vijesti/Hrvatska/Kinezi-nude-Hrvati-odbijaju-Europa-i-svijet-zapljusnuti-ulaganjima-a-mi-smo-dosad-bili-neodlucni?meta_refresh=true (Last Accessed: 9 December 2017).

51 | This paragraph is a direct quote from the author's work in Raditya-Lezaic, A. & Boromisa, A. (2018). China-Croatia Cooperation: Past, Present, and Future. *How the 16+1 Cooperation Promotes the Belt and Road Initiative China-CEEC Think Tank Book Series.* Huang, P. & Liu, Z. (Eds.). Paths International Limited.

52 | Stanzel, A., Krats, A., Szczudlik, J., & Pavlićević, D. (2016). *China's Investment in Influence: the Future of 16+1 Cooperation.* European Council on Foreign Relations. Retrieved from http://www.ecfr.eu/publications/summary/chinas_investment_in_influence_the_future_of_161_cooperation7204 (Last Accessed: 8 December 2017).

Compared to other Central and Eastern European (CEE) countries in the 16+1 mechanism, Croatia is lagging behind in attracting investments from China.[48] In 2016 alone, the value of newly signed Chinese contracting projects in the countries participating in the Belt and Road Initiative (BRI) reached US$126 billion. Contracts in BRI countries account for almost 50% of China's total turnover in overseas contracting projects[49]. While in Croatia, other than the residential and tourist resorts in Krapinske Toplice, no major investments had been made.

Back in 2012 before the BRI was launched, China had expressed similar interest to invest on the abovementioned railway route (Rijeka port to Budapest via Zagreb) but was unfortunately rejected by the Croatian government[50,51]. Back then, the Croatian government was expecting grants from the EU to finance such a big infrastructure project, once it joins the Union.

Lastly, one of the biggest challenges that face the BRI in Croatia is conforming to EU standards. It has been said that Croatia hesitated in accepting Chinese investments because the deal with China might not conform to the EU guidelines or could undermine EU policies[52]. The issue with the EU standards and regulations on public procurements and market competition has been one of the main reasons of the deadlock deals between Chinese investors and the Croatian Government and/or private businesses, especially when big infrastructural projects are involved in the negotiation.

5. Toward a Successful Belt and Road Initiative[53]

With reference to the opportunities and challenges brought by the BRI to Croatia, this part offers suggestions on achieving a mutual success for both China and Croatia by attempting to mitigate the barriers involved. In the end, it suggests ways which the BRI can incorporate the involvement of young people of Croatia into its overall framework.

5.1. Alternatives and Suggestions to the Belt and Road Initiative

2017 marked the 25th year since the establishment of diplomatic relations between the Republic of Croatia and the People's Republic of China.[54] During the meeting of the China-Croatia Joint Economic and Trade Commission held in Zagreb in March 2017, both parties highlighted the need for more intensive cooperation between the two countries. For a small country like Croatia, every opportunity for infrastructure projects is necessary for it to make use of its geo-strategic potentials.

However, due to the existing legal and political barriers for big transport infrastructure projects – such as port upgrades and railway construction – the first cooperation area within the MOU for BRI, Croatia and China should emphasise more of the focused sectors under the BRI's mechanism.

Trade and investment including cooperation between companies are, for example, an area of cooperation that both countries could further develop.[55] In 2016, trade exchange between Croatia and China was worth US$1.5 million, which was the highest in the history of their bilateral cooperation[56]. However, compared to the total trade exchange of China that reached US$3.6 billion the same year, the amount of trade exchange with Croatia is very small. Therefore, although trade is increasing, it is still significantly below its potential. Croatia's geographical position, knowledge, and local conditions in Southeast Europe make the country a partner worth serious consideration by China. As a member state of the EU, goods and services traded with Croatia are aimed not only at the Croatian market but could reach the entire EU with its population of 500 million people. Croatia, as an EU

53 | This section draws heavily from the author's work in Raditya-Lezaic, A. & Boromisa, A. (2018). China-Croatia Cooperation: Past, Present, and Future. *How the 16+1 Cooperation Promotes the Belt and Road Initiative China-CEEC Think Tank Book Series.* Huang, P. & Liu, Z. (Eds.). Paths International Limited.

54 | This paragraph includes direct quotes from the author's work in Raditya-Lezaic, A. & Boromisa, A. (2018). China-Croatia Cooperation: Past, Present, and Future. *How the 16+1 Cooperation Promotes the Belt and Road Initiative China-CEEC Think Tank Book Series.* Huang, P. & Liu, Z. (Eds.). Paths International Limited.

55 | This paragraph is a direct quote from the author's work in Raditya-Lezaic, A. & Boromisa, A. (2018). China-Croatia Cooperation: Past, Present, and Future. *How the 16+1 Cooperation Promotes the Belt and Road Initiative China-CEEC Think Tank Book Series.* Huang, P. & Liu, Z. (Eds.). Paths International Limited.

56 | HINA. (2017, March 28). *Hrvatska i Kina moraju produbiti suradnju, koristiti četiri platform.* Nacional. Retrieved from http://www.nacional.hr/hrvatska-i-kina-moraju-produbiti-suradnju-koristiti-cetiri-platforme/ (Last Accessed: 9 December 2017).

member, also has access to €1 billion of annual EU structural funds aimed at promoting entrepreneurship. Foreign investors, including Chinese ones, who set up business in Croatia have equal access to compete for this fund as a domestic entity.

Trade partnerships and cooperation between companies should not only involve big enterprises and multinational corporations. It should also involve small and medium-sized enterprises (SMEs). The 16+1 cooperation has recognised the potential of this kind of partnership. And the Budapest Guidelines for Cooperation between China and Central and Eastern Europe Countries (CEEC), which came out of the 6th Summit of China and CEEC, encourages SMEs in China and CEEC to enhance cooperation. It was decided in the Summit that Croatia will be the country to host the 16+1 SMEs centre.

To maintain economic and trade relations between Croatia and China, there is a mechanism called the Sino-Croatian Joint Committee for Economy and Trade that holds regular working meetings.[57] Several agreements have been signed, such as the Agreement on Economy and Trade, the Agreement on Reciprocal Promotion and Protection of investments, the Agreement on Avoiding Double Taxation and Prevention of Tax Evasion, the MoU on the Establishment of Croatian-Chinese Economic and Technological Zone, etc[58].

Referring to the previous part of the paper regarding the tourism sector in Croatia and its role in creating jobs, it shows tourism as an important sector that the BRI should focus on. Along with the transport infrastructures, tourism investments are much welcomed in Croatia. Within the 16+1 framework, the potential of tourism is recognised, and the growth of the sector is stimulated by conducting regular exchanges of experience, sharing best practices, engaging in joint research, organising promotional activities, networking between agencies, and developing regional tourism products. Among the 16 CEE countries, Croatia's potential in the tourism sector is highly recognised, which was proven by the fact that the 4th China-CEEC High-Level Conference on Tourism Cooperation was held in 2018 in Dubrovnik, Croatia[59].

57 | This paragraph is a direct quote from the author's work in Raditya-Lezaic, A. & Boromisa, A. (2018). China-Croatia Cooperation: Past, Present, and Future. *How the 16+1 Cooperation Promotes the Belt and Road Initiative China-CEEC Think Tank Book Series.* Huang, P. & Liu, Z. (Eds.). Paths International Limited.

58 | Croatian Ministry of Foreign and European Affairs. (2018). *Overview of Bilateral Treaties of the Republic of Croatia by Country — China.* Retrieved from: http://www.mvep.hr/en/foreign-politics/bilateral-relations/overview-by-country/china,66.html.

59 | Chinese Ministry of Foreign Affairs. (2017). *The Budapest Guidelines for Cooperation between China and Central and Eastern European Countries.* Retrieved from http://www.fmprc.gov.cn/mfa_eng/zxxx_662805/t1514534.shtml (Last Accessed: 10 December 2017).

5.2. Involvement of Young People in the Belt and Road Initiative

As mentioned, young people in Croatia have a very optimistic view of the BRI. Most respondents are sure that the BRI will bring positive impact by creating jobs for young people. When asked whether the BRI is relevant to the general needs of young people (see Appendix 1 – Part 5 Question no. 5), 46.66% responded yes, 46.66% responded maybe, and only 6.68% responded no. Some of the respondents commented on how the BRI will create a better investment climate which will lead to the creation of jobs. BRI is also expected to raise more investments in research and development, thus increasing the competitiveness of young Croatians.

The support that young people show for the implementation of the BRI and the cooperation between China and Croatia leads to the expectation that they will be highly involved in activities related to the BRI.

Other than job creation, young people might not be able to be directly involved in big infrastructure projects launched along the BRI. However, there is a focused area of cooperation within the MOU on the BRI between Croatia and China that could be highly beneficial to young people, which is cultural and people exchange in the sectors of education, arts, and culture. If the BRI stimulates people-to-people exchange, students, workers and young people in general can be highly involved. It was previously found that young people of Croatia believe that mobility brings a positive impact to them. In fact, 53.3% of the respondents had experienced student exchange, a semester abroad, an internship abroad, or international training themselves (see Appendix 1 – Part 2 Question no. 2). However, mobility for Croatian students and young people have so far been practised mostly within the EU member states. If extended to China on the basis of Croatia-China cooperation, the impact will be even higher.

5.3. Conclusion

Cooperation between China and Croatia is already well-established even though there is an asymmetry because Croatia is a very small country, which makes it hard to become a major Chinese partner. However, due to Croatia's unique geographical position and its being an EU member state, the partnership is worth consideration.

Croatia is ambitious to become a regional transportation hub, which is also grounds for participation in the BRI. However, various barriers and political issues have caused delays in the many interesting transportation infrastructure projects that China sees fit for Croatia. As such, the BRI should also focus on other areas of cooperation such as trade, cooperation between SMEs, and tourism, areas where Croatia exhibits high potential. Further cooperation in these areas will create jobs, increase the competitiveness of the workforce, increase research and development, and create a healthy climate for investments, which could, directly and indirectly, provide solutions to the current problems faced by young people, i.e. youth unemployment and massive emigration.

The optimistic and positive view that young people of Croatia have towards the BRI provides a solid ground for the possibility that they could engage in further involvement in the project. People-to-people relations, cultural and people exchange in the sectors of education, arts and culture, are the areas to develop if the BRI hopes to attain high youth participation.

Appendix: Questionnaire Form

This questionnaire is only in the English language, created for the chapter on the Croatian country report, which evaluates the opportunities and challenges faced by young people in Croatia. The answers are expected to give views on the level of satisfaction of the young people, their knowledge about existing opportunities, their professional and educational barriers and their opinions on the Government's policies to tackle these barriers. Please answer all of the questions. All answers will be used anonymously in the report. Your contribution is highly appreciated.

PART ONE: PERSONAL DATA

1. Please select your age group (highlight or underline):
   a. 20-24
   b. 25-29
   c. 30-34
   d. Over 35

2. Please select your highest degree acquired:
   a. High school or equivalent
   b. Bachelor's degree (*preddiplomski studij, bacc – všs prema starom sustavu*)
   c. Master's degree (*diplomski studij, mag – vss prema starom sustavu*)
   d. Non-doctoral post-grad degree (*mr. sc., post diplomski specijalistički studij*)
   e. Doctoral degree (*dr. sc.*)

3. Please select the major of degree you acquired:
   a. Arts and Humanities: arts, languages, literature, philosophy, etc.
   b. Social sciences: political sciences, economics, communication, history, journalism, etc.
   c. Health and medicine: medicine, dentistry, clinical laboratory, pharmacy, nursing, etc.
   d. Science, math and technology: ICT, engineering, math, statistics, architecture, etc.
   e. Business: accounting, finance, human resources, management, etc.
   f. Other, please specify: _____

PART TWO: MOBILITY IN EDUCATION AND TRAINING

Please answer with yes or no by filling in the right column with 'x'

NO.	QUESTION	YES	NO	MAYBE
1	Have you ever been involved in any projects or activities within the framework of the EU's Life-long Learning Programme, Youth in Action Programme, Erasmus+ Programme or Horizon 2020 Programme?			
2	Have you ever experienced student exchange, a semester abroad, an internship abroad or any international training during your study period (in high school or university)?			
3	Are you familiar with the Agency for Mobility and Programmes of the European Union and their work?			
4	Do you think mobility brings positive impact for Croatian students and workers?			
5	Do you think that Croatian accession to the European Union in 2013 has led to increasing numbers of students, researchers or workers who go abroad for exchanges and training?			

PART THREE: YOUTH UNEMPLOYMENT

Please answer with yes or no by filling in the right column with 'x'

NO.	QUESTION	YES	NO	MAYBE
1	Are you employed or self-employed full-time?			
2	Are you seeking employment?			
3	Have you ever used the help of governmental institutions (such as Croatian Employment Service (HZZ) – Zavod za zapošljavanje) to find a job?			
4	Do you think that the Croatian workforce is competitive in the EU market?			
5	Do you think the state provides enough assistance in regards to youth employment (such as help in finding a job, providing a personal consultancy with an advisor from the Croatian Employment Service (HZZ), providing free health insurance for unemployed youth, providing free workshops or training to enhance skills of the unemployed)?			

OPTIONAL QUESTION FOR PART THREE
(can be answered in English or Croatian)

What is your personal opinion on the work done by Governmental institutions regarding their assistance for unemployed youth, and do you have suggestions for policy that can be done to increase assistance for the unemployed youth?

_____

_____

_____

_____

_____

_____

PART FOUR: EMIGRATION OF THE YOUNG PEOPLE
Please answer with yes or no by filling in the right column with 'x'

NO.	QUESTION	YES	NO	MAYBE
	Did you move abroad and emigrate from Croatia?			
	Do you sometimes think of and/or consider moving abroad?			
	Do you think that the problem of massive emigration is directly linked with the problem of youth unemployment?			
	Do you think that Croatian accession to the EU in 2013 is connected with the increase in the number of young people moving abroad?			
	Do you think the state provides enough incentives to keep young people from moving abroad?			

OPTIONAL QUESTION FOR PART FOUR
(can be answered in English or Croatian)

In your opinion, what should the Government have done more in order to keep the young and educated people to stay and work in Croatia?

_____

_____

_____

_____

_____

_____

PART FIVE: POLICY AND BRI

Please answer with yes or no by filling in the right column with 'x'

NO.	QUESTION	YES	NO	MAYBE
1	Do you think that active measures for employment implemented by the Government (such as traineeship without working relations, support for self-employment or public works are good measures to reduce youth unemployment?			
2	Do you believe that there are policy measures (that are not yet implemented by the Government yet) that could tackle the problem of the massive emigration of young people?			
3	Do you believe that there is any correlation between the rise in Foreign Direct Investments that come into Croatia and the decline of youth unemployment and youth emigration?			
4	Do you think that Chinese investment within the framework of Belt and Road Initiative (or the new Silk Road) in Croatia could create employment for young people in Croatia?			
5	Do you think that that Belt and Road Initiative is relevant to the current needs of the young people?			

OPTIONAL QUESTION FOR PART FIVE
(can be answered in English or Croatian)

In your opinion, how could Belt and Road Initiative and Chinese investments positively or negatively impact the life of the young people in Croatia?

_____

_____

_____

_____

_____

_____

The old town square in Prague,
Czech Republic, at sunset.

CHAPTER 6

by   ALEXANDR LAGAZZI
University of Economics in Prague

JAROSLAV TON
University of Oxford

(Ordered by last name)

CZECH
REPUBLIC

CZECH
REPUBLIC

ESTONIA

LATVIA

LITHUANIA

POLAND

SLOVAKIA

HUNGARY

SLOVENIA

CROATIA

BOSNIA &
HERZEGOVINA

SERBIA

ROMANIA

MONTENEGRO

BULGARIA

NORTH
MACEDONIA

ALBANIA

1. Overview of the Political Structure of the Czech Republic

In 2018, the Czech Republic celebrated its 25th year of existence. It was also exactly 100 years since its democratic predecessor, Czechoslovakia, declared independence from the Austro-Hungarian Empire in 1918. In that 100-year period, Czechs experienced severe shifts in both the political direction and structure of the country before Czechoslovakia (peacefully) split in 1993 into the Czech Republic and Slovakia. Originally a democratic republic, Czechoslovakia lost its independence when it was annexed and re-forged into a Nazi protectorate from 1939 to 1945, only to become a *de facto* socialist republic and Soviet satellite until 1989.

This chapter will discuss the characteristics of the political structure of the Czech Republic and the opportunities present with regard to the Belt and Road Initiative (BRI). For the purposes of this chapter, the authors have deliberately excluded assumptions and/or speculations about the political structure of the various forms of states that existed before the official declaration of Czech independence in 1993. Thus, the years described shall be from 1993 until today.

1.1. Main Characteristics of the Czech Political Structure

Following the original Czechoslovak Constitution from 1920, the current Czech Constitution was adopted on 16 December 1992 (entry into force on January 1, 1993)[1] and it characterises the country as a 'sovereign, unitary, and democratic state governed by the rule of law, founded on respect for the rights and freedoms of man and of citizens'[2]. The Czech Republic is a pluralist parliamentary democracy that follows a continental civil law system, which incorporates human rights of its citizens as the main component.

The parliament is bicameral, consisting of the Senate (upper house) and the Chamber of Deputies (lower house) with 81 and 200 Members respectively. The president of the Czech Republic has limited powers, but most importantly he has the right to return bills to the parliament and to appoint the prime minister. The prime minister appoints the cabinet. The rest of the executive power is exercised by the Government,

1 | The full text of the 1920 Constitution is available at https://www.vlada.cz/en/media-centrum/aktualne/constitution-1920-68721/

2 | Constitute Project. (2002). *Czech Republic's Constitution of 1993 with Amendments through 2002.* Retrieved from https://www.constituteproject.org/constitution/Czech_Republic_2002.pdf.

which is responsible to the lower house of the parliament and has exclusive legislative initiative in terms of the state budget. Thus, considerable power is in the hands of the prime minister, who, given his mandate as the Head of Government, takes the helm of both domestic and foreign policies. The prime minister stays in office for as long as he/she retains the support of the majority of the Lower House members[3].

In terms of the legal order, the legislative process is as follows: when the Chamber of Deputies is presented with a bill, the lower house holds three readings of it. Then, if the bill is passed (by a majority of the present members), it is sent to the Senate. The upper house then has the power to veto it or send it (often with amendments) to the president, who formally signs it into law. Few major legislative changes have occurred since the 1993 Constitution was enacted with the exception of two changes – first, an amendment in the Constitutional Acts to present a legal framework for the country's European Union (EU) membership (2001)[4] and second, to allow a direct election of the president by popular vote (2012)[5]. Apart from those, a new Criminal and Civil Code was also entered into force in 2010 and 2014 respectively, updating the previous legislation.

When electing members to legislative bodies, the Czech Republic holds secret ballots by direct voting, electing the Lower House on the basis of proportional representation (a single party must obtain the threshold of at least 5% of the popular vote to gain seats or 10% in case of a two-party coalition), and the Upper House on the basis of a majority vote in a two-round runoff. The Czech Republic is divided into 14 constituencies and the D'Hondt method[6] is used as the calculation assuring proportional allocation of seats to the parties.

1.2. Czech Domestic Policy

In the electoral term that spanned from 2013 to 2017, the key document submitted in February 2014 by the Government was its Policy Statement. This was presented together with the

3 | Czech Government. (2017). *Members of the Government: Prime Minister.* Available at: https://www.vlada.cz/en/clenove-vlady/premier/premier-en-50677/.

4 | See Act No. 395/2001 Coll.

5 | See Act No. 71/2012 Coll. Available in English at: http://czecon.law.muni.cz/dokumenty/31586.

6 | For more information about the D'Hondt method, see the European Parliament Briefing on EU Democracy, Institutional and Parliamentary Law, available here: http://www.europarl.europa.eu/thinktank/en/document.html?reference=EPRS_BRI(2016)580901.

Coalition Agreement between the Czech Social Democratic Party (ČSSD), the ANO 2011 Movement[7] and the Christian Democratic Union – Czechoslovak People's Party (KDU-ČSL) that formed the Government of the Czech Republic (by gaining 20.45 %, 18.75 % and 6.78 % of votes respectively).[8]

The document presented the political priorities and outlined an agenda to fulfil them. The 2013-2017 Government, dubbed by then Prime Minister, Bohuslav Sobotka, first outlined the framework with five main priority points: 1) sustainable economic growth, increased competitiveness and efficient use of European funds; 2) rationalisation of state economic management with a focus on transparency; 3) eradication of corruption; 4) further development and universalisation of access to public services; 5) strengthening the rule of law; and active membership in the EU, NATO and the UN. Then, the document presented a 12-point programme that gave a detailed description of future policies:

1. In terms of economic development, the focus was clearly on science and research in technological advancements and how to achieve sustainable growth, transport modernisation and lowering energy intensity in the economy.

2. Budget and taxes-wise, a call was made for a rational central government budget and for enhancing efficiency in collecting taxes with regards to the newly adopted Civil Code from 2014.

3. In the agriculture field, an effort to increase competitiveness and keep up with the EU standards.

4. On climate change, a call for efforts to maintain a favourable environment and sustainable development through nature and landscape protection, waste reduction and preserving clean air standards.

5. For employment, there will be targeted maintenance of social cohesion through active employment policy and

7 | ANO. (2017). *Program hnutí ANO pro volby do Poslanecké sněmovny 2017 [Programme of the YES movement for the elections to the Chamber of Deputies 2017]*. Retrieved from https://www.anobudelip.cz/file/edee/2017/09/program-hnuti-ano-pro-volby-do-poslanecke-snemovny.pdf.

8 | The full result of the 2013 parliamentary elections in the Czech Republic is available at: https://volby.cz/pls/ps2013/ps2?xjazyk=CZ.

pushing for a return to regular pension increases linked to the rate of inflation.

6. In terms of healthcare, efforts will be made to improve its accessibility and quality.

7. For education, a call for greater affordability and quality whilst focusing on science, research, and innovation.

8. On culture, the government will strive to develop arts and support sport as important elements of civic and social life.

9. On the rule of law, an appeal to consistently adhere to the principle of separation of legislative, executive, and judicial powers and increase the citizens' engagement in public affairs with regard to combatting corruption and producing quality legislation.

10. For public administration, a call for high-quality, transparent, and corruption-resistant public administration as the cornerstone of the solid functioning of the state.

11. In terms of internal security, increased efforts to reinforce security priorities and capabilities.

12. On defence, a broadening of capacity to ensure a secure a stable environment[9].

During and after the course of Sobotka's mandate (2014-2017), which he (unlike previous governments) managed to finish in its entirety, the administration's successes and failures were pointed out. On the positive side, the ability to remain in power for the whole electoral period was possibly due to the fact that the goals presented were followed by concrete political and legislative actions to fulfil the Policy Statement. The programme, often dubbed as the *Socially Responsible Government*[10], offered a viable alternative for voters who were not satisfied with previous right-wing motivated policies

9 | Czech Government. (2014). *Policy Statement of the Government of the Czech Republic, Government of Bohuslav Sobotka, 2014,* pp. 24-25. Retrieved from https://www.vlada.cz/assets/media-centrum/dulezite-dokumenty/en_programove-prohlaseni-komplet.pdf.

10 | Dostal, V. (2017). *Sobotkova vláda paradoxů: Patří mezi nejúspěšnější, promrhala ale pověst v konfliktech [Sobotka's reign of paradoxes: He is one of the most successful, but he has lost his reputation in conflicts].* Retrieved from https://www.info.cz/cesko/sobotkova-vlada-paradoxu-patri-mezi-nejuspesnejsi-promrhala-ale-povest-v-konfliktech-20082.html.

11 | *Ibid.*

12 | The Visegrad Group, or V4, is a cooperative effort between the Czech Republic, Hungary, Poland, and Slovakia that is based on their shared cultural heritage and common interests in European integration. Reference: Visegrad Group. (n.d.). *About the Visegrad Group*. Retrieved from http://www.visegradgroup.eu/about.

13 | ČTK (Česka Tiskova Agentura). (2017). *Ze Sobotkovy vlády se sice ekonomika zvedla, ale chybí reformy, shodli se analytici [Although Sobotka's government has risen, the reforms are lacking, analysts agree]*. Retrieved from http://zpravy.e15.cz/volby/volby-2017/za-sobotkovy-vlady-se-sice-ekonomika-zvedla-ale-chybi-reformy-shodli-se-analytici-1340380.

14 | Kuchynova, Z. (2017). *Ekonomika se zvedla, ale chybí reformy, hodnotí analytici končící Sobotkovu vládu [The economy has risen, but reforms are lacking, say analysts ending Sobotka's government]*. Retrieved from http://www.radio.cz/cz/rubrika/udalosti/ekonomika-se-zvedla-ale-chybi-reformy-hodnoti-analytici-koncici-sobotkovu-vladu.

15 | Csornai, Z., Garai, N. & Szalai, M. (2017). *V4 migration policy: conflicting narratives and interpretative frameworks*. Barcelona Centre for International Affairs (CIDOB). Retrieved from https://www.cidob.org/content/download/65934/2018784/version/7/file/19-30_M%C3%81T%C3%89%20SZALAI%2C%20ZSUZSANNA%20CSORNAI%20AND%20NIKOLETT%20GARAI.pdf.

like lowering taxes and privatising public services. Sobotka's economic policy was a modest reaction to the post-2008 crisis environment in Europe, and as the unemployment rate fell, salaries continued to rise. Thus, the Czech Republic benefitted in those four years of rule due to proper management. Overall, the most substantive results were obtained in terms of social policy and tax collection[11].

On the other hand, the Government was also characterised for its 'inter-coalitional' rivalry, where the tensions in the relationship of the two coalition parties' leaders Bohuslav Sobotka and Andrej Babis (ČSSD's Prime Minister and ANO 2011's Finance Minister respectively) resulted in political disputes and, eventually, in voters switching party preferences and electing Babis as Prime Minister for the next term, causing a severe blow to ČSSD. Moreover, mixed signals were also sent to the EU, whereby the unity of the Visegrad Group's stance against migration quotas shed an unfavourable light on the Government, all being topped by the president's ambiguous comments on the topic.[12] Furthermore, concerns were raised over the effectiveness of achieving the Policy Statement – the construction progress of transport infrastructure was slow, the educational system continued to stagnate compared to other countries, and the digitalisation of state administration in the Czech Republic still lagged behind EU standards[13]. The Babis Government thus had to tackle reforms of pensions, healthcare, and the educational system[14], as those were seen as the major sectors that Sobotka's Government left behind.

1.3. Czech Foreign Policy

The Czech Republic's European Union membership, obtained in 2004, remains a central issue for the country and its Ministry of Foreign Affairs. However, as seen in the much-discussed migration quotas[15], Czechs together with their three closest neighbours – Slovakia, Hungary and Poland – have voiced their concerns over migration through the 'Visegrad' (V4) Platform. Beyond the V4 Group, the Czech Republic shares warm relations with Germany, Israel and other EU member states. In terms of foreign policy, the mandate of the Minister of

16 | In Czech politics, commonly mistaken even by the president itself, the role of the president is far less relevant to that of both the prime minister and foreign minister in terms of foreign policy.

17 | MFA (Ministry of Foreign Affairs of the Czech Republic). (2014). *Concept of the Czech Republic's Foreign Policy.* Policy Planning Department. Retrieved from http://www.mzv.cz/file/1574645/Concept_of_the_Czech_Republic_s_Foreign_Policy.pdf.

18 | *Ibid.*, p. 2.

19 | Czech Government. (2015) *Vláda schválila Koncepci zahraniční politiky České republiky [The Government approved the Concept of Foreign Policy of the Czech Republic].* Retrieved from https://www.vlada.cz/cz/media-centrum/aktualne/vlada-schvalila-koncepci-zahranicni-politiky-cr-132757/.

20 | *Ibid.*, p. 4.

21 | Institute for Economics & Peace. (2017). *Global Peace Index 2017: Measuring Peace in a Complex World.* Retrieved from http://visionofhumanity.org/app/uploads/2017/06/GPI17-Report.pdf.

22 | OECD DAC. (2016). *Official development assistance of the Czech Republic.* OECD DAC Statistical Reporting, 2016. Retrieved from http://www.mzv.cz/file/2102282/AJ_ODA2015.pdf.

23 | With an annual budget amounting to €2 million, the priority countries of the Transition Promotion Programme from 2015 are Burma/Myanmar, Belarus, Bosnia & Herzegovina, Georgia, Cuba, Kosovo, Moldova, Serbia and Ukraine. The full document is available at: http://www.mzv.cz/public/98/7c/e8/2239165_1648851_Human_rights_and_transition_promotion_policy_concept_of_the_Czech_Republic_.pdf

Foreign Affairs, together with the prime minister, is of primary importance[16].

The 'Concept of the Czech Republic's Foreign Policy' is an official document drafted by the Foreign Minister that provides a foreign policy framework for the interests of the country. The prime minister and his government are responsible for implementing it, together with the Foreign Ministry[17]. The document regards the Czech Republic as a 'small country in a global context and a medium-sized country on a European scale'[18] and is thus actively involved as a member of international institutions such as the EU, NATO, UN, and the Organisation for Security and Co-operation in Europe.

The Concept of Foreign Policy was formulated in 2014 and has been valid since 2015[19]. The document proposes the following objectives: 'security, prosperity and sustainable development, human dignity, including the protection of human rights, serving the people, and nurturing a good reputation abroad', noting that 'these objectives are not isolated but mutually interlinked and conditional upon each other'[20].

After the collapse of the Soviet Union, the dissolution of Czechoslovakia has domestically and internationally been regarded as one of the most successful cases of a peaceful transition to independence. After the transition, the level of domestic disorder has been minimal to non-existent: according to the rankings in the Global Peace Index, the Czech Republic maintained its 6th place worldwide in 2017[21]. Czech foreign policy thus also aims to share its unique experience with countries which are in peaceful transition towards democracy. Those efforts for promoting democracy are especially reflected in the Official Development Assistance (ODA) programme, in which the top five recipients of Czech aid are Moldova, Ukraine, Afghanistan, Bosnia & Herzegovina, and Ethiopia[22]. Moreover, other efforts include the Transition Promotion Programme, based on the Czech experience of social transition and democratisation, which targets a very similar list of countries[23] and co-finances Czech Non-Governmental Organisations working to achieve peaceful transitions.

24 | AMO (Asociace Pro Mezinarodni Otazky). (2017). *Velvyslanci České republiky 2017 – Infografika [Ambassadors of the Czech Republic 2017 – Infographics]*. Retrieved from https://www.amo.cz/wp-content/uploads/2017/08/AMO_velvyslanci-2017-infografika.pdf.

25 | Henley & Partners. (2017). *The Henley and Partners Visa Restriction Index 2017*. Available at: https://henleyglobal.com/files/download/hvri/HP_Visa_Restrictions_Index_170301.pdf.

26 | UNWTO (United Nations World Tourism Organisation). (2016). *Visa Openness Report 2015*. UNWTO, Madrid.

27 | MFA (Ministry of Foreign Affairs of the Czech Republic). (2017). *Bulletin 'Zahraniční politika České republiky' [Bulletin 'Foreign Policy of the Czech Republic']*. Retrieved from https://www.mzv.cz/jnp/cz/o_ministerstvu/organizacni_struktura/utvary_mzv/oazi_/mesicnik_zahranicni_politika_ceske.html.

Moreover, in the post-independence period, Czech foreign policy opened up both politically and economically, where the country established diplomatic relations with other countries. As of 2017, the Czech Republic had established relations with and was being diplomatically represented in more than 180 countries in the world and had 95 ambassadors on a mission in embassies worldwide[24]. Czechs could travel without a visa to 168 countries out of 218 listed by the Henley and Partners Visa Restriction Index[25], and the Czech Republic is ranked by the UN World Tourism Organisation as the 24th least-affected country by visa restriction worldwide[26].

Overall, the Czech Republic aims to maintain its course of foreign policy in line with the goals issued in 2015[27].

2. The Political Direction of the Czech Republic

From being a socialist regime for much of the second half of the 20th century to becoming a democracy in 1989, the Czech Republic has gone through significant political changes in the past 80 years. These changes have had profound effects on its foreign policy outlook. The difficult task of predicting its future political development, especially with respect to foreign policymaking, should therefore be rooted in historical analysis[28].

This chapter will analyse relevant foreign policy developments from a historical perspective, especially in relation to China. It will then examine recent political developments with an emphasis on the parliamentary elections in October 2017 and the presidential elections in January 2018. It will attempt to offer a non-conclusive prediction about the political direction of the Czech Republic, focusing mainly on its foreign policy,

28 | Vertzberger, Y. Y.. (1986). Foreign policy decisionmakers as practical-intuitive historians: Applied history and its shortcomings. *International Studies Quarterly*, 30(2), pp. 223-247.

especially towards China. The study will conclude that the Czech Republic remains and will remain a parliamentary democracy and its foreign policy will be concentrated on the EU. In relation to Czech-China relations and, in particular, to the Belt and Road Initiative (BRI), it remains to be seen how the parameters set by the priority of the EU membership will influence the Czech Republic's stance towards China in general and BRI in particular.

2.1. Czech-China Relations in Historical Perspective

Czechoslovakia became a socialist country in February 1948 and remained so until 1989. The regime, like many others in Central and Eastern Europe, was characterised by the ideological marks of Marxism-Leninism and emphasised a move towards a classless, communist society via a transitory period of socialist dictatorship[29]. Its foreign policymaking was heavily dependent on its position in the socialist bloc and its dependency on the Soviet Union[30]. This explains the persistent fluctuations in Czech-Chinese relations that will be outlined below.

Given the political and ideological context of the Cold War (CW), as well as clearly drawn power dynamics, Czechoslovakia's foreign policy was heavily oriented towards countries in the former socialist bloc, including China. The Czechoslovakian 'tradition of industrial exports and a great deal of political romanticism' in relation to China led to close ties between the two countries as well[31]. Between 1955-1957, following a series of 'bilateral treaties on trade, scientific, educational and cultural cooperation [...] the increasing volume of bilateral trade with China [...] ranked third', behind only the USSR and East Germany[32]. Consequently, the 1950s can be considered a 'heyday of Sino-Czechoslovak relations[33].

Nevertheless, in the 1960s, following the Great Leap Forward, Sino-Soviet relations deteriorated, and the formerly strong relations with the Central and Eastern European countries, including that of Czechoslovakia, were weakened[34]. The Cultural Revolution in China labelled the USSR leadership

29 | Verdery, K. (1996). Chapter 1 'What was Socialism and Why did it Fall?' in *What Was Socialism and What Comes Next?* Princeton University Press, pp.19-38; Munk, M.. (2000). Socialism in Czechoslovakia: What Went Wrong? *Science & Society*, 64(2), pp.225-236. Retrieved from http://www.jstor. org/stable/40403841.

30 | Staniszkis, J. (1986). *The Dynamics of Dependency*. East European Program, European Institute, the Wilson Center, 4.

31 | Furst, R., Pleschova, G.. (2010). Czech and Slovak Relations with China: Contenders for China's Favour. *Europe-Asia Studies*, 62(8), pp.1364-5.

32 | *Ibid.*

33 / *Ibid.*

34 | *Ibid.*

as not dedicated enough to the communist ideology, leading as a consequence of a further worsening of the relationship between China and Czechoslovakia[35].

Eventually, from the 1980s, the relationship warmed again, especially with the increased openness of China. A new wave of close political and economic co-operation followed, and some commentators even called it the 'best ever era for relations between Prague and Beijing'[36]. However, the events of 1989, both in China (Tiananmen Square movement) and in the Czech Republic (Velvet Revolution) threatened the warm ties as the Chinese regime hardened while Czechoslovakia started engaging in a democratic revolution[37].

The post-revolutionary Czech political scene focused on its liberal agenda, with special attention paid towards the betterment of human rights[38]. This was, to an extent, the personal political agenda of the first Czechoslovak President Vaclav Havel (later the first Czech president after the split into the Czech Republic and Slovakia in 1993). President Havel was personally invested in propagating a vision of global democratisation, which involved supporting the Tibetan movement (a movement led by Havel's personal acquaintance – the 14th Dalai Lama).[39]

These developments led to a critical view of China among Czech journalists. That critical stance of the media continues to this day, highlighting the problematic aspects of the relationship with China with respect to the domestic parameters of foreign policymaking[40]. The effect of the media on the general Czech population´s opinion of China and Chinese investments in the Czech Republic, however, seems surprisingly limited. Even in relation to sensitive investments in the field of nuclear energy, Czechs seem relatively well disposed to the Chinese presence, according to Turcsanyi[41].

Despite the high-profile meetings between the Dalai Lama and the former Czech president, the focus of Czech diplomacy was never on Tibet. Instead, both Chinese and Czech diplomats attempted to build on the rejuvenated and flourishing trade

35 | Ibid.

36 | Ibid.

37 | Sarotte, M.E.. (2012). China's Fear of Contagion: Tiananmen Square and the Power of the European Example. International Security, 37 (2), pp. 156-182.

38 | Furst, R., Pleschova, G.. (2010). Czech and Slovak Relations with China: Contenders for China's Favour. Europe-Asia Studies, 62(8), p. 1366.

39 | Ibid.

40 | Karaskova, I.. (2017). Obraz Číny v Česku. Politici mluví o sbližování, média ale ukazují hlavně negativa země [China's image in the Czech Republic: politicians talk about rapprochement, media focus on the negatives]. Hlídací Pes. Retrieved from https://hlidacipes.org/obraz-ciny-cesku-politici-mluvi-sblizovani-media-ukazuji-hlavne-negativa-zeme/.

41 | Turcsanyi, R.. (2017). Central European Attitudes Towards Chinese Energy Investments: The Cases of Poland, Slovakia, and the Czech Republic. Energy Policy, p. 101, pp. 711-722.

42 | Furst, R. & Pleschova, G..
(2010). Czech and Slovak Relations
with China: Contenders for China's
Favour. *Europe-Asia Studies*, 62(8),
pp. 1363-1381.

43 | Cadier, D.. (2012). Après le
retour à l'Europe: les politiques
étrangères des pays d'Europe
centrale [After the return to Europe:
the foreign policies of Central
European countries]. *Politique
étrangère*, (3), pp. 573-584.

44 | Henderson, K.. (2005). *Back
to Europe: Central and Eastern
Europe and the European Union*.
Routledge, iv.

45 | AMO (Asociace Pro
Mezinarodni Otazky). (2018). *Role
prezidenta v zahraniční politice po
zavedení přímé volby [Role of the
president in foreign policy after
the implementation of the direct
voting system]*. Retrieved from
http://www.amo.cz/wp-content/
uploads/2018/01/AMO_role-
prezidenta-v-zahranicni-politice-po-
zavedeni-prime-volby.pdf.

46 | *Ibid.*

47 | ČTK. (2015). *V Pekingu
proběhla největší vojenská přehlídka
v dějinách Číny. Přihlížel jí Zeman
i Putin [Largest military parade
in history of China took place in
Beijing. Attended by Zeman and
Putin]*. Retrieved from https://
www.lidovky.cz/v-pekingu-
probehla-nejvetsi-vojenska-
prehlidka-v-dejinach-ciny-prihlizel-
ji-zeman-i-putin-gv4-/zpravy-svet.
aspx?c=A150903_072457_ln_
zahranici_ELE.

48 | ČT24. (2016). *Maraton
prezidenta Si. Smlouva o spolupráci,
jednání se Sobotkou [Marathon
of President Xi. Cooperation
Agreement, Negotiations with
Sobotka]*. Retrieved from:
https://ct24.ceskatelevize.cz/
domaci/1738314-maraton-
prezidenta-si-smlouva-o-spolupraci-
jednani-se-sobotkou.

49 | *Ibid.*

and political relations from the mid-1980s[42]. These attempts continued throughout the 2000s.

Nevertheless, the main priority of Czech foreign policy was to shed its socialist heritage and integrate into Europe, or more generally into Western institutions[43]. Since the 1990s, Czech foreign policy efforts have been heavily focused on managing the Czech Republic´s membership in the EU as well as NATO[44]. Consequently, relations with China were not considered a priority and their importance faded.

2.2. Developments Post-2012

One of the most important recent developments in Czech politics was the change made to the voting system of the presidential election which came into effect in 2012. With a direct popular vote, the office of the president seemed to have gained more legitimacy and more interest from the general population[45]. The change gave the president more political power and thus more influence in directing the country's foreign policy[46]. President Milos Zeman was the first directly elected president in January 2013 and has been an important element in transforming Czech-China ties. In September 2015, Zeman´s attendance was keenly reported by the media because he was the only EU leader who attended the Victory Day parade in Beijing[47]. His attendance, along with his other efforts in harnessing ties with China was reciprocated when President Xi Jinping visited the Czech Republic in March the following year[48]. During this visit, another wave of non-binding treaties, mainly focused on trade and investment, were signed[49].

Other than the developments in the presidential election, one must also pay attention to the developments in the parliament. The victory of ANO in parliamentary elections in 2017 was labelled by some commentators as leading to a break from traditional foreign policy to one that would focus increasingly on Russia and China. If this proves to be the case, the government and the president would have finally agreed upon the increased importance of China for the Czech Republic, leading to an increased profile for China in the Czech foreign policy agenda. However, this perception is in clear opposition to ANO's programme, which states its emphasis on fostering ties in Europe and on advancing human rights[50].

However, with the importance of the Prime Minister, a pragmatic Andrej Babis, in the formulation of the Czech Republic's foreign policy, a clear break with EU-oriented foreign policy is not warranted. In relation to China, ANO´s foreign policy as of 2018 emphasised the 'respect of the rules of international trade in order to protect the Czech industry from unfair competition'[51]. Babis' pragmatism also makes him a firm ally of the EU. After all, his first foreign visit was to Brussels to participate in a negotiation with the European Council, suggesting that this will be the area of foremost attention in the foreign policy of his government[52].

2.3. Conclusion

As said above, Czech political history has been relatively turbulent in the last 80 years. Relations with China have been tumultuous, often as a result of Czech association with a powerful ally which complicated its relationship with China.

Moreover, as of 2018, it remained to be seen whether Czech policy towards China would become more conservative. ANO's agenda suggests that the Czech Republic is firmly in the EU camp – ideologically and otherwise – and its policy towards China might differ from that of the previous government, given the increasingly cautious European policies with regard to trade protection and foreign direct investments (FDI) screening on the entire EU-level[53]. Nevertheless, avenues of co-operation within the legal boundaries of the Lisbon Treaty offered space for further deepening of Czech-Chinese cooperation.

Another element influencing the relations with China would be the role of the Czech president. Being directly elected in a popular vote gives the president more legitimacy than the presidents elected before 2013 to set foreign policy agendas. President Milos Zeman has been active in promoting a relationship with China since his first mandate began in 2013,

50 | ANO. (2017). *Program hnutí ANO pro volby do Poslanecké sněmovny 2017 [Programme of the YES movement for the elections to the Chamber of Deputies 2017]*. Retrieved from https://www.anobudelip.cz/file/edee/2017/09/program-hnuti-ano-pro-volby-do-poslanecke-snemovny.pdf.

51 | *Ibid.*, p. 15.

52 | Bartonicek, R.. (2017). *Babiš krátce po jmenování premiérem zamíří do Bruselu, bude jednat o kvótách. Češi budou v Evropě slyšet, slibuje [Shortly after being named PM, Babiš will leave for Brussels to discuss quotas. Czechs will be heard in Europe, he promises]*. Retrieved from https://archiv.ihned.cz/c1-65980830-babis-kratce-po-jmenovani-premierem-zamiri-do-bruselu-bude-jednat-o-kvotach-cesi-budou-v-evrope-slyset-slibuje.

53 | Zalan, E.. (2017). *EU Defends New Rrade Rules After Chinese Criticism*. EU Observer. Retrieved from https://euobserver.com/eu-china/140392.

to the point where this could be seen as one of his important priorities. Moreover, his efforts can be understood as a strategy to create additional foreign policy goals, thus expanding the foreign policy beyond the prior, almost exclusive, focus on the EU and the United States[54]. It remained to be seen whether this policy would continue in his second mandate after his re-election in January 2018.

Again, Czech-Chinese relations should be seen in light of further developments in the EU as a whole, especially in relation to the closer FDI screening. The Czech Republic is a core member of the EU, and its national foreign policy decision-making will inevitably be restricted by the parameters commonly agreed upon in the EU structures. Developments on the EU level[55], however, suggest a more defensive stance towards Chinese trade and FDI. It remained to be seen whether the Czech stance would end up supportive of those defensive measures.

54 | Furst, R.. (2018). Czechia´s Relations with China: on a Long Road toward a Real Strategic Partnership? in: Weiqing Song (ed.): *China´s Relations with Central and Eastern Europe: From 'Old Comrades' to New Partners*. Routledge Contemporary China Series, pp. 117-136.

55 | European Commission. (2017). *Press Release: State of the Union 2017 – Trade Package: European Commission Proposes Framework for Screening of Foreign Direct Investments*. Retrieved from http://europa.eu/rapid/press-release_IP-17-3183_en.pdf.

3. Evaluation of the Current Opportunities and Challenges Faced by Young People in the Czech Republic

This section will elaborate on the characteristics of the Czech youth. It will then focus on the challenges and opportunities present in the country, especially as perceived by the youth. We will conclude this Part by observing that Czech youth are currently in a very good position with a stabilised economy marked by low unemployment, relatively high economic growth and rising wages. At present, challenges are not acute for the majority of Czech youth. Nevertheless, the near future may create additional challenges, especially with respect to digitisation, which will be briefly discussed as well.

3.1. Characteristics of Czech Youth

For the purposes of this Part, we define young people as those aged from 14, when they have finalised compulsory schooling, to the age of 30 years old, which is the age at which normally the processes of social maturation are finalised[56].

Only about 18% of the Czech population, or about 1.9 million people, were between the ages of 15-29, in 2013, which roughly corresponds to the definition set above. This stands in contrast to the year 2000, where youths made up 23.5% of the population. Therefore, Czech youths are experiencing a decrease in proportion to the total population over time. The absolute number of young people is steadily diminishing as well – since 2000, the number of young people fell by more than 500,000.

This reflects the trend of an ageing population as observed in European and Asian societies. An ageing population represents a challenge in itself – it increases pressure on the state finances, especially in generous welfare states. The inability of successive governments to engage in reform to tackle the challenge means that its consequences will have to be dealt with by the youth in the not-so-distant future.

Secondly, the general educational level of Czech youths is also an important factor to consider. 90% of young Czechs have completed some form of secondary education, which is above the European average of 80%[57]. However, Czechs lag behind in terms of higher education, with only 25.6% having completed a university degree compared to the European average of 36%[58]. Even though this does not currently represent an acute challenge, it may represent a hurdle in a transition towards a higher added-value based economy, an underlying feature of long-term prosperity.

Czech youth benefits from a relatively prosperous and macro-economically stable environment. In terms of material well-being and inequality, the Czech Republic does relatively well, with GDP purchasing power parity (PPP) Per Capita at nearly

56 | Linhart, J., Petrusek, M., Vodakova, A., & Marikova, H.. (1996). *Velký sociologický slovník [Large Sociological Dictionary]*. Praha: Karolinum; Zajic, J., et al.. (2014). *Bez Růžových Brýlí [Without Pink Glasses]*. The Centre for International Cooperation in Education, Prague.

57 | Zajic, J., et al.. (2014). *Bez Růžových Brýlí [Without Pink Glasses]*. The Centre for International Cooperation in Education, Prague.

58 | *Ibid.*

US$35,000, more than double that of China[59]. According to the OECD, the Gini index (a measure of income or wealth inequality) was at 0.26 in 2016, reflecting a socio-economic landscape that possesses relatively high equality of access to opportunities[60]. Such a society can give its members a prospect of a hopeful future, should adequate effort on their part be done. The Czech Republic exhibits features of such a society.

Overall, despite a decreasing number of young people in society and the related demographic pressures, the youth´s characteristics are rather positive. Czech youth attains a relatively high educational level and enjoys a relatively prosperous and equal societal environment.

3.2. Challenges and Opportunities Faced by the Youth

Analytically, we depart from the perspective that the challenges faced by youths are arguably best understood by the youths themselves. Out of the challenges that Czech youths identify as the most acute, we choose to emphasise the factors of unemployment, economic situation, and poor education as particularly relevant to Czech-Chinese relations and current and future co-operation within the BRI.

Despite being identified as some of the main challenges by Czech youth, in reality, unemployment rates and the economic situation had improved significantly in 2017 as compared to 2012 and 2009. Youth unemployment in 2016 was at 10.5% — even lower than before the global economic crisis from 2007 to 2008 when it was 10.7%[61]. Ranking amongst EU's best in youth unemployment, as shown in the table below, Czech youth were in a much better position than other member states in the EU, such as in Spain, Italy, or Greece, as illustrated below[62].

An improving economic situation also stands in contrast with it being identified as a key issue faced by Czech youths. The Czech economy has been experiencing relatively high growth, with a GDP growth rate of 4.3% in 2014, 2.3% in 2016 and an estimated average growth rate of 3.3% over the 2017-2019

59 | OECD. (2018). *Gross Domestic Product (GDP) (indicator)*. Retrieved from: https://www.oecd-ilibrary.org/economics/gross-domestic-product-gdp/indicator/english_dc2f7aec-en.

60 | OECD. (2017). Government at a Glance 2017. Retrieved from https://www.oecd.org/gov/gov-at-a-glance-2017-czech-rep.pdf.

61 | Eurostat. (2016). *Statistics Explained.* Retrieved from http://ec.europa.eu/eurostat/statistics-explained/images/6/6a/Youth_unemployment_ figures%2C_2007-2016_%28%25%29_T1.png.

62 | *Ibid.*

	YOUTH UNEMPLOYMENT RATE				YOUTH UNEMPLOYMENT RATIO			
	2007	2014	2015	2016	2007	2014	2015	2016
EU-28	15.9	22.2	20.3	18.7	6.9	9.2	8.4	7.7
EURO AREA	15.6	23.8	22.4	20.9	6.7	9.5	8.8	8.2
BELGIUM	18.8	23.2	22.1	20.1	6.4	7.0	6.6	5.7
BULGARIA	14.1	23.8	21.6	17.2	4.2	6.5	5.6	4.1
CZECH REPUBLIC	10.7	15.9	12.6	10.5	3.4	5.1	4.1	3.4
DENMARK	7.5	12.6	10.8	12.0	5.3	7.8	6.7	7.9
GERMANY	11.8	7.7	7.2	7.0	6.1	3.9	3.5	3.5
ESTONIA	10.1	15.0	13.1	13.4	3.8	5.9	5.5	5.8
IRELAND	9.1	23.9	20.9	17.2	5.1	8.9	7.6	6.7
GREECE	22.7	52.4	49.8	47.3	7.0	14.7	12.9	11.7
SPAIN	18.1	53.2	48.3	44.4	8.7	19.0	16.8	14.7
FRANCE	19.5	24.2	24.7	24.6	7.2	8.7	9.0	9.0
CROATIA	25.4	44.9	42.3	31.1	9.2	15.3	14.0	11.6
ITALY	20.4	42.7	40.3	37.8	6.3	11.6	10.6	10.0
CYPRUS	10.2	36.0	32.8	29.1	4.2	14.5	12.4	10,7
LATIVIA	10.6	19.6	16.3	17.3	4.5	7.9	6.7	6.9
LITHUANIA	8,4	19.3	16.3	14.5	2.3	6.6	5.5	5.1
LUXEMBOURG	15.6	22.3	16.6	19.2	4.0	6.0	6.1	5.8
HUNGARY	18.1	20.4	17.3	12.9	4.6	6.0	5.4	4.2
MALTA	13.5	11.7	11.8	11.1	7.3	6.1	6.1	5.7
NETHERLANDS	9.4	12.7	11.3	10.8	4.3	8.6	7.7	7.4
AUSTRIA	9.4	10.3	10.6	11.2	5.6	6.0	6.1	6.5
POLAND	21.6	23.9	20.8	17.7	7.1	8.1	6.8	6.1
PORTUGAL	21.4	34.7	32.0	28.2	8.6	11.9	10.7	9.3
ROMANIA	19.3	24.0	21.7	20.6	6.1	7.1	6.8	5.8
SLOVENIA	10.1	20.2	16.3	15.2	4.2	6.8	5.8	5.1
SLOVAKIA	20.6	29.7	26.5	22.2	7.1	9.2	8.5	7.2
FINLAND	16.5	20.5	22.4	20.1	8.8	10.7	11.7	10.5
SWEDEN	19.2	22.9	20.4	18.9	10.1	12.7	11.2	10.4
UNITED KINGDOM	14.3	17.0	14.6	13.0	8.8	9.8	8.6	7.6
ICELAND	7.1	10.0	8.8	6.5	5.6	7.7	7.1	5.4
NORWAY	7.2	7.9	9.9	10.9	4.4	4.3	5.5	6.1
SWITZERLAND								
TURKEY	17.2	18.0	18.6	19.6	6.3	7.3	7.7	8.2
UNITED STATES	10.5	13.4	11.6	10.4				
JAPAN	7.7	6.2	5.5	5.1				

Source: Eurostat (une_rt_a).

Figure 1: Youth unemployment rate[63] and youth unemployment ratio across the European Union, Iceland, Norway, Switzerland, Turkey, United States and Japan in 2007, 2014, 2015 and 2016.

period[64].

Nevertheless, some argue that strong pressures will begin to weigh down on the structure of the Czech labour market and economy in the future. According to the Czech government, a significant number of jobs may be lost in the next 10-20 years due to digitisation and automation. Moreover, these losses will be particularly felt in regions that face pre-existing structural disadvantages to begin with[65].

The analysis of the challenges and opportunities should therefore take into account the rapid technological changes that will influence the Czech economy in the near future. The evaluation of the prospects Czech youths face would not be complete without considering the probable drastic changes in their situation within the next 10-20 years.

Consequently, the next factor identified as a challenge by the Czech youth – a poor education quality – is the most important one to address. Broadly, the Czech educational system is unable to sufficiently provide the necessary skills for the country's present or future economic needs, but it does prepare the young population well in terms of soft skills, such as teamwork and communication – necessary for an increasingly digitalised future[66]. Additionally, the ranking of the Czech Republic in science, technology, engineering, and mathematics (STEM) has been consistently falling[67], though it does appear that this issue is being prioritised by the government. The governmental elementary education policy provides only a basic framework of reference on how to counteract the decline of science and mathematics and how to provide the soft skills necessary for the near future.

What opportunities do Czech youths have? As of 2017, the high economic growth and low unemployment combine to create significant upward pressure on wages. This is true of the public sector which wages have been raised by 10-15% (2016-2017 year on year), as well as in most areas of the private sector. A particularly lucrative employment can be found in informational technology (IT), which also seems to be

63 | The youth unemployment rate refers to the ratio of youth unemployed to total youth population.

64 | ČNB. (2018). *Prognóza HDP* (GDP prognosis). Retrieved from https://www.cnb.cz/cs/menova_politika/prognoza/index.html#HDP.

65 | Czech Government. (2015). *Dopady digitalizace na trh práce v ČR a EU [Impact of digitalisation on the labor markets in the Czech Republic and the EU].* Retrieved from *https://www.vlada.cz/assets/evropske-zalezitosti/analyzy-EU/Dopady-digitalizace-na-trh-prace-CR-a-EU.pdf.*

66 | For the current and future needs of the labour market, see PWC. (2017). *The Future of Workforce.* Available at: https://www.pwc.com/gx/en/services/people-organisation/workforce-of-the-future/workforce-of-the-future-the-competing-forces-shaping-2030-pwc.pdf. For Czech PISA score, see http://www.compareyourcountry.org/pisa/country/cze?lg=en.

67 | OECD. (2016). *PISA 2015 Results (Volume I): Excellence and Equity in Education.* PISA, OECD Publishing, Paris. Retrieved from https://doi.org/10.1787/9789264266490-en.

a comparative advantage of the Czech economy. In order to capitalise on the situation, Czech youth should be responsive to the demands of the market and dynamically adapt their skillsets, via formal education or self-study.

The future of automation, both in industry and services, will represent a challenge for Czech youths. The government should promote policies that will support educational changes and engagements related to key, high-tech and high value-added sectors of the economy in order to ensure prosperity and employability of Czech youth[68]. Some of the key sectors have also been prioritised by the Chinese government already, such as those stated in the China 2025 policy. Exchanging lessons of economic development, co-operating in science and research and promoting international private-public partnerships might be mutually beneficial and could create additional opportunities for Czech youths.

3.3. Conclusion

In this section, we have identified the characteristics of Czech youth. We have also identified the challenges perceived by the youth, relevant to Czech-Chinese co-operation, with some of these challenges being economic and educational. Co-operation in trade, science and research may counter some of these challenges, and create additional opportunities for Czech youths. In particular, co-operation with regard to education in a dynamically changing technological environment is desirable.

68 | For the role of the state in promoting entrepreneurship and growth, see Mazzucato, M. (2015). *The Entrepreneurial State: Debunking Public vs. Private Sector Myths* (Vol. 1). Anthem Press.

4. Evaluation of the Opportunities and Challenges Offered by the Belt and Road Initiative

In this section, we will analyse the opportunities and challenges that arise from the BRI. Indeed, the opportunities are manifold, relating to diverse areas such as research and development, civil aviation, and infrastructure. Broadly, the opportunities can be understood as arising from the goal of the Czech Republic to move up the global value chain and the ways in which the BRI can be conducive to this goal.

The challenges, from the European perspective, relate mainly to market access and perceptions of the BRI. The market access issue is one with deep roots and refers to the perception of unequal market access to China from the perspective of the EU. Second, the Chinese activities in the Czech Republic by Czechs themselves and their EU partners can be perceived as a threat. These perceptions, *inter alia*, have turned into a proposed policy of investment screening at the EU level[69], which may have an impact on BRI projects in the Czech Republic.

A useful theoretical framework in understanding the motivation and goals of the Czech Republic in participating in the BRI is that of international political economy (IPE). From the liberal perspective, especially since the end of the Cold War[70], wealth and prosperity have become increasingly important to the state[71]. There is historical evidence for this. One of the main political and economic goals of the Czech Republic post-1989 (the year of the end of socialism in the country) was to re-integrate with the West, i.e. the Euro-Atlantic structures, which represented prosperity and high quality of life[72]. Almost 30 years after the end of the socialist regime, this goal seems fulfilled only partially, and the additional goal of economic development is ever-present among Czechs[73].

Certain Czech political elites now recognise that China is one of the leaders in some industries, such as railways, and that it has great potential going forward in high technology and science[74]. Thus, both countries can benefit economically from co-operating. Therefore, there is an interest in exploring avenues of achieving economic upgrading beyond co-operation with traditional Euro-Atlantic partners, which in turn explains

69 | For a discussion on the reasons behind the proposal, see Bryan Cave Leighton Paisner LLP. (2018, September 16). *EU Screening of Foreign Direct Investments: European Commission Proposals Move Forward.* Retrieved from https://www.bclplaw.com/en-US/thought-leadership/eu-screening-of-foreign-direct-investments-european-commission-proposals-move-forward.html.

70 | Jackson, R. & Sorensen, G. (2016). *Introduction to International Relations: Theories and Approaches.* Oxford University Press.

71 | As opposed to traditional considerations, such as state survival. However, state survival and state wealth can be understood as two mutually reinforcing elements that affect state power.

72 | Cadier, D.. (2012). Après le retour à l'Europe: les politiques étrangères des pays d'Europe centrale [After the return to Europe : the foreign policies of Central European countries]. *Politique étrangère*, (3), 573-584.

73 | Interview conducted with a former senior Czech diplomat in spring 2018. Interview A. (2018). Prague.

74 | *Ibid.*

the interest of co-operation with China within the BRI. This theoretical perspective will be implicitly used throughout the chapter.

In the first section of this chapter, we will outline the key industries the Czech Republic aims to focus on. This will be conducive to the understanding of the strategic economic priorities of the country. We will also outline the areas of co-operation as presented in the BRI Memorandum of Understanding (MoU) between the Czech Republic and China. The second section will focus on some of the greatest opportunities that the economic partnership between the two countries presents. This mainly includes science, research and development, but also some other, more specific areas where the Czech Republic has the expertise and China has great demand, such as civil aviation. Generally, the strengthening of more equal trade ties underpins opportunities. In the third part, we will discuss the challenges to the BRI in the Czech Republic. The main challenges to the BRI are those of perception and of the investment screening policy at the EU level. Thus, we will conclude that the success of the BRI would be limited unless larger, structural issues are addressed.

4.1. Key Strategic Areas for the Czech Republic

In section 4.1, we will outline the key strategic industries of the Czech Republic. Indeed, these are the industries in which the Czech Republic enjoys a comparative advantage and that it seeks to upgrade in order to move up the global value chain.

The government of PM Bohuslav Sobotka (2013-2017) identified strategic industries for the Czech Republic. In other words, these are the industries in which the Czech Republic should aim to invest and specialise[75]. These are industries in which the Czech Republic traditionally enjoys a relative advantage and on which its economy heavily depends, such as the automobile and electrical engineering sectors. However, the governmental strategy specifies that the industry should be modified to fit the needs of the 21st century[76]. Thus, one can expect increased investment in this area.

75 | Czech Government. (2016). *Vláda schválila strategii inteligentní specializace, podpoří perspektivní odvětví vědy [The Government approved a strategy of intelligent specialization. It will support promising branches of science].* Retrieved from https://www.vlada.cz/cz/clenove-vlady/pri-uradu-vlady/pavel-belobradek/aktualne/vlada-schvalila-strategii-inteligentni-specializace--podpori-perspektivni-odvetvi-vedy--146681/.

76 | *Ibid.*

Indeed, the pace of technological change looms as a challenge to the automobile industry as we know it. The increasing pressure on the combustion engine, as well as the progress in autonomous vehicles, could radically transform the industry[77]. The task for the Czech Republic, in the automotive industry, for instance, is to adapt to a future of driverless cars. Therefore, technological upgrading of the automotive industry, as well as of auxiliary industries, is in order. That seems to be one of the economic priorities of the Czech government.

Indeed, the growth and success of other key industries identified by the government of Sobotka, such as electrical technical engineering, and IT, are closely related to the automobile industry. The priority role of these industries seems to be in line with the needs of a future of driverless vehicles alongside the integration of advanced technological and informational systems in the vehicles. Given that the Czech Republic enjoys a legal and commercial environment conducive to the development of self-driving automobiles and related high added-value industries, such as optics[78], the Czech automotive industry has the potential of being a global leader in the development of these advanced systems[79].

Despite being a strategic government priority, the automotive industry was not explicitly mentioned in the BRI MoU between the Czech Republic and China. As we will explain in the next section, however, there are areas closely related to the automotive industry that are objects of the BRI-related co-operation between the two countries. Moreover, there are other sectors that are in line with one of the overarching objectives of the Czech Republic – economic upgrading – that are present in the BRI MoU.

4.2. Opportunities for the Belt and Road Initiative in the Czech Republic

The route towards upgrading the economy in general leads through high added-value sectors, such as the upgraded automotive industry, which in turn can be achieved through the development of advanced technologies. In the case of

77 | McKinsey & Company. (2016). *Automotive Revolution – Perspective Towards 2030.* Retrieved from https://www.mckinsey.com/~/media/mckinsey/industries/automotive%20and%20assembly/our%20insights/disruptive%20trends%20that%20will%20transform%20the%20auto%20industry/auto%202030%20report%20jan%202016.pdf.

78 | Business Wire. (2018). *New CTA Scorecard Identifies the Most Innovative Countries.* Retrieved from https://www.businesswire.com/news/home/20180110006053/en/New-CTA-Scorecard-Identifies-Innovative-Countries.

79 | Czech Government. (2014). *Národní RIS3 strategie* (National RIS3 strategy).

the automotive industry, this can be technologies related to, for example, advanced optics, a vital component in the development and operation of driverless vehicles. Therefore, the priority of co-operation within BRI should be scientific and technological, from the perspective of the general economic and industrial upgrading of the Czech Republic. Sharing scientific discoveries, joining scientific programmes, and especially co-operating in applied research with practical results in the key industries may be of interest to the Czech Republic, and represents key opportunities for the BRI in the country.

Indeed, the strength of the Chinese scientific community has begun to draw a heightened interest from its potential partners, as the number of patents, 'often used as an indicator of technology innovation'[80], dramatically increased in a short period. In 2017, the number of filed patents from China stood at 1,338,503 compared to 1,101,864 in 2016, equivalent to an annual growth of 21.5%[81]. In terms of industrial designs, arguably the most relevant category of patents with respect to the Czech-Chinese economic co-operation, the number of Chinese patents stood at around 650,000 in 2016[82]. The quality of the patents is a different matter that is beyond the scope of this study, but the industriousness of Chinese scientists is undeniable and is reflective of the Chinese commitment to creating a highly productive national innovation system[83]. As such, it is potentially an interesting partner for a country, such as the Czech Republic, whose goal is an economic upgrade. The Czech Republic's scientific community is undeniably of great quality as well[84].

Nevertheless, strengthening of mutual trust, especially with respect to intellectual property rights, is a key condition to the solidification of scientific and technological co-operation. This key structural issue will be further discussed in section 4.3.

Apart from science and technology, the BRI MoU includes 19 other areas of potential co-operation. These include:

Infrastructure, Investment, Industry and trade, Energy resources, Finance, Transport and logistics, Medicine and

80 | Hu, A. G., Peng, Z., & Lijing Z.. (2018). *Patents and Innovation in China*. Innovation in the Asia Pacific. Singapore: Springer.

81 | WIPO. (2017). *World Intellectual Property Indicators 2017*. Geneva: World Intellectual Property Organisation.

82 | *Ibid*.

83 | Fu, X.. (2015). *China's Path to Innovation*. Cambridge University Press.

84 | Business Wire. (2018). *New CTA Scorecard Identifies the Most Innovative Countries*. Retrieved from https://www. businesswire.com/news/ home/20180110006053/en/ New-CTA-Scorecard-Identifies-Innovative-Countries.

health, Civil aviation, Norms and certificates, Agriculture, Culture, Tourism, Sport, Education, Environment, Regional co-operation, IT, Co-operation between think-tanks.

It is not within the scope of this study to analyse each of these areas. We will thus only briefly discuss those areas that we understand to be of strategic interest to the Czech Republic. If the Czech Republic were to further specialise in these areas, the country would move up the global value chain due to the high added value represented in these areas.

Specifically, an area that was further emphasised by Czech decision-makers was that of civil aviation[85]. The Czech civil aviation sector enjoys a comparative advantage in the design and construction of regional (i.e. small passenger capacity) aircrafts[86]. This kind of aircraft, *inter alia*, is increasingly sought after in China as demand for short-haul flights increases[87].

The co-operation in areas that would be conducive to Czech economic upgrading would at least partially correct the highly unequal balance of trade between the Czech Republic and China. In 2017, the balance stood at -418 billion Czech koruna (CZK) in 2017 (approximately US$19 billion) with Czech imports from China ten times larger than its exports to China[88]. Increased co-operation in areas with a high added value would allow the Czech Republic to develop these areas and lead to a desired economic development, should China open its market to Czech exports of this kind.

4.3. Challenges for the Belt and Road Initiative in the Czech Republic

This brings us to one of the main challenges of achieving sustainable success in the BRI. The Czech Republic, as a very active trading country, whose exports-to-GDP ratio stands at 80%[89], needs access to foreign markets to grow. China, with its increased demand for goods, such as small aircraft, is, therefore, a reasonable market for Czech high value-added exports. However, the contention is that China itself attempts to promote domestic high value-added production[90]. Should

85 | Interview A. (2018). Prague.

86 | *Ibid.*

87 | Xinhua. (2017). *China to Demand 6,103 New Passenger Aircraft Over 20 Years: AVIC.* Retrieved from http://www.chinadaily.com.cn/business/2017-11/24/content_34929102.htm.

88 | Czech Trade. (2018). *Čína: Obchodní a ekonomická spolupráce s ČR* (China: commercial and economic co-operation with the Czech Republic). Retrieved from http://www.businessinfo.cz/cs/clanky/cina-obchodni-a-ekonomicka-spoluprace-s-cr-19054.html.

89 | WITS. (2018). *Czech Republic Exports of Goods and Services % of GDP, 1988-2016.* Retrieved from https://wits.worldbank.org/CountryProfile/en/country/CZE/startyear/LTST/endyear/LTST/indicator/NE-EXP-GNFS-ZS.

90 | Fu, X.. (2015). *China's Path to Innovation.* Cambridge University Press.

it prioritise domestic producers and keep its markets relatively closed off to foreigners, it is argued that this would undermine the BRI in the long run. In order for it to be truly mutually beneficial, the BRI has to allow countries beyond China to reap the economic benefits that stem from the initiative.

Another challenge for the successful fruition of BRI projects in the Czech Republic comes from the cautious perception of Chinese projects by a part of the Czech political realm and by the EU.

Indeed, a segment of the Czech political sphere is not partial to the increase of Chinese investment in the country. Relatively large, centre-right parties such as TOP09 or the Civic Democratic Party (ODS), as well as parties with a strong focus on human rights in their foreign policy agenda, such as the Greens or the Pirates, are opposed to increased co-operation with China. The reasoning is that it would be against the fundamental principles – such as the respect for a wide scope of human rights – of the Czech Republic[91]. Should these parties win parliamentary elections and form a government in the future, there is a possibility that they would prioritise other projects than those related to Chinese investment, including investments that are a part of the BRI. At any rate, they will continue to affect the public debate about Chinese investment in the country.

Moreover, the Czech Republic is part of the EU, an EU that has been increasingly concerned about the nature of Chinese investments[92]. If Chinese investment is seen as targeting the acquisition of strategic industries and technology that form the cornerstone of European prosperity, the investment is likely to be curbed[93]. Respect for intellectual property rights, the commercial rather than political nature of the investment, as well as an increased opening up of the Chinese market to the EU are key in continuous and mutually beneficial relations between the EU, its member states, and China. It remains to be seen, therefore, to what extent BRI can be rolled out in the Czech Republic and in the EU in general, and to what extent this will be governed by economic versus political parameters.

91 | Interview conducted with a former senior diplomat. Interview B. (2018). Prague.

92 | European Commission. (2017). *Speech: President Jean-Claude Juncker's State of the Union Address 2017*. Retrieved from http://europa.eu/rapid/press-release_SPEECH-17-3165_en.pdf.

5. Suggestions and Alternatives for the Belt and Road Initiative

This section shall explore possible alternatives for the Belt and Road Initiative for the Czech Republic and present a synthesis of the previous parts findings.

As we explored in section 1, the Czech Republic is, after existing for only a fifth of a century, a sovereign state with very favourable economic conditions and an engaged member of the EU. Considering cases of internal political backlash and what can be perceived as political uncertainty in terms of foreign relations, the Czech Republic stands out in the Visegrad Group for its ability to respond to domestic and foreign issues with versatility. However, while statements and rivalry between high political representatives might be often misleading, there is little doubt that the primary focus shall be the Czech Republic's EU membership: full involvement within the European market and its development and liberalisation, energy union, digital single market, security and consumer protection and setting a concrete date for entering the Euro Zone provided that the Euro Zone reforms itself[94].

In terms of foreign relations and with regard to the EU, the Czech Republic is expected to revive the human rights agenda. However, for relations with China especially, they are dependent on the outcome of the presidential elections: Zeman's re-election suggests a continuation of the present, rather than warmer relations. Therefore, as pointed out in section 4, substantial importance is given to political decisions made on the EU level – as seen, for example, from the former President of the European Commission (2014-2019) Jean-Claude Juncker's proposal of FDI screening[95], which clearly responds to European concerns about Chinese investments into key sectors across the Union. Moreover, as presented during Juncker's 2017 State of the Union speech, 'trade is about exporting our standards, be they social or environmental standards, data protection or food safety requirements'[96], and a potential Czech approach to foreign policy with a focus on human rights would be clearly welcomed on the EU level.

Therefore, if Czech-Chinese relations are to be advanced, adherence to the legal boundaries of the Union will be the

93 | European Commission. (2017). *Press release: State of the Union 2017 – Trade Package: European Commission proposes framework for screening of foreign direct investments.* Brussels, 14. 9. 2017. Retrieved from http://europa.eu/rapid/press-release_IP-17-3183_en.pdf.

94 | ANO. (2017). *Program hnutí ANO pro volby do Poslanecké sněmovny 2017 [Programme of the YES movement for the elections to the Chamber of Deputies 2017].* Retrieved from https://www.anobudelip.cz/file/edee/2017/09/program-hnuti-ano-pro-volby-do-poslanecke-snemovny.pdf.

95 | European Commission. (2017). *Press release: State of the Union 2017 – Trade Package: European Commission proposes framework for screening of foreign direct investments.* Brussels, 14. 9. 2017. Retrieved from http://europa.eu/rapid/press-release_IP-17-3183_en.pdf.

96 | European Commission. (2017). Speech: President Jean-Claude Juncker's State of the Union Address 2017. Retrieved from http://europa.eu/rapid/press-release_SPEECH-17-3165_en.pdf.

stepping-stone towards mutual benefits. On that note, the Czech Republic is fit to engage within the BRI. However, since the sectors involved in the initiative are heavily politicised and usually under the scrutiny of both the public and media, the aisles of cooperation might be restricted. Focusing on the coordination between national authorities in assessing possible security concerns of sensitive takeovers is therefore essential.[97]

Overall, the BRI, if considered an investment project *per se*, is unique in its multilateral outlook and geopolitical range. However, on an EU-led multilateral level, the failures of agreements like the Trans-Pacific Partnership (TPP) or the Transatlantic Trade and Investment Partnership (TTIP) showed that setting a new standard for future deals can be very politically exhausting. As complicated as the dealmaking itself can be, the responsibility is rooted in US President Donald Trump's protectionism and in Brexit.

However, the EU seems to be gradually trying to take over the role of the US in terms of free trade, with Juncker calling the Union 'open for business'. By that, he was emphasising the successes of the Comprehensive Economic and Trade Agreement (CETA) with Canada and accelerating the process for EU trade-agreement making. In this regard, according to the Joint Communication to the European Parliament and the Council from July 2016, the document titled *Elements for a New EU strategy on China* clearly states that the EU must project a strong, clear and unified voice in its approach to China[98]. Therefore, a firm framework has been already set on the table, aiming to supplement the Comprehensive Agreement on Investment[99].

As opposed to the complex and often lengthy process of multilateral cooperation, the field of bilateral relations and investments is where the Czech Republic can truly benefit from the BRI. However, given the existence of EU rules and regulations on top of the common framework(s) to mutually approach the markets, there are already firmly established alternative sources of foreign investments available to the Czech Republic.

97 | Brunsden, J. (2017, September 9). Juncker to Lay Out Plans for Screening Foreign Takeovers in EU. *Financial Times*. Available at: https://www.ft.com/content/b59475aa-9701-11e7-b83c-9588e51488a0.

98 | European Commission. (2016). *Joint Communication to the European Parliament and the Council: Elements for a New EU Strategy on China*. Retrieved from https://eeas.europa.eu/sites/eeas/files/joint_communication_to_the_european_parliament_and_the_council_-_elements_for_a_new_eu_strategy_on_china.pdf.

99 | Proposed and introduced in the EU-China talks in 2013. See European Commission. (2016). *EU and China Agree on Scope of the Future Investment Deal*. Retrieved from http://trade.ec.europa.eu/doclib/press/index.cfm?id=1435.

If comparing the percentage of FDI to the Czech Republic by country, data available from the Czech National Bank (as of 31 December 2015) indicates that the first six countries – and the ones above 4.0 % – were all European: Netherlands (24.1 %), Austria (13.4 %), Germany (12.9 %), Luxembourg (12.2 %), France (7.7 %) and the UK (4.0 %)[100]. Moreover, the sectors that benefitted the most from investments were manufacturing, and financial and insurance activities (with 32.9% and 27.3% respectively)[101]. United Nations Conference on Trade and Development (UNCTAD) statistics show that the top five partners of the Czech Republic in 2016 were Germany, Slovakia, Poland, the United Kingdom and France[102].

The Czech Republic was the only country in the statistics that stood out in comparison to the rest of the Visegrad Group in terms of protection of investors – in all three indexes of 'manager's responsibility', 'shareholders' power' and 'investor protection' – by matching the German standards. Yet, it considerably lags behind Germany in the category of 'transaction transparency', being similar to that of Hungary[103].

Overall, the trade imbalance of Czech-Chinese economic relations can be observed from MIT's Observatory of Economic Complexity (OEC) dataset. In 2016, Czech exports that exceeded 5% of the total export capacity were to Germany (33%), Slovakia (8.4%), Poland (5.7%), the United Kingdom and France (both at 5.2%). Imports, however, came mostly from Germany (27%), China (13%), Poland (8.4%) and Slovakia (5.2%)[104]. The imbalance of Czech exports to China (compared to the overall Czech exports) and Chinese imports to the Czech Republic (compared to the overall Czech imports) are striking. Thus, the BRI faces serious competition in terms of bilateral cooperation in the Czech Republic, but both countries can benefit from concrete projects, should they be implemented in accordance to the EU law and target sectors conducive to Czech economic development.

100 | ČNB (Česka Narodni Banka) [Czech National Bank]. (2017). *Foreign Direct Investment in 2015*. Statistics and Data Support Department. Retrieved from https://www.cnb.cz/miranda2/export/sites/www.cnb.cz/en/statistics/bop_stat/bop_publications/pzi_books/PZI_2015_EN.pdf.

101 | *Ibid.*

102 | *Ibid.*

103 | Santander. (2017). *Trade Portal – Czech Republic: Foreign Investment*. Retrieved from https://en.portal.santandertrade.com/establish-overseas/czech-republic/foreign-investment.

104 | OECD. (2018). *Gross Domestic Product (GDP) (indicator)*. Retrieved from: https://www.oecd-ilibrary.org/economics/gross-domestic-product-gdp/indicator/english_dc2f7aec-en.

Tahkuranna Parish in Estonia
(photo taken by the author).

CHAPTER 7

by   MIKK RAUD
Stanford University

ESTONIA

LATVIA

LITHUANIA

POLAND

CZECH
REPUBLIC

SLOVAKIA

HUNGARY

SLOVENIA

ROMANIA

CROATIA

BOSNIA &
HERZEGOVINA

SERBIA

MONTENEGRO

BULGARIA

NORTH
MACEDONIA

ALBANIA

ESTONIA

This chapter offers a unique overview of Estonia and its relevance to the Belt and Road Initiative (BRI) — a topic that has previously attracted little attention in Estonia or anywhere else despite the BRI's international coverage. Before exploring the opportunities that the BRI creates for Estonia and the next generation of Estonians, the first sections of this paper will provide an insight into Estonia's political structure and attempt to predict the dynamics of the country's political future. After evaluating the current opportunities and challenges faced by Estonian youth, the focus will turn to analyse the BRI framework and its potential benefits and shortcomings from an Estonian perspective.

1. Political Structure of Estonia

On 24 February 2018, the Republic of Estonia celebrated its 100th birthday. The nation's political framework is rooted in a century characterised by dramatic geopolitical change. The international chaos that emerged towards the end of World War I created a window of opportunity for Estonians, who had lived under foreign rule for many centuries, to seek national sovereignty and independence. The Estonian War of Independence was fought against the newly established Soviet Russia, who controlled the region at the time, and it ultimately resulted in Estonia's founding. The building of a new nation-state formally began in February 1920 after the signing of the Treaty of Tartu with the Russians. However, the freedom of self-governance was short-lived and came to a violent end in 1940, when the Soviet Union forcefully annexed Estonia and brought it into the communist bloc towards the end of the Second World War.

While half a century of Soviet rule took a heavy toll on Estonia, the nation refused to give up its strong identity – a sentiment that laid the groundwork for its eventual shift to independence on 20 August 1991. Unwilling to remain a satellite of Russia, unlike many other former Soviet states, Estonia set its sights on the Western framework through harsh, but extremely necessary political and economic reforms with a clear goal in mind: become a liberal, democratic European state. The historic lesson of having already lost independence once acted as a constant reminder not to take it for granted. Thus, despite serious socio-economic problems in the early 1990s, securing independence was considered a priority.

Choosing the path of international neutrality prior to World War II had proved unsuccessful for Estonia during its first attempt at independence. Learning from history, the country's young leaders at the end of the 20th century knew that the best way to secure Estonia's sovereignty in the new Europe was to integrate with Western powers and their allies. When addressing the Swedish Institute on Foreign Affairs in September 1995, President Lennart Meri stated: 'For Estonia, Europe's future is determined by two entities: the European Union and NATO. These two organisations embody our aspirations: prosperity and security. By 2002, Meri was a former president and had become convinced that 'Estonia's political, economic and cultural independence can only be granted by NATO and the European Union.' These statements encapsulate the reasoning behind Estonian efforts to gain access to the EU and NATO during the first decade of its independence. The country finally achieved membership to both supranational bodies in 2004. The prerequisites of joining these organisations assume stable and democratic governance, as well as a smoothly functioning economy. Working towards these landmark goals drove the general development of the country.

Today, independent Estonia operates in accordance with the constitution adopted in 1992, which grants the people the highest power of governance. Estonian citizens elect the 101-member parliament, called the Riigikogu, every four years. The parliament is unicameral and exercises the country's legislative power. Other than fulfilling legislative duties, the parliament elects a president every five years. If no candidate achieves a two-thirds majority within three rounds, the electoral process will be offloaded to an election committee, consisting of members of parliament and representatives of local governments. If the election committee also fails to elect the president, the election process will be shifted back to the parliament. This happened during the 2016 presidential elections when Kersti Kaljulaid became the first female President of Estonia. While largely acting as the moral and symbolic representative of the Estonian people, the president also typically makes a proposal to the head of the election-winning party to form the government, which holds executive powers.

Under the principle of separation of powers, legislative and executive authorities are supplemented by an independent judiciary, which has three levels – the highest of which is the Supreme Court of Estonia (Riigikohus). Similar to most European countries, Estonia follows the civil law system as opposed to the common law system practised in the United Kingdom and many of its former colonies.

2. Political Direction of Estonia

Having provided a basic overview of Estonia's statehood and governmental structure, the following chapter will make predictions and assessments about the Republic's political future. Outlined first is a brief explanation of how Estonia became the innovation powerhouse it is recognised as today. The bulk of this section, however, breaks down the country's most pressing issues into three main categories: 1) International tensions with Russia and domestic challenges with the Russian speaking population, 2) Internal demographics, including migration, and 3) The populist movement challenging not only Estonia but a number of other European countries.

2.1. Technological Innovation

Estonia's development in the first decade of reclaimed independence was largely characterised by the foreign political goals of achieving membership of the EU and NATO projects, which naturally involved the introduction of domestic reforms that took the country forward. After memberships were granted in 2004, the country needed to set its sights on a new national project. For a long time, public discourse was characterised by the search for the 'Estonian Nokia' – a term that President Meri had coined in 1999, referring to the neighbouring Finnish mobile phone giant whose success on the global stage brought the Nordic country into the limelight[1]. It soon became clear that rather than pushing for a single product or company to create international recognition for Estonia, the country itself instead had the potential to grow as a global brand. Driven by the constraints arising from a small population and lack of natural resources, Estonians have realised that the best way to overcome such limitations is through innovation. In 2001, Estonia launched the X-Road, a platform that provides and integrates public and private e-services through interactive databases. As a corollary, Estonia became the first country in the world to allow e-voting in the 2005 local elections and has gradually brought a variety of other public services operations online. At the same time, Estonia's official documentation and archives have been digitised.

Estonians have launched world-famous companies like Skype, GuardTime and TransferWise, demonstrating that innovation is not only limited to the government level. Today, newspaper headlines often refer to Estonia as 'the preview of tech future', 'most digital country in the world', or 'most entrepreneurial European country'[2,3,4]. Thus, Estonia has found its own Nokia – operating an effective government while championing entrepreneurship and innovation throughout society, thereby

1 | Veidemann, R.& Korv, N. (2017, February 23). *Sõnad, mille saime eelmistelt presidentidelt ehk 'tädi Maalist', 'vabakonnani' [Words we received from the former presidents, i.e. from 'tädi Maali' to 'vabakond']*. Retrieved from https://arvamus.postimees. ee/4023537/sonad-mille-saime-eelmistelt-presidentidelt-ehk-tadi-maalist-vabakonnani.

2 | Arets, M. (2017, April 6). *How Estonia Became the Most Digital Country in the World*. Retrieved from http://smartcityhub.com/ governance-economy/how-estonia-became-the-most-digital-country-in-the-world/.

3 | Gray, A. (2017, March 16). *Europe's Most Entrepreneurial Country? It's Not the One You May Expect*. Retrieved from https://www.weforum.org/ agenda/2017/03/europes-most-entrepreneurial-country/.

4 | Walt, V. (2017, April 27). *Is This Tiny European Nation a Preview of Our Tech Future?* Retrieved from http://fortune.com/2017/04/27/ estonia-digital-life-tech-startups/.

becoming a beacon of digitisation for other countries across the world.

All of this would not have happened without appropriate policies from the highest level of government, creating an environment in which new businesses and pioneering ideas are allowed to flourish, no matter how unconventional. In the World Bank's report on entrepreneurship, entitled *Doing Business 2018*, Estonia ranked 12th among 190 countries, ahead of close neighbours such as Finland, Latvia and Lithuania. This scoring is based on the ease of opening a business, obtaining loans, conducting cross-border trade and enforcing contracts, among other standards[5].

What explains the materialisation of such open-minded policies is the fact that the Reform Party, the most liberal party along the Estonian political spectrum, led the government from 2005 to 2016. Over a decade, the Reform Party's government managed to grow the economy, take Estonia successfully through the global financial crisis, create the preconditions for becoming a technologically 'smart' nation, and reinforce Estonia's position in the EU and NATO. However, the later years of the Reform Party's rule were plagued by political scandals and subsequent loss of public support. The growing negative perception of the party in the wake of political controversy led to the administration's stagnation[6]. Thus, in November 2016, the reformists' coalition government collapsed, and they were forced into opposition and replaced by the Centre Party.

While two of the three coalition parties were unaffected by the change in government – the Social Democratic Party and the Pro Patria and Res Publica Union Party – the shift in leadership from liberals to centrists was a significant one. Some have labelled the move as a leftist turn, but it is important to look at the reasons behind such a shift. While the Reform Party-led government was suitable for entrepreneurs and the already-affluent, many blamed their policies for neglecting social issues faced by the financially disadvantaged, as well as farmers and workers.

5 | Äripäev. (2017, October 31). Eesti ärikeskkond ikka tugeval kohal [Estonian business environment still performing well]. *Äripäev*. Retrieved from https://www.aripaev.ee/uudised/2017/10/31/eesti-arikeskkond-ikka-tugeval-kohal..

6 | Postimees. (2016, October 21). Juhtkiri: stagnatsioon Eesti poliitikas [Op-ed: Stagnation in Estonian politics]. *Postimees*. Retrieved from https://arvamus.postimees.ee/3880547/juhtkiri-stagnatsioon-eesti-poliitikas.

Within a year of holding office the Centre Party's government initiated several reforms, which included healthcare and pensions, re-establishing farmers' financial support, and expanding free public transportation from the capital to include inter-city lines and trains[7]. Even though many of these changes have been criticised by the new reformist opposition for being overly populist and economically infeasible, Prime Minister Jüri Ratas's approval rating was above average – 3.23 out of 5 in October 2017, compared to his predecessor Taavi Rõivas's rating of 2.58 in September 2016, shortly before he left office[8]. This indicates that the people were indeed tired of the Reform Party's rule and are open to hearing different voices from the top.

From an economic point of view, 2017 was a more successful year for Estonia than 2016 – GDP grew 5.7% in the second quarter of the latter year as compared to the year before, primarily due to the increased activity in construction, science and technology, and logistics.[9] Such figures allowed the Centre Party to claim that they are able to tackle social problems, as well as ensure Estonia's economic growth and continued entrepreneurial mindset, adding legitimacy to their policies.

2.2. Security, Demographics and Populism

Any party's future success will depend on their ability to deal with three issues facing Estonia and many other European countries. These include security, demographics (including migration and integration), and populism. Regarding security, Estonia's eastern neighbour, Russia, has become increasingly expansionist – best illustrated by its annexation of Crimea and continuing armed conflict in Eastern Ukraine. While Russian President Vladimir Putin has voiced his ambitions to create a Eurasian Union headed by Russia, much resembling the Soviet Union, it is important for small border countries like Estonia to deter Russia from attempting any advances at their sovereignty's expense. Therefore, as long as Russia's administration retains its aggressive rhetoric, any party leading Estonia must ensure security is the ultimate priority.

7 | Täna täitub peaminister Ratase koalitsioonivalitsuse esimene aasta [Today marks the first-year anniversary of Prime Minister Ratas's coalition government]. *Valitsus*. Retrieved from https://www.valitsus.ee/et/uudised/tana-taitub-peaminister-ratase-koalitsioonivalitsuse-esimene-aasta.

8 | Jaagant, U. (2017, October 23). *Peaminister Ratase reiting on langenud [Prime Minister Ratas's rating has decreased]*. Retrieved from http://epl.delfi.ee/news/eesti/peaminister-ratase-reiting-on-langenud?id=79895774; Tamm, M. (2016, September 29). *Hinnang Taavi Rõivase tööle kukkus tugevalt [Rating on Taavi Rõivas's performance fell significantly]*. Retrieved from http://epl.delfi.ee/news/eesti/hinnang-taavi-roivase-toole-kukkus-tugevalt?id=75752831.

9 | Rudi, H. (2017, August 31). Eesti majanduskasv jätkub erakordse hooga [Estonia's economic growth continues at remarkable speed]. *Postimees*. Retrieved from https://majandus24.postimees.ee/4228073/eesti-majanduskasv-jatkub-erakordse-hooga.

Estonia has hosted an increased number of allied troops, as well as building infrastructure critical for receiving physical support from NATO. Despite its appeal to the Russian-speaking community and its previous leader's tight connections with Putin's party, the Centre Party did not stray away from Estonia's basic security principle – more integration with the West, both militarily and politico-economically, for better protection of independence.

The demographics in Estonia is another key area concerning the political direction of the country. Specifically, attention will be drawn to the ethnic composition of the country. Firstly, as one of the most palpable results of the Soviet occupation, Estonia's population of 1.3 million comprises 25 per cent of ethnic Russians[10]. Poor integration policies in the early 1990s left Russian speakers largely isolated from the rest of the population. The majority of them live in a single county bordering Russia (Ida-Virumaa) as well as in particular districts of the capital Tallinn. They also attend schools specifically made for Russian speakers, so it unsurprising that the average Estonian does not socialise with many Russians and the communities remain largely separated. There are, of course, exceptions and the current youth are much better positioned to close the gap between the communities than their parents were.

However, longstanding ethnic tensions between the two groups still exist, well demonstrated by the so-called April Riot of 2007 when hundreds of Russian speakers took to the streets against a repositioning of a Soviet-era statue, showing their distrust towards the Estonian government. This relatively social issue received renewed attention after the annexation of Crimea in 2014, with media discussion focusing on Ida-Virumaa as a potential next target for Putin, precisely because of its social 'distance' from the rest of Estonia and the population's supposed pro-Russian political orientation. Fortunately, the Russian government has not yet tested the region's loyalty, but the mere prospect illustrates how a poorly handled social issue may quickly escalate into a national security issue.

Another problem concerning demographic composition

10 | Statistics Estonia. (2017, June 1). *Rahvaarv rahvuse järgi [Size of population according to ethnicity]*. Retrieved from https://www.stat.ee/34267.

touches on the European refugee crisis, fuelled by domestic instability in several Middle Eastern and North African countries. In seeking a solution to the crisis, the EU decided to relocate the refugees reaching the southern member states, who were bearing the heaviest burden, more evenly across all EU member states. Naturally, that plan included Estonia. As of July 2017, Estonia had received 161 refugees, which is not a large number comparing with the 160,000 subject to the quota-based allocation[11]. Significantly, widely circulated stories from Sweden, France and Germany made Estonian people highly suspicious about opening their doors to refugees who are often thought of as so-called comfort, or economic, migrants in search of improved standards of living instead of asylum seekers fleeing from persecution and threats of conflicts. At worst, locals feared that an influx of asylum seekers from war-torn Middle Eastern countries would allow terrorists to cross the border. The collective memory of how Soviet-era mass immigration threatened the survival of a small country's culture also acted as a cautionary reminder to policymakers in Estonia.

Regarding the refugee crisis, the Estonian Conservative People's Party, resembling the UK Independence Party and other populist movements in Europe, has opted to push the anti-immigration agenda in favour of preserving Estonian culture. Other mainstream parties advocate an approach more in solidarity with the EU and the need to solve the crisis together.

A final important observation about Estonia's political direction concerns the EU and its weakening popularity among several European nations, driven by populist movements best illustrated by the Brexit campaign in the UK, or Marine Le Pen's campaign during the French presidential election. Indeed, the Union-wide economic difficulties and social problems have made some member states believe that rather than facing these issues as a unified European community, it is more effective to take an isolationist approach and stand alone. Similar sentiments have not only arisen among the old Western European democracies but also within the former communist bloc, which less than three decades ago made great efforts to become a part of

11 | Government of Estonia. (2017, September 7). *Pagulasküsimus [The question of refugees]*. Retrieved from https://www.valitsus.ee/et/pagulased.

unified Europe. Isolationist movements in small European countries like Hungary and Poland are driven by the feeling that their cultures and welfare are more threatened by tight international cooperation than larger, more advanced states. Concerns that arise from issues like the refugee crisis and terrorist attacks help fuel those sentiments and lead to criticism towards supporters of a liberal Europe.

As the Estonian President Kersti Kaljulaid explained in a speech in 2017, Europe is currently facing the threat of self-occupation, in which what would usually be outlandish ideas, such as right-wing nationalism, become more attractive among the population and start to gradually dominate and suppress others. It is at this moment that democratic values, which are traditionally treasured in the EU, become endangered.

The future of Estonia will depend on how it will position itself vis-à-vis the EU – either continue to be a vocal supporter of unity or allow the Conservative People's Party to rally enough support to gradually change the country's course towards more isolationist politics. According to a survey released in February 2017, 77 per cent of Estonia's population supports the country's membership in the EU, marking a slight decrease from the year before. Scholars suggest this is the result of frequent coverage of the refugee crisis, as well as Brexit[12]. Nevertheless, nationwide support is high enough to maintain the status quo and not lead to a radical change overnight. Unless significant events beyond Estonia's borders and out of its control occur, either economic or political, Estonia's current dynamics within the EU is unlikely to see a dramatic change.

Overall, Estonia has come a long way from an underdeveloped post-Soviet state to become a global frontrunner of digital governance and innovation. The country has found its niche but is simultaneously facing political and social challenges, which, if not properly addressed, may derail Estonia from its progressive course. A survey investigating people's support for mainstream parties as of November 2017 showed that the Reform Party was leading the race with 29 per cent, followed by the Centre Party's 26 per cent. The Social Democrats and

12 | Rozbaum, M. (2017, February 1). *Uuring: Eesti elanike toetus Euroopa Liidule on stabiilselt kõrge [Survey: Estonian residents' support for the EU is continuously high]*. Retrieved from https:// www.ohtuleht.ee/785083/uuring-eesti-elanike-toetus-euroopa-liidule-on-stabiilselt-korge.

13 | Kantar Emor (2017, November). *Erakondade toetusreitingud [Support ratings of parties]*. Retrieved from http://www.emor.ee/erakondade-toetus/.

the Conservatives remained at 15 per cent and 14 per cent respectively[13]. With these figures in mind, the Reform and Centre Parties were best positioned to determine Estonia's political future in the challenging years ahead.

3. Challenges and Opportunities Faced by Estonian Youth

In the following part, the analysis will shift from the state level to the day-to-day experience of Estonian people, more specifically the country's youth who will one day shape its future. While the three large-scale challenges mentioned in the previous section are all bound to have an influence on the lives of young people, their concerns differ significantly from the older generations and require further inspection. Higher education and unemployment are the most pressing challenges facing 18-25-year-olds in Estonia. These issues will be introduced below, as well as the opportunities afforded to youths by the country's thriving business environment and its increasing integration with the European markets.

The education system in Estonia is certainly among the strongest in Europe. In the 2016 Programme for International Student Assessment (PISA) test – conducted every year among 15-year-old students around the world – Estonians performed the best in Europe. The results came as no surprise as Estonia has consistently been among the top PISA performers since its first participation in 2006.

Technological innovation is partly responsible for the success of the education system. 2012 saw the launching of ProgeTiger, a programme aimed at preschool, primary and vocational education to integrate technological education into the curricula. It resembled the 1996 Tiigrihüpe ('Tiger's Leap') programme, which modernised school networks by giving them internet access and introducing computer education courses[14].

However, while these examples highlight the excellent developments in primary and secondary education systems, higher education in Estonia is not meeting the same standard. Namely, Estonia's top university, the University of Tartu, was ranked 314th in the world at the QS World University Rankings 2017, while the second-best, Tallinn University of Technology, positioned 601-650th [15,16]. Despite the fact that Estonian universities have gradually improved their positions year by year, they remain far from the top spots. Surprisingly the apparent substandard especially applies to engineering and technology courses – key areas for the world's leading digital state – as no Estonian universities came within the top 500 in these categories. The universities are good enough on a local, and even regional, level but cannot provide Estonian youth with competitive qualifications comparable with the world's top educational institutions, leaving them at a disadvantage in global job markets. World-class education relies on world-class teaching, but unfortunately, a lack of funding directed to universities prevents them from offering globally competitive salaries to incoming professors. According to salary data from the University of Tartu (2016)[17], the average professor's salary in 2016 was €1,835, which is well above the Estonian average but cannot compete with salaries offered in the world's leading institutions. As a result, the institutions and academics struggle to balance good teaching with conducting ground-breaking research.

Indeed, statistics show that more and more young people have decided to travel abroad for their university degrees to improve their career prospects. While half a decade ago, the number of people joining Dream Foundation, a programme helping Estonian youth to study abroad, was only around twenty, the number is now in the hundreds[18]. The downside of encouraging Estonian students to explore the world and thereby increase the country's global presence is that it

14 | ProgeTiger Programme. (2017). *Information Technology Foundation for Estonia.* Retrieved from http://www.hitsa.ee/it-education/educational-programmes/progetiger.

15 | QS Top Universities. (n.d.). *University of Tartu.* Retrieved from https://www.topuniversities.com/universities/university-tartu/undergrad.

16 | QS Top Universities. (n.d.). *Tallinn University of Technology.* Retrieved from https://www.topuniversities.com/universities/tallinn-university-technology-taltech.

17 | University of Tartu. (2016). *Tartu Ülikooli 2016. aasta palgaandmed [University of Tartu Salary Data for 2016].* Retrieved from https://www.ut.ee/sites/default/files/www_ut/ulikoolist/tu_palgastatistika_2016.pdf.

18 | Käämer, L. (2015, July 3). *Üha enam Eesti noori läheb välismaale ülikooli [More and more Estonian youth are studying abroad].* Retrieved from http://www.pealinn.ee/newset/uha-enam-eesti-noori-laheb-valismaale-ulikooli-n147478.

becomes difficult to attract the youths to return home after their studies. Some of the most ambitious and brightest young people are effectively being 'lost' from the country, as illustrated by the brain drain problem.

A second, more complex, concern for young people in Estonia is employment. Namely, there is an overproduction of university degrees and underproduction of vocational degrees, leading to both unemployment among those with a degree, and unsatisfied demand for skilled workers. Unemployment figures are more severe among those without a secondary or university education, with around 20% jobless compared to 8.8% for university degree holders[19]. Work experience requirements also leave young people struggling after they leave higher education. Employers tend to hire those with previous experience in a workplace environment, which many students naturally do not have. This generates a vicious cycle of joblessness. The lack of practical experience combined with the questionable quality of many degrees from domestic universities with a lack of highly qualified staff (in the eyes of an employer) only amplifies the problem. While such a situation is not unique to Estonia, many other developed countries combat the work experience problem through increased emphasis on internship and traineeship programmes during university study. Unfortunately in Estonian society, this is not prioritised, leaving graduates disadvantaged. While more proactive students do indeed seek summer placements during their studies, most of these positions tend to be unpaid and thereby favour those from more economically advantaged families.

Other than the aforementioned difficulties acquiring work opportunities, the other reason why young Estonians search for careers abroad is low salaries and a relatively high cost of living in their home country. This issue intensified after Estonia adopted the euro in 2011. In 2017, the average salary in Estonia was €945 (after taxes) per month, well below the EU average of €1,520 per month and more than three times lower than Denmark's €3,095 per month which is leading the ranks. These figures leave Estonia ranking 19th of the 27 EU member states.[20] What demonstrates an incompatibility

19 | Mägi, M. (2017, March 1). *Vähese kogemusega noored tajuvad tööturul lootusetut olukorda [Youth with little work experience feeling helpless in the job market].* Retrieved from https://tarbija24.postimees.ee/4030361/vahese-kogemusega-noored-tajuvad-tooturul-lootusetut-olukorda.

20 | Fischer, R. (2017, April 22). *Average Salary in European Union 2017.* Retrieved from https://www.reinisfischer.com/average-salary-european-union-2017.

between salaries and price indexes is the fact that Estonia ranks sixth from bottom in terms of actual individual consumption, essentially meaning that an average Estonian's purchasing power with the salary available to them is one of the lowest within the EU[21]. In general, it is difficult for the average Estonian young person to find a suitable job that offers remuneration sufficient to afford a comfortable lifestyle amidst the rising living costs.

Having broadly highlighted the difficulties facing young Estonians, it is important to investigate the ways they can be overcome. Indeed, it is no easy task to bring changes to the higher education system or solve the issues of unemployment and low salaries. And the existence and understanding of these problems throughout Estonian society exacerbate problems in other related areas, such as career path choice. There is an awareness that a degree in humanities, as opposed to, say, applied sciences will make it harder to find a suitable job and is therefore likely to influence a high-school graduate's decision making. Of course, not everyone has the necessary passion and ability to study physics or engineering, but highlighting the benefits of such strategic thinking would certainly make some people reconsider their career choices.

As mentioned, many university degrees are not considered valuable by employers, yet the Estonian society still seems to value university education over a vocational one. Vocational programs equip a person with a specific skill demanded by society, thereby increasing the likelihood of finding a stable job and income. In fact, graduates who receive the highest salaries in Estonia (€2,100 per month) come from the Information Technology College, which prepares its students for work in information systems analysis, development and management. The IT College merged with the Tallinn University of Technology in August 2017 and hands out diplomas with a value equivalent to university degrees. So it appears to actually be transcending its traditional status as an institution dedicated to higher vocational education.

21 | Postimees. (2017, June 13). Eesti on elanike ostujõult ELi vaesemate seas [Estonia among the poorest in the EU in terms of people's purchasing power]. Retrieved from https://majandus24.postimees. ee/4144965/eesti-on-elanike-ostujoult-eli-vaesemate-seas.

Other opportunities come from the fact that Estonia provides an easy business environment where startups thrive, with the third-highest amount of startups per capita in Europe (31 for every 100,000 people), according to the e-Estonia initiative (2017)[22]. More and more young people are opting for untraditional career paths and becoming their own bosses – all one needs is a bright idea and the ability to pitch it well to the investors. For example, Estonian-born Taxify, a taxi-hailing app operating in Europe, Australia, Africa and Central America, is one of taxi giant Uber's biggest competitors today. Run by Markus Villig, a young man born in 1993, the company (which was rebranded as Bolt in 2018) secured an investment from Didi Chuxing, the world's largest ride-sharing platform from China, further paving the way to undiscovered markets around the globe[23].

Again, it is unreasonable to expect everyone to want to be an entrepreneur, but it is important to underline that such opportunities exist, and are more available, in Estonia than in most other countries in the world. Therefore, the strategically-minded young, equipped with the courage and willingness to shape their own future, have the best chances in the Estonian job market.

In addition, while Estonia may be disadvantaged by its lack of world-class universities and competitive salaries in many professions, young people aspiring to study in top academic institutions or work abroad now have a better chance to do so than ever before in Estonia's history. As EU citizens, the bureaucratic hurdles subject to Estonians when studying or working in other EU countries are small. For example, for studying in British universities, which are some of the best in the world, Estonians can pay local fees thanks to their EU citizenship, much lower than those paid by students from non-member states. Of course, it remains to be seen how Brexit will affect the status quo.

When it comes to employment, the free movement principle enshrined in Article 45 of the Treaty on the Functioning of the EU allows member state citizens to work and reside in another EU country with no work permit, while enjoying equal treatment with the destination country's nationals in all

22 | E-Estonia. (2017, June). *Estonia Ranks Third in Europe Regarding the Highest Number of Startups per Capita.* Retrieved from https://e-estonia.com/estonia-is-ranked-the-third-in-europe-regarding-the-highest-number-of-startups-per-capita/.

23 | Russell, J. (2017, August 1). *China's Didi Invests in Taxify, An Uber Rival Operating in Europe and Africa.* Retrieved from https://techcrunch.com/2017/08/01/chinas-didi-invests-in-taxify/.

regards[24]. Again, it is therefore up to an individual to go and explore the opportunities offered outside the motherland. By doing so, each individual tacitly acts as an ambassador for Estonia, raising awareness abroad about both opportunities and challenges the country is facing.

In all, the problems which the Estonian youth face in terms of education and employment are surely not specific to Estonia and are undoubtedly common in other European countries. While it is up to the state to adjust its policies in order to tackle these issues, membership in the EU, as well as the domestic business environment Estonia has developed, create plenty of opportunities for the youth to succeed.

24 | European Commission (2017). *Free Movement – EU Nationals*. Retrieved from http://ec.europa.eu/social/main.jsp?catId=457&langId=en.

4. The Belt and Road Initiative – Opportunities for Estonia

China is not the first nation that comes to mind when talking about strategically important partners for Estonia. However, China's rising prominence on the global stage, as well as Estonia's increasing ability and interest in reaching out to countries beyond its borders (highlighted by the e-Residency initiative, for example), has pushed the two countries closer over the past two years. Indeed, the former Estonian Ambassador to China, Marten Kokk, said he believes the relationship between Estonia and China is currently at its historic peak. Estonia opened its new chancery building in Beijing in 2015, followed by a series of ministerial-level meetings, cultural exchanges, and trade deals.

Despite all of this, Estonia's reaction to the BRI has been lukewarm, a contrast to the immense attention the project has received domestically in China, as well as among foreign observers. In fact, Estonia, along with Lithuania and Slovenia were some of the last European countries to officially agree to align with the Initiative in November 2017[25]. Despite being touted by some as one of the most important projects of the 21st century, it is rare to see Estonian reporters picking up news or developments regarding the BRI. Potentially because of vague understanding of the BRI's underlying goals, the project has clearly not been given sufficient attention in Estonia. The following section will aim to bridge this gap and provide an overview of several opportunities the BRI presents to Estonia.

Before moving on with the specific implications of the BRI, it is useful to have a look at some indicators of the current relationship between Estonia and China. The two countries have established several economic treaties. These include agreements on trade and economic cooperation in 1993, favourable investments in 1994, and avoiding double taxing in 1999. Trading agreements in 2015 allowed easier access for Estonian fish and dairy products to enter the Chinese market. In total numbers, the Republic imported €555.8 million worth of goods from China, while exporting only $168.4 million back in 2016, amounting to a trade deficit of more than €387 million[26]. Positively for Estonia, the first three quarters of 2017 alone marked a 38% increase in Estonian exports to China compared to 2016 and comparing 2016 to 2015 also showed a 24.2% increase[27].

Outside the economic figures, the number of Chinese visitors coming to Estonia has increased significantly, reaching 17,000 in 2016, in line with the global growth of Chinese tourists. Estonian musicians and performers, including the Estonian National Symphony Orchestra and the Estonian Philharmonic Chamber Choir, have also been successfully touring China. The Shanghai Opera House performed at the Saaremaa Opera Days in summer 2018, demonstrating a meaningful two-way cultural interaction. Relations between Estonia and China have been improving on several fronts, but there is still room for further, more enhanced cooperation that perhaps the BRI could play a role in developing.

25 | Hu, Y. (2017, November 29). *3 Final CEE Countries Align with Belt, Road Initiative*. Retrieved from http://www.chinadaily.com.cn/world/cn_eu/2017-11/29/content_35117825.htm.

26 | Foreign Ministry of Estonia. (2017, March 6). *Hiina [China]*. Retrieved from http://vm.ee/et/riigid/hiina?display=relations.

27 | Kokk, M. (2017, November 29). Personal Interview; *Ibid.*

4.1. The Central and Eastern Europe 16 Plus China Framework

Foreign countries' level of involvement in the BRI will be dependent on various factors, including proximity to the Chinese-proposed trade routes, political relations with China and their own ability and will to utilise the BRI framework for their own advantage. In April 2012, China agreed with 16 European countries to launch the Central and Eastern Europe 16 plus China framework – the 'China-CEEC' or simply '16+1'. While all 16 countries share similarities due to their communist past, as well as relatively small size compared to China, they also differ significantly in terms of culture, geopolitical posturing (11 are EU members, 5 are non-members), and the level of politico-economic engagement with China. Despite being a member of the framework from the start, Estonia's commitment has so far been rather limited. Namely, after regaining its independence from the Soviet Union, Estonia has tried to shake off the 'post-Soviet' or 'post-communist' image. Having successfully integrated into the Euro-Atlantic economic and security community with memberships in the European Union and NATO, the country is looking to brand itself as a Nordic or Northern European country. Hence, placing Estonia in the same category with Central and Eastern Europe does not necessarily resonate well with Estonians' self-perception and this could affect willingness to engage in the project. Furthermore, the Estonian leadership has been hesitant about the China-CEEC framework's practicality as well as feasibility[28].

Putting pseudo issues aside, it is time for Estonia to undertake a more proactive role in the China-CEEC. As mentioned, each country's successful involvement in the BRI depends on its own ability to make use of the tools and platforms provided – the 16+1 is certainly one of the main instruments Estonia currently has for maximising the benefits of its partnership with China.

In one of the few articles published in the Estonian media on the BRI and China-CEEC in 2016, Liisi Karindi, an expert on Chinese social and economic policies, and Urmas Varblane, an economist, drew attention to five reasons why Estonia should increase its involvement in China-CEEC[29].

28 | Karindi, L.& Varblane, U. (2016, September 8). *Hiina äri koju kätte [Business with China right to one's doorstep]*. Retrieved from https://arvamus.postimees.ee/3828039/liisi-karindi-ja-urmas-varblane-hiina-ari-koju-katte.

29 | *Ibid.*

Firstly, it offers Estonia a direct pathway to the highest levels of the Chinese government, which would otherwise be complicated given Estonia's small size. Connections established through deeper high-level interactions have, for example, led to agreements on fish and dairy product exports.

Secondly, participating in the framework enables Estonia to reach out to a wider Chinese audience through events and forums organised by the Chinese on local government levels.

Thirdly, as there are 15 other countries involved in this branch of the framework, expanding connections does not necessarily have to be limited to China – meetings and events held within the 16+1 group also open new avenues of cooperation with other countries on the European continent, while optimising resources.

Fourthly, it is important not to underestimate how committed the Chinese are on developing the trade and transport corridor between Asia and Europe, in which Central and Eastern Europe serve as an important link, potentially bringing benefits for all the participating countries.

Finally, by increasing involvement in the framework, Estonia can further the Republic's national interests, as well as proactively engaging with opportunities that naturally offer potential benefits for the country. As the latest positive example of the China-CEEC framework's usefulness, the then Estonian Minister of Entrepreneurship and IT, Urve Palo, signed three cooperation agreements in the China-CEEC high-level meeting in Budapest in November 2017, concerning e-commerce, ICT (Information and Communications Technology) and logistics. The move effectively signalled Estonia's alignment with the BRI and its overall goals[30].

30 | Kralla, A. (2017, November 28). *Urve Palo allkirjastas Hiinaga kolm majanduslepet [Urve Palo signed three economic agreements with China].* Retrieved from http://www.logistikauudised.ee/uudised/2017/11/28/urve-palo-allkirjastas-hiinaga-kolm-majanduslepet.

4.2. Estonia's Role as a Smart Transportation Hub

Having established that the 16+1 can serve as a useful platform for Estonia to experience meaningful engagement in the BRI, as well as reap other benefits, the following chapters will focus more deeply on the specific areas that carry the most potential for Estonia – transportation and technology, and the fusion of the two.

The main reason Estonia has been historically conquered by numerous foreign powers is its geographical location. An extensive coastline, long rivers, and flat surfaces made it easy to invade, but they now offer an appealing environment for overseas trade. Moreover, some describe Estonia as the easternmost corner of European civilisation, facing off the Orthodox Russians and more oriental cultures farther East. Controlling such key territory is therefore highly important not only for economic but also for geopolitical purposes. While the independent Estonia of today has clearly sided with Western civilisation, it does not mean the country should not benefit from its location at the crossroads of East and West. While the role of transit hub and trade intermediary is already familiar to Estonia, the BRI may enable the country to take its involvement to the next level.

The proposed infrastructure of BRI consists of the land-based Belt route and the maritime Road route. The maritime routes passing by Estonia offer natural trading ease. Estonia operates nearly 30 well-functioning ports, which are some of the closest EU ports to China. As well as the potential to enhance the already existing port infrastructure, the BRI can bring even bigger improvements to railway infrastructure. Feeding more cargo to transhipment ports, new railway connections may help Estonia to consolidate its position as a major trade intermediary. Indeed, Estonia is one of the geographically closest EU members to China, and the EU market is one of the key destinations for Chinese goods. Efficiently passing Chinese goods on to other EU countries and vice versa seems to be one of the best options for Estonia to profit from the BRI. A couple of examples help illustrate the situation.

Hoping to expand its market reach towards the West and hit its two billion customer goal, Alibaba, one of the most valuable Chinese companies, is catering to the West with a separate, English-language online retail platform AliExpress. Recognising the strategic location of Estonia, Alibaba has contracted Post11, a joint venture of Omniva (the Estonian postal service) and SF Express (one of the largest Chinese postal service companies), to distribute goods ordered through AliExpress to the Russian and European markets. While primarily focusing on 10 countries in Northern and Eastern Europe, Post11 also delivers about 20% of all Russian-placed orders from AliExpress. Admittedly, Post11 is competing with postal service firms from many other European countries, including parcel giants like the German DHL. Therefore, any new business opportunity must be grabbed as soon as possible, as noted by Omniva's leaders[31].

Another example that demonstrates Estonia's good positioning vis-à-vis Europe and China is the Swedish furniture producer IKEA's agreement with Estonia's EVR Cargo (which became Operail in 2018) to get its Chinese manufactured products to Western European markets through the potential China-Estonia railway connections. The connection would provide a more cost-effective and significantly faster avenue than the previously used sea transportation, taking 10-12 days – down from 45 days[32]. There are plans for a direct train link between Changchun, a Chinese city in Jilin province, and Estonia[33]. These two cases show how Estonia's role in the BRI can and should be used to inject hope for larger projects, such as establishing logistics centres for prepositioning goods near ports and railway hubs.

31 | Ristikivi, H. (2017, February 3). *Omniva vaatab ootusärevalt tulevikku [Omniva looking to the future filled with hope]*. Retrieved from http://www.logistikauudised.ee/uudised/2017/02/03/omniva-vaatab-ootusarevalt-tulevikku.

32 | IKEA Shipping China Made Furniture via Rail to Estonia for European Markets. (2017, November 9). *Silk Road Briefing*. Retrieved from https://www.silkroadbriefing.com/news/2017/11/09/ikea-shipping-china-made-furniture-via-rail-estonia-european-markets/.

33 | Kokk, M. (2017, November 29). *Personal Interview*.

What may help Estonia and other Baltic countries improve their position is the projected high-speed railway connection from Tallinn to Warsaw, effectively linking the Baltic states to the dense European railway network. While facing certain legal and environmental obstacles, Rail Baltic's construction is expected to be completed by mid-2025. Even though plans were drawn up long before the BRI even existed, what best describes Rail Baltic's potential relevance in terms of the BRI is a futuristic yet realistic aspiration, which envisions shipping Chinese goods through the Arctic Sea to Finland, and transporting them with trucks to Helsinki, which, in the future, will be connected via an undersea tunnel to Tallinn. The construction is expected to start in the early 2030s[34].

It is, therefore, no wonder that the China Railway Construction Corporation Limited has shown interest in bidding for Rail Baltic[35]. Such an initiative is welcomed given the more affordable price of Chinese expertise compared to other international bidders, but Estonia will likely approach any deal with caution. While the Chinese have already committed to building the Land-Sea Express Route connecting the Greek port of Piraeus with Budapest in a crucial part of the BRI's European section, they have discovered that merely injecting money is not enough given the complicated regulatory environment of the EU. Namely, the construction of the Hungarian section of the proposed high-speed railway has come under EU Commission investigation because the country failed to offer a public tender[36]. To avoid similar issues, the Estonian and Chinese counterparts will need to make sure that complying with the EU regulations have to be prioritised, should any projects be agreed upon in the future.

Even though Rail Baltic will certainly benefit Estonia, it will not necessarily reduce competition with other countries, which obviously aim to make similar gains from facilitating Chinese intervention. Over the years, a network of more than 40 lines of direct cargo trains between China and Europe has emerged, kicking off with the Chongqing-Duisburg line in 2012[37]. Concerning Estonia's immediate neighbours, Finland's rail terminal Kouvola has been receiving trains with freight cargo from China, joining Latvia's Riga and Poland's Lodz[38]. It seems clear that the Chinese would like to see a junction point for their cargo's distribution from Eastern Europe to the rest of Europe.

Political tensions with Russia must also be taken into account. Despite having the advantage of sharing the same railway width as Russia and thus not having to shift goods from one railway to another, Estonia's eastern neighbour has signalled its preference to send goods through Belarus or ask a considerably higher price for the transfer. Russia has been modernising the ports of Ust-Luga and Primorsk to supplement the already capable

34 | Ibid.

35 | Postimees. (2017, May 31). Hiina raudteehiid on huvitatud Rail Balticu arendamisest [Chinese railway giant is interested in developing Rail Baltic]. Postimees. Retrieved from https://majandus24.postimees.ee/4130619/hiina-raudteehiid-on-huvitatud-rail-balticu-arendamisest.

36 | Belt and Road Advisory. (2017, November 11). Legal Quagmire Blocks Belt and Road Initiative in CEE? Belt and Road Blog. Retrieved from https://beltandroad.ventures/beltandroadblog/2017/11/11/legal-quagmire-blocks-belt-and-road-initiative-in-cee.

37 | Shepard, W. (2017, September 13). How China's Belt And Road Sparked A Renaissance Of Transportation Innovation. Retrieved from https://www.forbes.com/sites/wadeshepard/2017/09/13/how-chinas-belt-and-road-just-sparked-a-renaissance-of-technological-innovation/#1f4a3cb138f7.

38 | IKEA Shipping China Made Furniture via Rail to Estonia for European Markets. (2017, November 9). Silk Road Briefing. Retrieved from https://www.silkroadbriefing.com/news/2017/11/09/ikea-shipping-china-made-furniture-via-rail-estonia-european-markets/.

Saint Petersburg and Kaliningrad ports and avoid using the services of its international neighbours, especially with regard to oil shipments[39]. Therefore, it remains crucial for Estonia to maintain and improve the quality of its own ports so that it can continue to offer effective and attractive transhipment hubs for Chinese goods. Notably, while the Chinese have shown interest in investing in the port of Tallinn, no concrete agreements have been signed.

Amidst the competition, it is important to bear in mind that China subsidises the majority of the freight cargo shipments already occurring between China and European countries. However, even the funds allocated for the BRI have their limits, and it is likely that Chinese money will end up with those able to best sell their service. Here, Estonia seems to have an advantage with its advanced technological capacity. Namely, Chris Devonshire-Ellis from Dezan Shira & Associates, an Asia-focused corporate advisory firm, predicts that the long-term winners in this competition to facilitate Chinese transhipments to European markets will be those able to adopt technologies like blockchain and 5G the fastest. This is to deal with the gap in data sharing that currently exists between the EU and non-EU countries, leading to demand for a more effective service[40].

With perhaps the most advanced digital governance in the world, Estonia's so-called 'smart society' infrastructure is well-positioned to achieve the required breakthroughs using technical know-how and innovation ingrained in the culture. Perfect for supporting e-commerce and e-customs, nearly all of the country's commerce is already IT-driven. Indeed, it is not only the physical infrastructure that matters, but also the speed of processing documentation and transactions, which prevents the emergence of major bureaucratic bottlenecks that often cause problems with railways at the borders between two countries.

As the frontrunner of the Digital Single Market within the EU, Estonia has been promoting the principle of 'once only', which means that once a product's documentation has been cleared in one location, it can also be seen and verified in

39 | Statistics Estonia. (2016, January 29). *Sadamate ja raudtee kaubamaht oli viimase kümnendi väikseim [Trading figures of ports and railways lowest in the past decade].* Retrieved from https://blog.stat.ee/tag/sadamad/.

40 | Devonshire-Ellis, C. (2017, July 20). *China Eyes Estonia as Smartest and Nearest Port for EU Access.* Retrieved from https://www.silkroadbriefing.com/news/2017/07/20/china-eyes-estonia-smartest-nearest-port-eu-access/.

other countries – a capacity that is clearly desirable in the BRI project. While Estonia and a handful of other countries already facilitate this innovation that significantly increases the efficiency of delivery processes, its maximum potential can only be achieved if all countries along the BRI trading routes adopt a similar system. After highlighting its own competence in 'clearing' goods in an efficient manner through digitised logistics centres, Estonia's niche could be found in developing a similarly harmonised network within the BRI. Providing such a convenient business environment would be extremely welcome from the Chinese perspective, whose billion-strong e-consumer market needs similar innovations to facilitate its trading with the EU[41].

Estonia is thus well positioned to rely on the already existing digital infrastructure and develop it further to fit the demands arising from increasing trade and transhipment needs internationally. Thus, both parties can benefit from a partnership optimising Estonia's digital capacity.

4.3. Technologies Improving Sino-Estonian Cooperation

One of the most important areas that could potentially characterise Estonia's involvement in the BRI concerns the country's technological advancement. Often described as the 'smartest' digital nation in the world, Estonia has a global reputation for recognising how innovation and flexible policies, such as those driving digital governance, can enhance people's livelihood. China, too, has made technological leaps, becoming the global leader in developing and adopting disruptive technologies – demonstrated by the rapid rise of fintech (financial technology). The focus on digitisation in both Estonia and China shows a shared interest in positively embracing technological change, and this sets a strong basis from which cooperation can develop. There is also evidence that certain avenues of cooperation in this field already exist, shown by the recent launch of the 'Digital Belt and Road', the relevance of which will be discussed in the later stage of this section. In what follows, further opportunities from the BRI relating to technology will be explored and analysed.

41 | *Ibid.*

Perhaps the most important initiative to introduce first is a pilot project launched by the Estonian government in 2014 called e-Residency. Simply, it is a government-issued ID scheme available for anyone in the world. It enables online companies to benefit from operating under a centralised system committed to improving business and entrepreneurship. Participants, or so-called e-residents, can access the entirety of the EU market, do their business banking, sign documents digitally, and use a host of other digital services offered by the Estonian government. The project, which is one of Estonia's flagship initiatives, gives those involved a chance to achieve location independence while doing business by effectively breaking down international borders.

As of 2018, more than 30,000 individuals owned e-Residency ID cards. Applicants for e-residency hail from 154 countries and have so far established more than 4,300 companies[42]. Initially proposed as an idea to benefit the Estonian economy, it has now become clear that e-Residency marks an attempt to create a new borderless digital nation. Even though the BRI has been primarily viewed as a large-scale infrastructure project, it is clear that the Chinese are looking for ways to get rid of this sentiment and expand the understanding of BRI to include business and cultural exchange between China and the countries along the Belt and Road. The e-Residency scheme might be able to help in this regard and benefit both the Chinese and Estonians, as well as the rest of Europe.

As already said, e-Residency can provide an easy route to doing business within the EU, serving as an attractive and convenient platform for Chinese entrepreneurs hoping to enter the European market. So far, the Estonian Embassy in Beijing has already issued over 500 cards, while the total number of applicants from China has surpassed 1,200. Moreover, Estonia's start-up ecosystem itself is one of the busiest in Europe and operates within an encouraging regulatory environment that favours new businesses. The culture encourages innovation-driven individuals to explore even the craziest sounding ideas.

E-Residency could also become a gateway between Estonian

42 | Rang, A. (2017, November 30). Who are Estonia's E-residents? *Medium*. Retrieved from https://medium.com/e-residency-blog/who-are-estonias-e-residents-f8c2ba2bee3d.

and Chinese tech entrepreneurs by offering a platform for Chinese ideas to reach Europe quicker, overcome certain legal challenges, and promote technological cooperation between the two countries on a higher level. One starting point could be launching a collaboration with certain Chinese cities that share relevant interests. While innovation-driven megacities like Shenzhen and Hangzhou may be suitable candidates, it is likely that approaching relatively unknown cities would attract even more interest. This idea, too, comes with certain challenges, however. For example, as e-Residency servers are physically far from China, the provision of services for cardholders in China may not be as smooth as for Europeans. Moreover, despite the potential of the Estonian idea, executing it alone on a scale large enough to make a worldwide impact will be very difficult. Thus, the BRI framework could be used to bring more attention to the project and enhance its potential. It could feasibly promote e-Residency within non-EU nations along the Belt and Road as well as demonstrate that small countries like Estonia are not merely recipients of Chinese investments or trade facilitators but can contribute remarkably to the BRI's overall success.

A more specific form of cooperation could be developed around the aggressively expanding blockchain – a technology already extensively used in Estonian digital government services, including e-Residency, as well as increasingly researched and applied in the Chinese tech community. Both China and Estonia have been toying with the idea of issuing a cryptocurrency – 'estcoin' is being trialled in Estonia and the People's Bank of China's has produced its own prototype – to make financial services more efficient and widely available. While neither of the countries has formally executed their plans in this domain, the experimentation shows how aligned both nations are with their interest in embracing financial innovation. National cryptocurrency, issued by either Estonia or China, would help scale up e-commerce and thereby generate more trade and business for the BRI infrastructure facilities.

Blockchain technology itself can be applied to build an efficient and secure trade finance platform, facilitating the whole

BRI – an idea suggested by James Lau, Hong Kong's then Secretary for Financial Services and the Treasury[43]. The China Federation of Logistics & Purchasing is already cooperating with the Shenzhen Digital Singularity, a technology firm, to bring blockchain into the logistics business[44]. As innovation is limitless, Estonian scientists and entrepreneurs with long years of expertise and experience in blockchain could potentially get a hold of this idea and work together with the Chinese. Other than improving the overall effectiveness of BRI, joint development of such a platform could mark the beginning of tight tech-based cooperation, which can be extended to Research and Development (R&D) on 5G – crucial for successfully developing 'smart' cities, which both China and Estonia are attempting to create.

Finally, cooperation within the technology sphere can also enhance cultural understanding between Estonia and China. If Chinese entrepreneurs with ambitions in the EU market make use of the opportunities arising through e-Residency, and, say, set up a company requiring local personnel, an initial impact would see more jobs available for the local people. While the current image of Chinese culture in the eyes of the average Estonian is still centred around exotic food and cheap consumer goods, introduce the tech-savvy Estonian to a handy smartphone app or a high-quality gadget designed by Chinese entrepreneurs, and the rebranding of China may begin.

While the Estonian tech entrepreneurs are surely aware of the success of their Chinese counterparts, the physical presence of Chinese tech enthusiasts in Estonia would certainly help launch more efficient cooperation as well as boost cultural interactions between the two nations.

Likewise, encouraging Estonians to study and engage with China serves the same purpose. Perhaps a step in the right direction was the agreement between the Tallinn University of Technology and Chinese tech giant Huawei, the latter of which has decided to support ICT students in Estonia and take them on a study tour to China under the Future Seeds Scholarship[45].

43 | Dale, B. (2017, November 14). *Hong Kong Official Touts Blockchain for China's 'Belt and Road' Plan.* Retrieved from https://www.coindesk.com/hong-kong-trade-finance-belt-and-road/.

44 | Shenzhen Digital Singularity. (2016, December 20). *Chinese Logistics Industry Enters Blockchain Era with the New Blockchain Application Sub-Committee.* Retrieved from https://www.prnewswire.com/news-releases/chinese-logistics-industry-enters-blockchain-era-with-the-new-blockchain-application-sub-committee-300382342.html.

45 | Embassy of the People's Republic of China in the Republic of Estonia. (2016, June 14). *Huawei Technologies Signed Agreement with Tallinn University of Science and Technology on Future Seeds Scholarship Program.* Retrieved from http://ee.china-embassy.org/eng/dtxw/t1372008.htm.

Moreover, Robotex, an Estonian-driven robotics competition, is cooperating with a network of robotics clubs in China to bring the competition to the Far East, essentially establishing a partnership allowing young Estonian engineers to visit China and meet their Chinese colleagues[46]. A further educational collaboration of a similar kind to attract young talents to study and work in Estonia is a good idea – getting to know the local environment through studies will increase the likelihood of Chinese youth staying in Estonia and potentially starting their own business. The already established Work in Estonia programme – aimed at attracting bright foreign minds within the Chinese tech contingent – is another avenue for the Chinese to bring their talent to Estonia.

The advanced technological thinking shared by both Estonia and China clearly offers plenty of room for cooperation between the two nations. It remains to be seen if and how these ideas materialise, and how much so within the BRI framework. As mentioned at the beginning of the chapter, the 'Digital Belt and Road' idea, launched in the 4th Internet Conference in Wuzhen in 2017, could offer the first tangible platform for cooperation. The proposed idea includes expanding access to broadband networks, helping e-commerce cooperation and increasing investments in the IT sector[47]. While Estonia is currently not part of the new initiative, these areas fall largely in line with Estonia's overall direction. Therefore, getting involved is surely something to consider to better execute the ideas discussed above.

46 | Äripäev. (2017, September 27). *Robotex läheb Hiina [Robotex is going to China]*. Retrieved from https://www.aripaev.ee/uudised/2017/09/27/robotex-laheb-hiina.

47 | Chen, Q. (2017, December 3). *Consensus Grows at Internet Conference*. Retrieved from http://www.globaltimes.cn/content/1078509.shtml.

5. Estonian Youth and the Belt and Road

The sections above discuss a variety of topics including

Estonia's political system and future direction, problems and opportunities for its youth, as well as the country's potential avenues for participation in the BRI. The final part of this paper will address how the Estonian youth can connect with the BRI and explore ways to improve elements of the Chinese initiative so that it engages more young people in Estonia.

Firstly, it is useful to consider how the BRI could present solutions for the two major problems facing 18-to-25-year-olds discussed above – access to world-class education and issues regarding employment and salaries. It is no secret that as well as expanding its economic influence, China is also interested in increasing its soft power – best done through culture and education. The Confucius Institutes have been around for a long time to facilitate Chinese culture and language learning in foreign countries. But in 2015 China launched the University Alliance of the Silk Road, bringing together more than 130 universities across the five continents. Led by the Xi'an Jiaotong University (located in Xi'an in Shaanxi province), the aim of this programme is to promote educational cooperation, training, and cross-cultural understanding between the universities[48]. Unsurprisingly, no Estonian universities are part of the initiative.

Even though there are certain concerns about the influence of Chinese soft power extending into the political sphere, the opportunities offered by the BRI for educational institutions ought not to be overlooked. One possibility is that the Asian Centre of the University of Tartu could lead the university to join that initiative. This would provide Estonian students with a chance to engage with a variety of top universities abroad, while also expanding their knowledge on the BRI – something evidently very limited among Estonians.

While it was argued above that the best chance for success in the Estonian job market is for those who have studied in applied sciences or gained concrete skills, the gap in knowledge about the BRI also needs filling. If the BRI becomes an influential vehicle in the European region, Estonia will need experts familiar with the Chinese culture, language, business style, and of course the mantra of the BRI itself. Those

48 | OBOReurope. (2017, August 4). *Education Is a Priority of the Belt and Road Initiative.* Retrieved from http://www.oboreurope. com/en/education-priority/.

recognising the potential of the BRI today will thus be ahead of the game in five- or 10-years time when the project's potential has materialised more fully.

It is understandable that the Soviet-era generation in Estonia has a more cautious view of the world and is perhaps unable to think globally. The current youth, however, are world citizens unafraid of foreign cultures and are in a prime position to take part in global initiatives such as the BRI. The key in bringing this message home to today's youth is in communication – raising awareness of opportunities and urging those courageous enough to experiment, engage and learn to get involved with long-term goals in mind.

As the Digital BRI demonstrates, China's flagship project will not only be about infrastructure. Taking the proposed ideas on Sino-Estonian cooperation on the logistics and technology front into account, it is likely that there will be demand for people with technical know-how. Combining this with the ability to talk to and do business with the Chinese will give any individual a strong starting position. As a positive example, academic institutions like Mustamäe Gymnasium and Kuressaare Gymnasium are already teaching Chinese – courses that complement the technical capabilities derived from the ProgeTiger, which already sets a good foundation for a youngster to get ahead of others in this field. Thus, while there is currently limited interest in Chinese culture and language among the majority of Estonian youth, the BRI and the corresponding professional and educational opportunities should make this a much more attractive avenue of exploration and social investment.

Of course, with regard to attracting the interest of Estonian youth towards the BRI and China more generally, there is still a lot of work to do. Ask an average high school graduate in Estonia if they know what Belt and Road is and the majority of them will answer 'no.' Similar situations are likely to occur with university students. This is all due to insufficient communication and a deficiency in international thinking, an area in which China excels as one of the world's most influential players.

In order to overcome this hurdle, both Estonians and Chinese are responsible for driving change. From the Estonian side, Enterprise Estonia, an institution promoting business and entrepreneurship in Estonia and with international partners, has recently increased lobbying efforts to make Estonian governmental organisations realise the potential of Chinese partnerships, the China-CEEC, and the BRI. Constant communication is therefore extremely necessary, especially when it comes to promoting initiatives that do not come to people's mind by default.

Similar communication-based approaches must be adopted to engage the youth. China naturally has its hands tied with regard to undertaking such activities in a foreign country, yet it is still important for the Chinese to make themselves more visible in Estonia. This can be achieved through the Chinese Embassy organising promotional events or lectures on the BRI. Seminars exposing opportunities to travel to China with academic initiatives and adverts about learning Chinese domestically could also achieve this. By working with local universities and promoting discussion on the BRI and its relevance to Estonia, China can help students to recognise their home country as an attractive location for foreign investment that can potentially offer employment. Positively, the Chinese diplomatic delegation in Estonia has started organising gatherings for Estonians in order to build a network of like-minded individuals who are already experienced with Chinese affairs or are simply interested in China. These kinds of initiatives are important to build bottom-up connections between the two nations and may one day lead to bigger achievements in Sino-Estonian relations.

As a general point, it seems that while the opportunities arising from the BRI to Estonian youth are far from public discussion, there are clear indications that cooperations in education as well as on the digital front are most likely to materialise. That requires interest and curiosity from both sides, and the best way to achieve this is through communication, both on the national and day-to-day levels.

Aerial panorama of the Old Town in Tallinn, Estonia.

Taking into account Estonia's political organisation and the society's stage of development, this paper found that Estonia has plenty of potentials to successfully engage with the BRI. Wider engagement on the state level through cooperation in logistics and technology can create new opportunities for Estonian youth both through education and employment. However, the BRI or even the rise of China are very much abstract concepts in the eyes of the majority of Estonians. In order to overcome this problem, the level of communication on all fronts must be increased and intensified. Only in this way can the BRI become a mutually beneficial vehicle for both Estonia and China.

Wine cellar in Tokaj, Hungary.

CHAPTER 8

by ADRIENN LUKÁCS
Assistant Lecturer at the
University of Szeged

HUNGARY

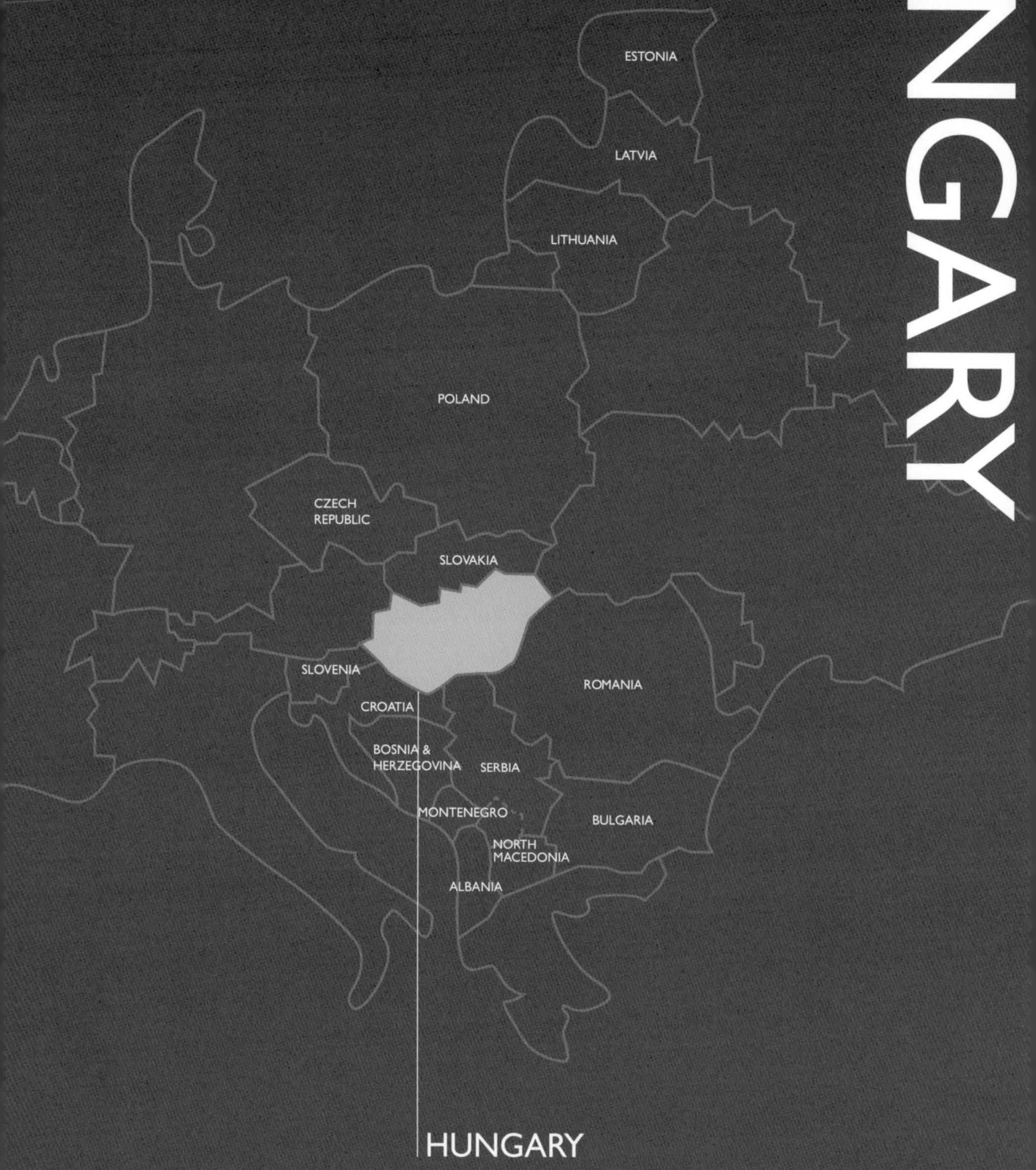

HUNGARY

ESTONIA

LATVIA

LITHUANIA

POLAND

CZECH
REPUBLIC

SLOVAKIA

SLOVENIA

ROMANIA

CROATIA

BOSNIA &
HERZEGOVINA

SERBIA

MONTENEGRO

BULGARIA

NORTH
MACEDONIA

ALBANIA

The chapter discusses the opportunities and challenges arising from Hungary's participation in the Belt and Road Initiative (BRI), with special regard to the economic situation of the country and young people in Hungary. The main focus of the article is on the foreign economic policy of the country and the situation of young people. The aim of the article is to enumerate the possibilities and risks represented by the BRI, starting by analysing the current situation in Hungary.

The chapter is composed of five parts. Part 1 presents the political overview of Hungary. It will detail the political structure of the country and the economic policies of the ruling government. Part 2 evaluates the contemporary political landscape by analysing the (economic) programmes of the political parties. Part 3 presents the opportunities and the challenges that young people face in Hungary. Part 4 analyses what opportunities or risks the Belt and Road project could mean to Hungary. Finally, Part 5 enumerates the possibilities that the Initiative can offer to young people.

1. Overview of the Political Structure of Hungary

Part 1.1. discusses the general political structure of Hungary. Hungary is a parliamentary republic, meaning that the power comes from the citizens but is exercised through elected representatives. This system consists of two major organs: the parliament, which is the supreme body of popular representation, and the government, which is the general means of executive power. These two bodies are closely connected and cooperative since the government is a more active body while the parliament's role is to limit and control the government[1]. Part 1.2. will discuss the main foreign

1 | Chronowski, N., & Drinóczi, T. (Eds.). (2007). *Európai kormányformák rendszertana*. Budapest: HVG-ORAC, p.22.

economic relations of the ruling party, with special regard to Hungary's status as an EU Member State, and the government's 'Opening to the East' policy.

1.1. General Overview of the Political Structure of Hungary

The parliament is the legislative body, the central decision-making unit of Hungary. It is an important body of political communication, which has different roles and responsibilities connected to the functioning of the government[2]. During the parliamentary elections of 2014, nine different political parties won seats in the parliament[3]. The democratic party *Fidesz – Magyar Polgári Szövetség* (Alliance of Young Democrats – Hungarian Civic Alliance) in close alliance with the party *KDNP* (Christian Democratic People's Party) obtained 133 mandates. The biggest opposition was formed by the alliance of the following left-wing parties: *MSZP* (Hungarian Socialist Party), *Együtt* (Together), *DK* (Democratic Coalition), *PM* (Dialogue for Hungary) and *MLP* (Hungarian Liberal Party), who altogether obtained 38 mandates. The extreme right party *Jobbik* (Movement for a Better Hungary) won 23 mandates, while the green party *LMP* (Politics Can Be Different) obtained 5.

The supreme body of executive power is the government. Hungary has a parliamentary system of government, more precisely governance by the prime minister. It means that the prime minister's person has key importance, regarding both the composition of the government and the determination of the government's policies[4]. The Fundamental Law lays out the most important regulations with regard to the government. Concerning the composition of the government, Subsection (3) of Article 16 states that '[t]he Prime Minister shall be elected by parliament on a recommendation by the president of the Republic' and Subsection (7) of Article 16 states that '[m]inisters shall be appointed by the president of the Republic upon recommendation by the prime minister.' The Fundamental Law also states that 'the ministries shall be listed in an act'[5]. After the parliamentary elections of 2014, the alliance of the Fidesz and KDNP parties obtained the majority of mandates

2 | Az Országgyűlés és az Országgyűlés Hivatala [The Parliament and the Office of the Parliament]. (n.d.). *Bevezetés, jogi szabályozás [Introduction, legal regulation]*. Retrieved from: https://www.parlament.hu/bevezetes-jogi-szabalyzas1.

3 | Az Országgyűlés és az Országgyűlés Hivatala [The Parliament and the Office of the Parliament]. (n.d.). *Bevezetés, jogi szabályozás [Introduction, legal regulation]*. Retrieved from: https://www.parlament.hu/bevezetes-jogi-szabalyzas1.

4 | Trócsányi, L., Schanda, B., & Csink, L. (Eds.). (2016). *Bevezetés az alkotmányjogba: az Alaptörvény és Magyarország alkotmányos intézményei (5th ed.) [Introduction to constitutional law: the Basic Law and the constitutional institutions of Hungary]*. Budapest: HVG-ORAC.

5 | Subsection (1) of Article 17 of Fundamental Law of Hungary (2011). Section 1 of the Act XX of 2014 on the listing of ministries in Hungary enumerates 10 ministries in total: Cabinet Office of the Prime Minister, Ministry for National Economy, Ministry of Agriculture, Ministry of Defence, Ministry of Foreign Affairs and Trade, Ministry of Human Capacities, Ministry of Interior, Ministry of Justice, Ministry of National Development and the Prime Minister's Office. Among the different ministries, two could have relevance to the subject of this paper: the Ministry for National Economy and the Ministry of Foreign Affairs and Trade. It is a Hungarian 'speciality' that the portfolios of foreign affairs and foreign trades are the competences of the same minister.

in the parliament[6]. It was Fidesz and KDNP's second consecutive term as governing parties as they also won the parliamentary elections of 2010.

1.2. The Foreign Economic Policy of the Government in Power[7]

Hungary is a Member State of the European Union, which naturally affects the country's foreign economic policy – due to the EU common trade policies – and economic relations. However, an opening towards the eastern regions could be observed in Hungary, manifested especially in the strategy 'Opening to the East'.

Hungary acceded to the EU on 1st May 2004. The country's membership in the EU highly influences its foreign economic relations (and the mobility of citizens), especially through the four fundamental freedoms of the EU – freedom of movement of persons, goods, capital and services[8]. The current circumstances also shape Hungary's economic relations: the common membership and the geographical proximity facilitate trade and commerce between the EU's member states.

Out of Hungary's 10 most important export partner countries in 2015, 10 were from the EU, and out of the 10 most important import partner countries, nine were EU member states (the 10th was Russia)[9]. The country's most important economic partner is Germany. More than one-quarter of both the export (five times bigger than with the second most important partner, Romania) and import (four times bigger than with the second most important partner, Austria) of goods from and to Hungary was realised with Germany in 2015[10]. The reliance of Hungary on trade with fellow EU states evidences the importance and impact of its membership on the national economy.

6 | Fidesz defines within its core values inter alia the strengthening of families, the honouring of labour, the increase of the cohesive force of the nation and the enforcement of the traditions and Christian values. KDNP defines itself as a modern party with Christian ideas, originating from the Hungarian and European historical Christian traditions and the mentality of the modern western European Christian democratic parties. Sources: Fidesz. (2013). *A Fidesz – Magyar Polgári Szövetség Alapszabálya [Statutes of the Fidesz - Hungarian Civic Association]*. Retrieved from http://static.fidesz.hu/newsite/documents-file/l/ya/1391092271-fidesz-magyar-polgari-szovetseg-alapszabalya.pdf; Kereszténydemokrata Néppárt. (n.d.). *A Kereszténydemokrata Néppárt politikájának alapelvei [Principles of Christian Democratic People's Party policy]*. Retrieved from https://kdnp.hu/celjaink/alapelvek.

7 | As of autumn 2017.

8 | These four freedoms are components of the common market and were already defined in the Treaty of Rome, 1957.

9 | Központi Statisztikai Hivatal [Hungarian Central Statistical Office]. (2016, July). Jelentés a külkereskedelem teljesítményéről, 2015 [Foreign Trade Performance Report, 2015]. Retrieved from https://www.ksh.hu/docs/hun/xftp/idoszaki/kulker/kulker15.pdf.

10 | Központi Statisztikai Hivatal [Hungarian Central Statistical Office]. (2017, January 13). *Statisztikai Tükör - A Magyarország és Németország közötti kapcsolatxei [The crisis in the European Union and the difficulties of opening up in the East]*. Retrieved from https://www.ksh.hu/docs/hun/xftp/stattukor/mo_nemet_kapcs.pdf

Moreover, the European Union's budget and funding provide the possibility of realising several investments or participating in different projects (e.g. creating infrastructure, renovating buildings, research projects, etc.), thus enabling the development of the country. For example, in 2014, the funding from the EU constituted 6.57 % of the gross national income (€6.62 billion), which was the highest percentage among EU member states' [11].

As early as 2010, Fidesz acknowledged Hungary's double 'identity', of being part of Europe but at the same time the border to the East[12]. To compensate for the economic difficulties arising within the European Union, in 2011, the government announced the strategy of opening up to the East and then, in 2015, the strategy of opening up to the South[13]. This article will focus on the former.

The strategy of opening up to the East basically consists of export development towards the dynamically developing countries, through which Hungary can benefit from their growing import need[14]. One of the main objectives of the strategy is to obtain market possibilities for Hungarian small and medium-sized enterprises (SMEs) through the establishment of trading houses[15]. This strategy is aimed at several countries which are also part of the Belt and Road Initiative (such as India, South Korea or Vietnam). Therefore, Hungary's participation in the Belt and Road project can also contribute to closer economic cooperation with these countries. Among these countries, the focus will be on Hungary's relations to Russia and China, as these are the two most important foreign economic powers – besides the United States – with regard to cooperation in the field of commerce and investments[16].

Chinese-Hungarian economic relations are becoming increasingly important. Outside of Europe, China

11 | European Commission. (n.d.). *EU Expenditure and Revenue.* Retrieved from https://ec.europa.eu/budget/graphs/revenue_expediture.html.

12 | Országgyűlés [House of the National Assembly]. (2010, May 22). *A Nemzeti Együttműködés Programja [National Cooperation Programme].* Retrieved from http://www.parlament.hu/irom39/00047/00047.pdf.

13 | Lentner, C. (2016)Vázlat a magyar gazdaságpolitika nem konvencionális eszközrendszeréről és eredményeiről – Magyar Nemzeti Bank centrikusan [Outline of the non-conventional instruments and results of Hungarian economic policy - Magyar Nemzeti Bank centrally]. *Hitel, 29(9), 3-18.* Retrieved October 28, 2017, from http://www.hitelfolyoirat. hu/sites/default/files/pdf/01-lentner.pdf. The strategy of opening up to the South was suggested by the Hungarian foreign affairs minister in 2015 and focuses on Africa and Latin-America. See more in: MTI. (2015, March 5). Szijjártó Péter meghirdette a déli nyitás stratégiáját [Péter Szijjártó announced the strategy of opening in the south]. *Origo.* Retrieved from https://www.origo.hu/itthon/20150305-szijjarto-peter-meghirdette-a-deli-nyitas-strategiajat.html.

14 | Nemzetgazdasági Minisztérium. (2011, December). *Magyar Növekedési Terv - Konzultációs anyag [Hungarian Growth Plan - Consultation material]. (2011, December).*Retrieved October 28, 2017, from http://www.innovacio.hu/download/allasfoglalas/2011_12_28_mnt_konzultacios_anyag.pdf.

15 | Szretykó, G. (2013). *Az Európai Unió válsága és a keleti nyitás nehézségei [The crisis in the European Union and the difficulties of opening up in the East].* Retrieved October 28, 2017, from http://kgk.sze. hu/images/dokumentumok/kautzkiadvany2013/makropenzugy/szretyko.pdf.

16 | Becsey, Zs. (2014). A keleti nyitás súlya a magyar külgazdaságban [The weight of opening to the East in the Hungarian foreign economy]. *Polgári Szemle, 10* (1-2), pp. 428-443.

17 | In 2015 imports with China constituted 5.7 % of the imports of Hungary, while the total Chinese working capital exceeded US$4 billion in 2016. See Központi Statisztikai Hivatal [Hungarian Central Statistical Office]. (2016, July). *Jelentés a külkereskedelem teljesítményéről, 201xa [Statutes of the Hungarian Socialist Party]*. Retrieved October 28, 2017, from KSH website: https://www.ksh.hu/docs/hun/xftp/idoszaki/kulker/kulker15.pdf.

18 | Xinhua News Agency. (2015, June 7). China Signs Cooperation Document with Hungary over Belt and Road Initiative. *Belt and Road Portal*. Retrieved from https://engyidaiyilu.gov.cn/zchj/zcjd/1847.htm.

19 | Kong, T. (2015, December 14). *The 16+1 Framework and Economic Relations Between China and the Central and Eastern European Countries*. Retrieved October 28, 2017, from http://councilforeuropeanstudies.org/critcom/161-framework-and-economic-relations-between-china-and-ceec/.

20 | For example, Act II of 2014 on the promulgation of the agreement between the Government of Hungary and the Government of the Russian Federation on the cooperation regarding the use of nuclear energy for peaceful purposes.

21 | For more information see: Act XXIV of 2014 on the promulgation of the agreement between the Government of the Russian Federation and the Government of Hungary about the state loan provided to the Hungarian Government in order to finance the construction of the nuclear power plant in Hungary.

22 | It is an investment of significant volume. The agreement was subject to heavy attack from the media, reflecting the viewpoint that by these agreements Hungary makes a step back towards the Soviet era and will be dependent on Russia. See for example: Stubnya, B. (2017, August 27). *LMP: Eszement időutazás részei vagyunk [LMP: We are part of a crazy time journey]*. Retrieved October 28, 2017, from http://index.hu/belfold/2017/08/27/lmp_putyin_paks_2_kossuth_ter_demeter_szell_vago/; Nyüzsi. (2013, January 07). *Orosz gyarmat leszünk, pokoli hosszú időre [We will be a Russian colony for a hell of a long time]*. Retrieved November 06, 2017, from http://hvg.hu/velemeny.nyuzsog/20130107_Orosz_gyarmat_leszunk_pokoli_hosszu_idore.

23 | Gazprom Export. (n.d.). *Hungary*. Retrieved from http://www.gazpromexport.ru/en/partners/hungary/.

is Hungary's most significant commercial partner, with a growing number of exports and of Chinese companies investing in Hungary[17]. The Belt and Road Initiative (BRI) represents many opportunities for Hungary. On 6th June 2015, Hungary's and China's foreign affairs ministers signed a memorandum of understanding, which makes Hungary the first European country participating in the Initiative.[18] Hungary – along with the other countries in the Central European region (covered by the so-called 16+1 framework) – plays an important role for China, as they can constitute a bridgehead to advance into the European market[19]. A detailed discussion of the BRI in relation to Hungary is contained in Part 4 of this paper.

Russia has a key role in Hungary's opening towards the eastern countries. Russia is a crucial partner in Hungary's energy sector, and Hungary has signed various energy agreements with it[20]. In 2014, the Russian and Hungarian governments signed an agreement in which Russia shall grant a loan (up to a maximum of €10 billion) for the purpose of expanding Hungary's nuclear plant (the so-called Paks investment) [21]. The Paks investment was perceived by the government as a comprehensive economic development strategy, while the opposition – as the next chapter will point out – did not share this viewpoint.[22] Russia also plays an important role in Hungary's gas supply: from 2021, a new long-term agreement with Russian energy multinational Gazprom regulates the issue of gas supply[23].

2. The Future Political Direction of Hungary: With Special Regard to the Economic Programmes of Parties Running in the Parliamentary Elections of 2018[24]

In the spring of 2018, parliamentary elections were held in Hungary, during which the citizens of the country elected members of the parliament for the period of 2018-2022. Several parties ran at the elections, and this part will present their political programmes regarding the most relevant elements related to the subject of this paper[25]. Part 2.1. will briefly examine the parties that were present in the parliament before the elections, while Part 2.2. will deal with the parties that gained relatively considerable support according to polls. Table I helps to review these parties and their main orientations.

2.1. Parties Present in the Parliament Between 2014-2018

This Part will discuss the political programmes of the parties that were present in the parliament during this period; namely, the programmes of Fidesz-KDNP, which are the governing parties, the coalition of the left-wing parties (MSZP, Együtt, DK, PM, MLP), the extreme right party, Jobbik, and the green party, LMP.

I have previously presented the economic programme of the governing parties, *Fidesz-KDNP*. As the parties had not yet published their political programme for the elections, it was presumed that in case they won the elections of 2018, they wished to continue their existing political and economic programme.

It was not yet clear how exactly the left-wing parties would form an alliance. The *MSZP* defined itself as a social-democratic party[26]. It clearly stated that it wishes to strengthen Hungary's cooperation with the European Union and it envisaged the country's future as a Member State in the European Union[27]. At the time of writing the study, the party had not published its detailed political programme.

The *Együtt* party aimed to restore constitutional democracy, to promote economic and social development and national

24 | The content of this part was written as of autumn 2017.

25 | At the time of writing not all of the parties had released their official political programme for the elections.

26 | Magyar Szocialista Párt. (n.d.). *A Magyar Szocialista Párt Alapszabálya [Statutes of the Hungarian Socialist Party]*. Retrieved from http://mszp.hu/sites/default/files/a_magyar_szocialista_part_alapszabalya.pdf.

27 | Magyar Szocialista Párt. (n.d.). *Új Társadalmi Szerződés Magyarországért -Ajánlat a Nemzetnek [New Social Treaty for Hungary - Offer to the Nation]*. Retrieved from http://mszp.hu/page/download?ct=doc&cid=213&dt=atch&did=459.

NAME OF THE PARTY	THE PARTIES' OWN DEFINITION OR MAIN ORIENTATION[28]
PARTIES CURRENTLY PRESENT IN THE PARLIAMENT	
FIDESZ (Alliance of Young Democrats – Hungarian Civic Alliance)	Democratic party
KDNP Christian Democratic People's Party	Modern democratic party with Christian ideas
MSZP (Hungarian Socialist Party)	Social democratic party
Együtt (Together)	Aims to restore constitutional democracy, to promote economic and social development and national identity, and it supports European integration
DK (Democratic Coalition)	The progressive, western-oriented, centre-left party
PM (Dialogue for Hungary)	Left-wing, green political community
MLP (Hungarian Liberal Party)	Liberal party
Jobbik (the Movement for a Better Hungary)	Value-based, conservative, radical national Christian party
LMP (Politics Can Be Different)	Represents left-wing, liberalist and community-based conservative ideas and has an ecological and radical democratic approach
PARTIES CURRENTLY NOT PRESENT IN THE PARLIAMENT	
Momentum	New political generation, aiming to promote equal opportunities and solidarity and enhance the national cooperation
MKKP (The Hungarian Two-tailed Dog Party)	Joke party

Table 1 – Political Parties Running at the 2018 Parliamentary Elections

identity, and it supported European integration[29]. It expressed its views of supporting the European Union in its political programme for the elections of 2018. Regarding foreign relations, Együtt wished to lay new foundations for political and economic relations with Russia. It also expressed that the conditions of asserting Hungary's interests should be strengthened, and its vulnerability should be decreased.[30]

The *DK* – a progressive, western-oriented, centre-left party – was committed to the European Union. The party suggested a renewal in foreign policy, and the strengthening of ties to the European Union and to the European countries in the region[31]. The DK also wanted to terminate the Paks contract with Russia. Various measures were aimed at the young generation to help improve their chances of entering the labour market and to facilitate their housing situation[32]. Párbeszéd (previously referred to as PM) – a left-wing, green political community – was committed to the European Union and wished to strengthen the EU. Párbeszéd also wanted to improve the economic situation of Hungary. Concerning the housing problems of young people, the party envisioned a ceiling for rent and the creation of low-rent social apartments and dormitories[33]. *MLP* is a liberal party, which had not issued its political programme for the elections at the time of writing.

Jobbik, which is an extreme-right party, defined itself as a 'value-based, conservative, radical national Christian party.'[34] Although their political programme for the 2018 elections had not been released, according to Jobbik's founding document, Hungary's membership in the European Union is unacceptable. In their programme of 2014, they advocated either withdrawal from the European Union or the re-negotiation of the relationship with the EU. In this document, they stated that they wished to channel foreign economic relations towards the eastern economic region. They deemed it necessary to establish a Central-European axis from Poland to Croatia, which could constitute the fundaments of regional cooperation, and these countries could enforce their interests in a uniform manner. Jobbik desired a change of paradigm in Hungarian foreign policy and wished to explore the possibilities in the dynamically

28 | On the exact sources of the Parties' own definitions, see the founding documents referred to in section 2.2.

29 | Field, R. (2014, June 22). *Together 2014 Issues New Political Manifesto*. Retrieved from https://budapestbeacon.com/together-2014-issues-new-political-manifesto/.

30 | *Ibid.*

31 | Új Köztársaságért Alapítvány & Demokratikus Koalíció. (2017). *Sokak Magyarországa - A Demokratikus Koalíció Programjavaslata Vitaanyag 2.0 [Hungary of Many - Program Proposal of the Democratic Coalition Discussion paper 2.0]*. Retrieved from https://ujkoztarsasagert.hu/wp-content/uploads/SOMA.pdf.

32 | *Ibid.*

33 | Párbeszéd Magyarországért. (n.d.). *Mindenki számít! Karácsony Gergely, a Párbeszéd miniszterelnök-jelöltjének programja [Everyone matters! Program of Gergely Karácsony, Prime Minister-designate for Dialogue]*. Retrieved from https://parbeszedmagyarorszagert.hu/files/public/mindenki_szamit_2.0.pdf.

34 | Jobbik Magyarországért Mozgalom. (n.d). *Alapító Nyilatkozat [Founding Statement]*. Retrieved from https://www.jobbik.hu/alap%C3%ADt%C3%B3-nyilatkozat.

developing Chinese, Russian, Indian, Turkish and Central-Asian markets. Germany, Russia and Turkey had a central role in Jobbik's foreign policy. According to them, Turkey could be a possible important intermediary towards the Middle East. They also wished to develop good relations with China and South-East Asia on a realpolitik basis[35]. Jobbik's programme contained dispositions aimed at young people: they wished to promote the return of young expatriates to Hungary. To achieve this goal, they intended to introduce a state-financed programme for the construction and purchase of homes and rented dwellings, and they wished to create more jobs for, and prevent the indebtedness of, young people[36]. Jobbik aimed to provide students with the possibility to gain experience abroad by renewing exchange student programmes and scholarships. The party also recognised the power of receiving foreign students as they can contribute to the development of economic, cultural and commercial relations and to cooperation between the countries[37].

The *LMP*'s political ideology was inspired by different ideologies, so it reflected left-wing, liberalist and community-based conservative ideas, and had an ecological and radical democratic approach[38]. The party visualised Hungary's future within the European Union. Together with the European green parties, it aimed to assume an initiator role in solving the EU's political crisis, promoting cooperation within Europe, especially within the Central-European region and shaping the European Union's foreign and commercial policy[39].

2.2. Other Parties in the Public Mind in Hungary

According to certain polls, two other political parties might obtain considerable support: the Momentum party and the MKKP[40].

Momentum first appeared in the public forum with their 'NOlimpia' campaign, in which they called for a local referendum against Hungary's application for the 2024 Olympic Games. The party, which is mostly composed of young people, aims to create a new political generation by promoting equal

35 | Jobbik Magyarországért Mozgalom. (2014). *Kimondjuk. Megoldjuk. - A Jobbik országgyűlési választási programja a nemzet felemelkedéséért [Jobbik's parliamentary election program for the rise of the nation].* There is no direct link to a pdf, but it is downloadable from this link: https://adoc.pub/queue/kimondjuk-megoldjuk-a-jobbik-orszaggyelesi-valasztasi-program.html.

36 | *Ibid.*

37 | *Ibid.*

38 | LMP - Lehet Más a Politika. (n.d.). *A Lehet más a politika kezdeményezés alapító nyilatkozata [The founding statement of the policy initiative may be different].* Retrieved from https://lehetmas.hu/wp-content/uploads/2018/03/371939699-Alapito-nyilatkozat-pdf.pdf.

39 | LMP - Lehet Más a Politika. (n.d.). *Alapító Nyilatkozat és Kezdeményezésünk céljai. [Founding Statement and Objectives of Our Initiative].* Retrieved from: https://lehetmas.hu/alapito-nyilatkozat/.

40 | Publicus. (2017, October 01). Pártok támogatottsága és politikusok népszerűsége, 2017 szeptember [Party support and popularity of politicians, September 2017]. *Publicus.* Retrieved from http://www.publicus.hu/blog/partok_tamogatottsaga_es_politikusok_nepszersege_2017_szeptember/.

41 | Momentum. (2017). *Momentum 2018 - Indítsuk be Magyarországot! [Momentum 2018 - Let's start in Hungary!]*. Retrieved October 28, 2017, from Momentum's website: https://program.momentum.hu/static/pdfs/momentum-program-2018.pdf.

42 | *Ibid.*

43 | According to the Momentum party, the details of the project are not clear, and it does not come with significant advantages to the Hungarian economy compared to the high costs of the development. Source: Momentum Blog. (2017, May 06). A belgrádi vasút lesz Orbánék négyes metrója [The Belgrade railway will be Orbánék's fourth metro]. *Greenfo*. Retrieved from https://greenfo.hu/hir/a-belgradi-vasut-lesz-orbanek-negyes-metroja/.

44 | *Ibid.*

45 | *Ibid.*

46 | Lencsés, K. (2016, October 04). *Félre a tréfát: a kétfarkúak indulnak a 2018-as választásokon [Set aside the joke: the two-tailed will run in the 2018 election]*. Retrieved October 28, 2017, from http://nol.hu/belfold/felre-a-trefat-a-ketfarkuak-indulnak-a-2018-as-valasztasokon-1634663.

47 | Pálfi, R. (2014, September 08). *Bejegyezték a Magyar Kétfarkú Kutya Pártot [The Hungarian Two-Tailed Dog Party was registered]*. Retrieved October 28, 2017, from http://24.hu/belfold/2014/09/08/bejegyeztek-a-magyar-ketfarku-kutya-partot/.

48 | Althoff et al. (2018). Us vs. Them in Central and Eastern Europe Populism, the Refugee Other and the Re-Consideration of National Identity. *Friedrich-Ebert-Stiftung*. Retrieved from https://library.fes.de/pdf-files/bueros/budapest/14599.pdf.

49 | During the previous referendums, the number of invalid votes barely passed 2%. Source: MG. (2016, October 2). Megközelítőleg sem szavaztak soha ennyien érvénytelenül [Nor have they ever voted so invalid]. *Index*. Retrieved October 28, 2017, from http://index.hu/belfold/2016/10/02/ervenytelen_szavazatok/.

50 | Nelson, S.S. (2018, April 7). Hungary's Satirical 'Two-Tailed Dog' Party Will Debut In Sunday Elections. *NPR*. Retrieved from https://www.npr.org/sections/parallels/2018/04/07/599928312/hungarys-satirical-two-tailed-dog-party-will-debut-in-sunday-elections.

opportunities, solidarity, and enhancing national cooperation[41]. In May 2017, they announced that they intended to run at the 2018 parliamentary elections. In their political programme they stated that instead of opening towards the east, they aimed to reinforce Hungary's position in the European Union and to promote investments in the Central European region[42]. Momentum wished to terminate the construction of the Budapest-Belgrade railroad[43] and the Paks investment[44]. Despite these desired terminations, Momentum also wished to cooperate and to establish prosperous relations with China and Russia in the future[45]. Several of their proposed plans were aimed at young people, by improving the availability of education and affordable housing.

MKKP (The Hungarian Two-tailed Dog Party) is a joke political party that declared its intent to run at the 2018 elections[46]. Although MKKP started its activity as a joke party, it became an officially registered party in 2014[47]. In October 2016, during the so-called migrant quota referendum, they encouraged their followers to express their disagreement and boycott the referendum by casting invalid ballots. Partly due to this campaign, they succeeded in mobilising voters, 6.14 % of whom cast invalid ballots[48], which is considerably higher than the invalid votes of the previous referendums[49]. The analysis of its political programme and economic programme would not contribute to this article, as it can basically be summarised by the phrase 'free beer plus eternal life'.[50]

In conclusion, it can be stated that the foreign economic policy of Hungary could be influenced depending on whether a party other than Fidesz won the elections or whether the parties in opposition could effectively cooperate. The parties in opposition expressed their disagreement towards Russia, especially the Paks investment, but such vehement opposition was not present regarding Chinese-

Hungarian relations (except for Momentum). The parties were not explicit on this subject, so it remained to be seen whether the parties emphasising Hungary's identity as an EU Member State would do so at the expense of opening to the East. However, it should be noted that according to polls made in September of 2017, Fidesz was still the most supported party[51].

Since this article's original submission (autumn 2017) and publication, a considerable amount of time has passed; a brief update is necessary in order to take into the consideration the results of the parliamentary elections held on 8 April 2018[52]. According to the results[53] published by the National Election Office, FIDESZ-KDNP obtained two-thirds of the mandates (67.34%) of the parliament, thus winning the election. The biggest opposition was formed by Jobbik, having 12.56 % of the mandates. Then followed MSZP-PM with 10.05 %, DK with 4.52 %, LMP with 4.02 % and Együtt with 0.5 %. Momentum and MKKP could not get into the parliament as they did not obtain enough votes[54].

51 | Publicus Research. (2017, October 01). *Pártok támogatottsága és politikusok népszerűsége, 2017 szeptember [Party support and popularity of politicians, 2017 September]*. Retrieved from http://www.publicus.hu/blog/partok_tamogatottsaga_es_politikusok_nepszersege_2017_szeptember/.

52 | The update would consider only the results, but not the alliances between the parties or analysis of the political programmes published in the meantime.

53 | Based on the 98.96 % of the votes, as at the time of updating the article, the votes cast by Hungarians staying abroad were not yet counted.

54 | Nemzeti Választási Iroda. (2018, April 10). *Országgyűlési képviselők választása, 2018. április. 8. Tájékoztató adatok az országgyűlés összetételéről [Election of Members of Parliament, April 2018. 8. Informative data on the composition of the National Assembly]*. Retrieved April 10, 2018 from http://www.valasztas.hu/dyn/pv18/szavossz/hu/l50.html.

3. The Current Challenges and Opportunities Faced by Young People in Hungary

This Part presents what difficulties young people face in Hungary, and what their prospects are. Part 3.1. discusses the challenges, with special regard to the high youth unemployment rate, low wages, the difficult housing situation, and the increasing number of young people deciding to leave Hungary. Part 3.2.

presents the opportunities, such as the Youth Guarantee Plan, the EU financed mobility programmes, dual training and the opportunities created by new investments. The BRI could also contribute to increasing the opportunities of young people in Hungary, which will be further discussed in Part 5.

3.1. Challenges Faced by Young People in Hungary

According to a 2015 survey conducted by Youthonomics, a think tank aiming to help generations Y and Z, the so-called Youthonomics Global Index was created. This index uses 59 different criteria (such as youth unemployment, quality and cost of education, the ability of young people to afford housing and save for the future, public deficit, access to technology, etc.), and ranks 64 countries 'according to whether they are creating the conditions that will allow youth[55] to flourish and prosper'[56]. Hungary was ranked 38th, becoming the second-worst country in the Index from the European Union (Croatia is the last EU country, at rank 51). The study brings attention to the fact that if young people are overlooked, and their basic needs are not ensured, they can – in an age of unprecedented international mobility – leave their country of origin for better conditions[57]. This migration of young people poses a concern for Hungary.

Youth unemployment is a serious issue in the European Union, with a much higher rate compared to the unemployment rate for all ages. More precisely, in August 2017 the youth (aged between 15 and 24) unemployment rate was 16.7 % in the EU, while the overall unemployment rate was 7.6% for the same date[58]. In Hungary, according to the Central Statistical Office, the youth (aged between 15 and 24) unemployment rate was 11.4 % (with 39,700 unemployed persons) for the period of June-August 2017, while the overall unemployment rate was 4.2%[59].

Another problem that (young) people have to face is low real wages. In 2017, the Hungarian monthly minimum wage was 127,500 Hungarian forints (HUF)[60] (which is approximately €420), making the Hungarian minimum wage the fifth-lowest among the 22 EU member states that have a national minimum

55 | A difficulty regarding this part of the paper is that the sources cited define different age groups as young people. It will be indicated what the concerned source means by 'youth'. In the case of the Youthonomics Global Index, young people are people aged from 15 to 29.

56 | Youthonomics. (2015). *Global Index 2015. Putting the Young at the Top of the Global Agenda.*

57 | *Ibid.*

58 | Eurostat. (2017). *Unemployment Statistics Monthly Data.* Retrieved from http://ec.europa.eu/eurostat/statistics-explained/index.php/Unemployment_statistics#Youth_unemployment_trends.

59 | Központi Statisztikai Hivata. (2017, September 27). *Gyorstájékoztató. Munkanélküliség, 2017. június–augusztus [Quick guide: Unemployment, June – August 2017].* Retrieved from https://www.ksh.hu/docs/hun/xftp/gyor/mun/mun1708.html.

60 | There is another type of minimum wage called guaranteed wage minimum, with a higher amount. (HUF 161,000, approximately €517). In order to receive the guaranteed wage minimum, the job has to require minimum secondary-level qualifications.

wage[61]. For example, in 2014, Hungary had one of the lowest median gross hourly earnings: €3.59. The EU average was €13.14 , and in the most important migration destinations of Hungarians (Germany, the United Kingdom and Austria) this sum was also higher[62].

The housing situation of young people is also a growing issue. With the increase in rental and real estate prices, young people can hardly afford their own apartment. The price of real estate has increased (especially in the capital, Budapest), therefore it takes much longer to save enough money to buy property. For example, the average monthly wage in Hungary is HUF 173,500 (approximately €557), while the average sale price of one square metre in Budapest is HUF 489,811 (approximately €1,573), and the average rent price for a 50-square-metre residential property is HUF 127,920 (approximately €411)[63]. The rent prices have also increased, resulting in the fact that a huge part of income is spent on rent, which makes it even harder to purchase real estate[64]. According to a survey conducted by Fundamenta, more than half of the Hungarians in their twenties, and one-third of them in their thirties, still live with their parents[65]. This phenomenon can, in part, be traced back to the difficult housing situation.

These factors have led to the situation that many (young) Hungarians have become desperate and do not see a promising future in Hungary, so they decide to leave the country hoping to live in better conditions abroad. Therefore, migration is a growing issue. A survey in 2012 revealed that the number of Hungarians who consider moving abroad (either for a shorter or a longer period or also for emigration) has grown: one out of five Hungarian adults consider leaving the country. This number is the highest among young people (people aged between 18-29): every second young person considers leaving

61 | Conversely, in Germany the statutory minimum wage is €8.84 per hour (approximately €1,414 per month, for a full-time job), and €8.8 in the United Kingdom (approximately €1,408 per month). Source: Fric, K. (2017, February 09). Statutory Minimum Wages in the EU 2017. *Eurofound*. Retrieved from https://www.eurofound.europa.eu/observatories/eurwork/articles/statutory-minimum-wages-in-the-eu-2017.

62 | Median gross hourly earnings were EUR 15.67 in Germany, EUR 14.02 in Austria and EUR 14.81 in the United Kingdom. Source: Eurostat. (2017). Median Hourly Earnings, All Employees (Excluding Apprentices) by Sex. *Eurostat*. Retrieved from http://appsso.eurostat.ec.europa.eu/nui/show.do?dataset=earn_ses_pub2s&lang=en.

63 | Torontáli, Z. (2017, February 16). *A fiatalok lassan lemondhatnak a sajátlakás-álomról [Young people can slowly give up the dream of owning a home]*. Retrieved October 29, 2017, from http://hvg.hu/gazdasag/20170216_fiatalok_sajat_lakas_vasarlas_inflacio.

64 | *Ibid.*

65 | Fundamenta. (2017, March 01). *Harminc felett mamahotelben: aki tehetné, költözne [Over thirty in a mommy hotel: whoever could do it would move]*. Retrieved October 29, 2017, from https://www.fundamenta.hu/sajtoszoba/-/asset_publisher/nU7PtnNCR0Pb/content/harminc-felett-mamahotelben-aki-tehetne-koltozne.

66 | Sik, E. (2012). *Csúcson a migrációt tervezők aránya [The proportion of migration planners is at its peak]*. Retrieved October 29, 2017, from http://www.tarki.hu/hu/news/2012/kitekint/20120523_migracio.html.

67 | Blaskó, Zs., & Gödri, I. (2015). A Magyarországról kivándorlók társadalmi és demográfiai összetétele [Social and demographic composition of emigrants from Hungary]. In Zs. Blaskó & K. Fazekas (Eds.), *Munkaerőpiaci tükör* (pp. 59-71). Budapest: MTA Közgazdaság- és Regionális Tudományi Kutatóközpont Közgazdaság-tudományi Intézet.

68 | In 2014 the population of Hungary was 9,877,365.

69 | Gödri, I. (2015). Nemzetközi vándorlás [International migration]. In J. Monostori, P. Őri, & Z. Spéder (Eds.), *Demográfiai portré 2015 [Demographic portrait 2015]* (pp. 187-211). Budapest: KSH Népességtudományi Kutatóintézet, 206-207.

70 | Source: Section 48/A of Act CCIV of 2011 on national higher education. The government raised these measures to the level of constitutional status by amending the Fundamental Law stating that '[a]n act of Parliament may set as a condition for receiving financial aid at a higher educational institution the participation in, for a specific period of time, employment or enterprise that is regulated by Hungarian law'. Source: Section (3) of Article XI of the Fundamental Law (2011).

71 | Nyírő, Zs. (2017). *Külföldi továbbtanulás a legjobb hazai gimnáziumok tanulói körében – 2017 [Continuing education abroad among the students of the best domestic grammar schools - 2017]*. Budapest: MKIK Gazdaság- és Vállalkozáskutató Intézet. Retrieved October 29, 2017, from http://gvi.hu/files/researches/514/gimi_2017_elemzes_170717.pdf.

72 | Article 30 of the Fundamental Law (2011). The detailed dispositions are to be found in Act CXI of 2011 on the Commissioner of Fundamental Rights.

the country[66]. The most popular destinations are the United Kingdom, Germany and Austria. Among those who have left, the overrepresented groups are: young people, people with a high professional qualification or a higher education degree, and single people (men are also slightly overrepresented)[67]. It is hard to estimate how many people have left the country. According to certain statistics, 330,000 Hungarians lived abroad[68] at the beginning of 2014[69].

The Hungarian government introduced a change in 2013 – still in force today – trying to prevent young people from leaving after finishing their state-financed studies in higher education (so-called 'binding to soil'). The new dispositions state that as a condition to receive financial aid at a higher educational institution, the beneficiary must complete his/her studies in a certain period of time, and after the completion of the studies, he/she has to work in, or start, an enterprise in Hungary for a period equalling the amount of time spent pursuing studies within 20 years. In the case of non-compliance with these dispositions, he/she is obliged to pay back the tuition fee provided by the state.[70] However, it can be observed – though as a phenomenon it is mostly limited to the most talented students of elite high schools – that a growing number of students decide to leave Hungary right after finishing high school and start their higher education studies abroad, completely by-passing Hungarian higher education[71].

3.2. Opportunities Available for Young People in Hungary

In the legal sphere, several documents aiming especially at young people have been adopted. At a very general level, the Fundamental Law must be mentioned. It created the Deputy of the Commissioner of Fundamental Rights: the deputy protecting the interests of future generations.[72] In

2009, the parliament adopted the National Youth Strategy in order to provide equal opportunities for young people in the fields of education, employment, housing and family planning[73]. In 2011, the government also issued a complex programme targeting young people, the Programme for the Future of New Generations, which became the government's youth policy framework programme[74].

Despite the challenges previously examined, young people have several opportunities available to compensate for these difficulties. For example, some dispositions of the Job Protection Act provide the employer with tax allowances if he/she employs a person younger than 25[75]. Furthermore, the EU (co-)financed Youth Guarantee Implementation Plan targets young people under the age of 25. The programme aims to provide these individuals with adequate and quality employment opportunities, internships, and educational opportunities to decrease youth unemployment and to improve young people's life circumstances[76].

In Hungary, there are four components of the Youth Guarantee Plan, financed by EU sources: (1) the Youth Guarantee, (2) Youth Guarantee active labour market programme, (3) young people becoming entrepreneurs, and (4) internship for career starters[77]. For example, the Young People Becoming Entrepreneurs programme offers different opportunities to young people, such as help in creating a business plan or providing financial aid for starting an enterprise[78]. By promoting internships, young people can gain professional experience, which will later increase their chances to find a job. Within this project, businesses who employ young people can receive wage and contribution subsidies.

International mobility can represent an opportunity for young Hungarians, especially for those pursuing

73 / Emberi Erőforrások Minisztériumának. (2009). *Nemzeti Ifjúsági Stratégia 2009-2024 [National Youth Strategy 2009-2024]*. Retrieved from http://www.ifjusagitanacs.hu/docs/nis_091109.pdf.

74 | Új Nemzedék Központ. (2012). *Az új nemzedék jövőjéért program - A Kormány ifjúságpolitikai keretprogramja [Program for the Future of the New Generation - The Government's Youth Policy Framework Program]*. Retrieved from http://www.ujnemzedek.com/uploads/static_pages/attachments/AttachmentText_2/%C3%BAj_nemzed%C3%A9k_j%C3%B6v%C5%91j%C3%A9%C3%A9rt_program_netes.pdf

75 | Section 462/B of Act CLVI of 2011 on the amendment of tax laws and other related regulations.

76 | *Magyarország Ifjúsági Garancia Akcióterve [Hungary Youth Guarantee Action Plan]*. (2014, April 11). Retrieved October 29, 2017, from http://ngmszakmaiteruletek.kormany.hu/download/8/4c/c0000/Ifj%C3%BAs%C3%A1gi%20Garancia%20Akci%C3%B3terv.pdf.

77 | Ifjúsági Garancia. (n.d.). *Az Ifjúsági Garancia Európai Uniós forrásból megvalósuló programjai Magyarországon [Youth Guarantee programs implemented in the European Union in Hungary]*. Retrieved October 29, 2017, from http://ifjusagigarancia.gov.hu/europai-unios-finanszirozas.

78 | OFA Nonprofit Kft. (n.d.). *Fiatalok vállalkozóvá válásának támogatása országos programme [National programme to support young people to become entrepreneurs]*. Retrieved October 29, 2017, from http://ofa.hu/regio.

studies in higher education. Spending a certain period abroad, pursuing higher education studies or internships can not only increase students' chances in the labour market but can also provide a unique possibility to get to know another culture and gain everlasting experiences in a multicultural environment. These programmes focus mainly on other European countries. The most well-known international mobility programme was the Erasmus (EuRopean Community Action Scheme for the Mobility of University Students) programme, which made it possible for European students to study in another European country[79]. However, an opening towards non-European countries can be observed, as in 2014 this programme was re-baptised as Erasmus+ allowing students to study in non-European countries.

There are other programmes at Hungarian students' disposal, such as the CEEPUS (Central European Exchange Programme for University Studies), or the Campus Mundi programme (intended especially for Hungarians). In the frame of the Campus Mundi programme, students can also apply to non-European countries. Higher education is not the only field promoting international mobility: young people (17-30) can also do voluntary work in the framework of the European Voluntary Service's (EVS) programmes. The Belt and Road Initiative can also provide further possibilities in the field of international mobility, as will be seen in Part 5.

In 2015, the government introduced the dual training initiative, which enables students to gain true working experience during their studies by providing them with the chance to work at a company while still in school. The theoretical training takes place in the higher educational institute, while the hands on experience is provided at a company in cooperation with the educational institute. The trainee gains real professional experience (and receives a financial contribution for the length of the studies), while the cooperating company will not have to train new employees if the students are chosen to be hired, as they were trained during this internship[80].

Certain investments can also enhance young people's

79 | European Commission. (n.d.). *Erasmus+*. Retrieved from http://ec.europa.eu/dgs/education_culture/repository/education/library/statistics/2014/hungary_en.pdf

80 | Duális Diploma. (n.d.). *Mi Az A Duális Képzés [What is Dual Training]*. Retrieved from: http://www.dualisdiploma.hu/mi-az-a-dualis-kepzes.

opportunities in the concerned regions. Two cases, both from the Southern-Plain region, from the county seat, in the city of Szeged best illustrate this point. The first one is the establishment of the ELI-ALPS Research Institute,[81] which was in a huge part financed by the EU, and opened in May 2017 and which is estimated to create 400 jobs and aims to allure Hungarian researchers back from abroad[82]. The second example is British Petrol, which opened its business service centre in Szeged in 2017. This operation was projected to create 500 jobs until 2020, targeting young career starters[83].

81 | The research facility was built within the framework of the ELI (Extreme Light Infrastructure) project, which is simultaneously implemented by Hungary, the Czech Republic and Romania, realizsd with trans-European cooperation and the worldwide scientific community.

82 | Nemzeti Fejlesztési Minisztérium, Kommunikációs Főosztály [Ministry of National Development, Communication Department]. (2013, July 5). Várhatóan 400 új munkahely, több új egyetemi képzés, hazatelepülő kutatók - Lakossági fórum az ELI-ről Szegeden [400 new jobs expected, several new university courses, returning researchers - Citizens' Forum about ELI in Szeged]. Kormanyportal. Retrieved from https://2010-2014.kormany. hu/hu/nemzeti-fejlesztesi-miniszterium/fejlesztes-es-klimapolitikaert-valamint-kiemelt-kozszolgaltatasokert-felelos-allamtitkarsag/hirek/varhatoan-400-uj-munkahely-tobb-uj-egyetemi-kepzes-hazatelepulo-kutatok-lakossagi-forum-az-eli-rol-szegeden.

83 | MTI. (2017, September 27). Megnyílt a British Petrol szegedi üzleti szolgáltatóközpontja [British Petrol's business service center in Szeged has opened]. Autopro. Retrieved from https://autopro.hu/szolgaltatok/megnyilt-a-british-petrol-szegedi-uzleti-szolgaltatokozpontja/189840.

4. Opportunities and Challenges Offered by the Belt and Road Initiative to Hungary

The BRI could offer several opportunities in Hungary. This Part will present the advantages of the Initiative that the Hungarian government hopes to benefit from. The role of the Central and Eastern Europe Countries, with special regard to Hungary, will be addressed. Also, the benefits and risks offered by infrastructure development will be further presented through the example of the Budapest-Belgrade railroad.

The Belt and Road Initiative represents a great opportunity for European countries, including Hungary as an EU Member State, to reinforce European-Chinese relations. China is already the European Union's second-biggest trading partner[84], and its relations with China could be further developed through the BRI. As one of (international online news magazine) *The Diplomat's*[85] articles pointed out, the Belt and Road Initiative could connect to the adopted EU-China 2020 Strategic Agenda for Cooperation[86] in several regards. The fields especially concerned are trade and investment, peace and security, and people-to-people exchange, defined also by the Strategic Agenda. Connecting China to Europe through the development of infrastructure can enlarge and accelerate the transfer of goods between China and Europe. For example, the existing rail link between the city of Chongqing (in China) and the city of Duisburg (in Germany) has already largely contributed to the transfer of goods since its opening in 2011[87]. By means of the enhanced cooperation between China and Europe, security issues could be faced together and these two regions could create deeper cooperation. The BRI can also constitute a bridge between the two cultures – just as the ancient Silk Road contributed to the cultural exchange, and served as a passage, between the two – and promote intercultural understanding and trust[88].

For China, the countries in Central and Eastern Europe have key importance. The Belgrade Guidelines for Cooperation between China and Central and Eastern Europe Countries (2014) outlines the principal areas of cooperation.[89] The objectives of the Guidelines state, amongst others, enhanced cooperation and connectivity, promotion of trade and investments, expansion of financial, scientific, and technological

84 | European Commission. (n.d.).*Countries and Regions - China*. Retrieved from https://ec.europa.eu/trade/policy/countries-and-regions/countries/china/.

85 | Yan, S. (2015, April 09). Why the 'One Belt One Road' Initiative Matters for the EU. *The Diplomat.* Retrieved October 29, 2017, from https://thediplomat.com/2015/04/why-the-one-belt-one-road-initiative-matters-for-the-eu/.

86 | The EU-China 2020 Strategic Agenda for Cooperation is a strategic development plan jointly adopted by the two parties, defining the main objectives and areas of future cooperation.

87 | Yan, S. (2015, April 09). Why the 'One Belt One Road' Initiative Matters for the EU. *The Diplomat.* Retrieved October 29, 2017, from https://thediplomat.com/2015/04/why-the-one-belt-one-road-initiative-matters-for-the-eu/.

88 | *Ibid.*

89 | Ministry of Foreign Affairs of the People's Republic of China. (2014, December 17). *The Belgrade Guidelines for Cooperation between China and Central and Eastern European Countries.* Retrieved from: http://www.fmprc.gov.cn/mfa_eng/wjdt_665385/2649_665393/t1224905.shtml.

cooperation, and deepening people-to-people exchanges[90].

As previously mentioned, Hungary was the first European country to join the BRI. The Hungarian government expressed on several occasions that Hungary can profit greatly from the Initiative and that they would like to take an active part in the project. The Hungarian Prime Minister stated that Hungary – along with other Central European countries – would be an ideal pillar for the BRI and expressed his intent to cooperate with the East. The prime minister also referred to the Eastern origins of Hungarians, making Hungarians even more interested in cooperation with the East[91]. The Hungarian Minister of Foreign Affairs and Trade envisaged Hungary as the Western end-point of the modern Silk Road[92] which fits with Fidesz's previously mentioned conception of Hungary's double 'identity'. By being part of Europe and at the same time the border to the East, Hungary is a crucial member of the Belt and Road[93].

The Hungarian Minister of Foreign Affairs and Trade defined three concrete goals with relation to the BRI: (1) the application of Hungarian technology during the developments; (2) the realisation of infrastructure projects in Hungary; and, (3) increasing investments in Hungary.

First, regarding the application of Hungarian technology during the developments, the Minister primarily meant the fields of water management, city management, and IT. Second, the construction of as many infrastructure developments connecting Europe to China via Hungary as possible would create the potential to increase Hungarian exports to the East. Finally, investments are facilitated by tax cuts and tax benefits for research and development, and the Digital Hungary Programme, introduced by the government[94].

These objectives can be well integrated into Hungary's Opening to the East policy, adopted in 2012, particularly regarding the objectives of increasing exports and investments[95]. The Minister noted that the 'European Union has been losing a lot of competitiveness recently, we understand that the best way to regain the competitiveness is to build up practical cooperation

90 | *Ibid.*

91 | Cabinet Office of the Prime Minister. (2017, June 01). *Hungary is an ideal pillar of the One Belt, One Road Initiative.* Retrieved from http://www.miniszterelnok.hu/hungary-is-an-ideal-pillar-of-the-one-belt-one-road-initiative/.

92 | HKTDC. (2017, June 17). *Hungary: Leading the Way in BRI Co-operation.* Retrieved from https://beltandroad.hktdc.com/en/insights/hungary-leading-way-bri-co-operation.

93 | Országgyűlés [House of the National Assembly]. (2010, May 22). *A Nemzeti Együttműködés Programja [National Cooperation Programme].* Retrieved from http://www.parlament.hu/irom39/00047/00047.pdf.

94 | Daily News. (2017, May 15). *Hungary's foreign minister: Europe must seize One Belt, One Road opportunities.* Retrieved October 29, 2017, from https://dailynewshungary.com/hungarys-foreign-minister-europe-must-seize-one-belt-one-road-opportunities/.

95 | Hidvéghi, B. (2012, September 28). *Külgazdaság új megközelítésben: a „keleti nyitás – nyugati tartáspolitikája [Foreign economy in a new approach: the policy of 'opening up to the east - keeping it to the west'].* Speech presented in Eger (Hungary). Retrieved October 29, 2017, from http://www.mkt.hu/docs/2012-10-02-18-54-52-Hidvegi_Balazs.pdf.

with this part of the world (Asia)'.[96]

The Hungarian government has expressed its disagreement with the European Union, especially in relation to the migrant crisis. As an illustrative example, the government initiated a National Consultation and billboard campaign entitled 'Stop Brussels' in which it asked Hungarian citizens whether they agree with certain decisions made by Brussels interfering with Hungary's sovereignty[97].

However, as Max Gebhardt pointed out in an article, it should be kept in mind that in spite of any opening to the East, above all Hungary is '[…] chained to the EU by economic and social reality'[98]. Moreover, other countries' approach towards China can pose a problem, as certain countries regard China's aims and strategic purposes with suspicion.[99] Therefore, despite aiming at closer cooperation with the East, Hungary should not undermine its relations with the European Union.

One of the developments in the framework of the BRI that directly concerns Hungary is the modernisation of the Budapest-Belgrade railway. The railroad would connect Budapest to the Chinese-run Piraeus port (in Greece) through Belgrade (in Serbia) and is expected to make faster connections between Europe and China[100]. According to the Hungarian government, the modernisation of this line could contribute largely to the strengthening of Hungary's strategic position, and would make Hungary the most important transit country, in the region[101].

Hungary would connect to the Silk Road in two regards: first, through the southern railway route, and second, the Maritime Silk Road would connect to this southern route through the port of Piraeus[102]. However, there are also risks related to this project. Certain researchers expressed their concerns

96 | Qianhui, Z. (2017, May 14). *Hungary aims for bigger role in Belt and Road Initiative*. Retrieved October 29, 2017, from http://europe.chinadaily.com.cn/world/2017-05/15/content_29342870.htm.

97 | On the questions and the European Commission's reaction to them, see: European Commission. (2017). *Facts Matter - European Commission responds to Hungarian National Consultation*. Retrieved from https://ec.europa.eu/commission/sites/beta-political/files/commission-answers-stop-brussels-consultation_en.pdf.

98 | Gebhardt, M. (2017, August 05). *China's Belt and Road Initiative – Hungary at a Crossroads*. Retrieved October 29, 2017, from https://english.atlatszo.hu/2017/08/15/max-gebhardt-chinas-belt-and-road-initiative-hungary-at-a-crossroads/.

99 | Chang, F. K. (2016, October 03). Who Benefits from China's 'One Belt, One Road' Initiative?.*Foreign Policy Research Institute*. Retrieved fromhttps://www.fpri.org/2016/10/benefits-chinas-one-belt-one-road-initiative/.

100 | Xinhua News Agency. (2017, February 8). *China-Europe land-sea fast transport route opens*. Retrieved October 29, 2017, from https://eng.yidaiyilu.gov.cn/info/iList.jsp?tm_id=139&cat_id=10058&info_id=6593.

101 | MTI. (2016, September 09). A Budapest-Belgrád vasútvonal Közép-Európa legnagyobb infrastrukturális projektje [The Budapest-Belgrade railway line is the largest infrastructure project in Central Europe]. *Demokrata*. Retrieved from https://demokrata.hu/vilag/a-budapest-belgrad-vasutvonal-kozep-europa-legnagyobb-infrastrukturalis-projektje-91086/.

102 | PAGEO. (2017, March 20). *Hungary – A Key State on the Silk Road*. Retrieved October 29, 2017, from http://www.geopolitika.hu/en/2017/03/20/hungary-a-key-state-on-the-silk-road/.

103 | Kratz, A. (2015). The geopolitical roadblocks. In *'One Belt, One Road': China's Great Leap Outward* (pp. 8-10). European Council on Foreign Relations. Retrieved October 29, 2017, from http://www.ecfr.eu/page/-/China_analysis_belt_road.pdf.

104 | Kynge, J. (2017, February 20). EU Sets Collision Course with China Over 'Silk Road' Rail Project. *Financial Times*. Retrieved September 1, 2018, from https://www.ft.com/content/003bad14-f52f-11e6-95ee-f14e55513608.

105 | Xinhua. (2015, March 15). Air China to Start Budapest-Beijing Flight. *China Daily Europe*. Retrieved from http://europe.chinadaily.com.cn/world/2015-03/15/content_19812612.htm.

over the broad scope of the Initiative and China's relatively weaker competitive advantage in the field of the Maritime Silk Road[103]. Administrative and executive difficulties can arise since Hungary's European Union membership means that the country must comply with EU rules. For example, in 2017 the European Commission began an investigation into possible infringement of EU public tender requirements in the case of the Budapest-Belgrade railway.[104]

The connectivity between the two countries is realised not only through the railway but also through direct flights. In 2015, a direct flight was established between Beijing and Budapest, resulting in increased tourism and exports[105]. With successful cooperation, the connectivity of the two countries might be further strengthened; therefore, the establishment of direct flights to other destinations might be considered.

5. The Belt and Road Initiative in Relation to Young People in Hungary

Part 5 explores what challenges and opportunities young people face in Hungary. Although the Belt and Road Initiative itself will not solve their problems, it can have positive effects on young people and contribute to improving their situation. This Part will discuss the possibilities of facilitating young people's employment, the promotion of cultural and people-to-people exchange, and cooperation in research and connectivity between the two countries.

One of the main objectives of the BRI is to enhance economic cooperation and to promote infrastructure developments and investments. Foreign investments realised in the framework of the Initiative could contribute to the creation of employment opportunities in Hungary. Explicit attention should be paid to encouraging the employment of young people. To facilitate their employment, either existing benefits (such as tax benefits) could be used more effectively or new ones could complement them. For example, financial support could be awarded, provided by the BRI, after satisfying a certain quota of employing young people from Hungary. Or – depending on the type of investment – paid internship programs could help young people gain real professional experience. The creation of these new companies could be integrated into the dual training programme. Measures like these might contribute to the amelioration of the deficiencies in the labour market for young people in Hungary.

In accordance with Hungary's opening to the East policy, to promote young people's employment (and mobility), a bilateral Working Holiday Scheme was established between China and Hungary for young people. The scheme allows up to 200 young people (aged between 18 and 30) to stay and work in the country for a definite period of time[106]. The creation of such mobility programmes can contribute to the realisation of the Initiative's aims, such as enhancing people-to-people exchanges and building a bridge between different cultures.

The involvement of multiple countries naturally leads to people-to-people exchange and the promotion of different cultures. The ancient Silk Road was not only a way to boost commerce, but also a way of spreading cultures alongside the Road: the Initiative also goes beyond connecting the economies of the countries and also aims to connect people[107]. It is especially open-minded young people with suitable educations and backgrounds who are most likely to have the opportunity to benefit from the Initiative[108]. Among the core areas of cooperation, the Education Action Plan for the Belt and Road Initiative (issued by the Ministry of Education of the People's Republic of China) defines closer people-to-people exchanges, supporting talent, and achieving common development.[109] The realisation of these objectives could provide a large opportunity for young Hungarians.

106 | *Ibid.*

107 | Chow, V. (2017, September 13). *China's Silk Road Initiative Could Have Far-Reaching Consequences for Entertainment Industry.* Retrieved October 29, 2017, from http://variety.com/2017/biz/news/silk-road-infrastructure-initiative-1202555743/.

108 | Wong, D. (2016, May 26). *Belt and Road offers HK young people enormous opportunities.* Retrieved October 29, 2017, from http://www.chinadaily.com.cn/hkedition/2016-05/26/content_25471377.htm.

109 | Belt and Road Portal. (2017, October 12). *Education Action Plan for the Belt and Road Initiative.* Retrieved from https://eng.yidaiyilu.gov.cn/zchj/qwfb/30277.htm.

The Education Action Plan emphasises the importance of breaking language barriers among the participating countries.[110] The spread of Chinese as a foreign language has been observed recently in Hungary; the Chinese language is one of the most popular language majors at Hungarian linguistic universities[111]. There are several Confucius Institutes in Hungary (Budapest – 2006, Szeged – 2012, Miskolc – 2013, Pécs – 2015) and education centres can be found in more cities, making the learning of the Chinese language more accessible to citizens. Also, several grammar schools and universities provide Chinese language courses. This is said to be the influence of the Belt and Road Initiative[112]. If enhanced cooperation with China is realised, the knowledge of the Chinese language could contribute to better marketplace potential for young people or be further used in international mobility. Therefore, the work of Confucius Institutes should be reinforced to promote Chinese culture in Hungary more successfully. Exchange trips (between schools or institutions) could promote closer cultural cooperation and could contribute to networking.

People-to-people exchange is a key objective of the Initiative, and the Education Action Plan aims to create bilateral and multilateral agreements between the Belt and Road Countries.[113] The strengthening of educational cooperation and relations between various institutes would make it easier for students to benefit from international mobility. This form of educational cooperation could be similar to the EU-run Erasmus and EVS programmes and could provide the possibility of studying abroad as an exchange student, gaining professional experience in the course of an internship, or performing voluntary work.

Studying in China could be a great possibility for Hungarian students, not only because of the rich cultural heritage and the beautiful landscape of China, but because Chinese universities have a growing international reputation which could also be a great asset in the labour market[114]. This cooperation could be developed not only between China and Hungary, but with several Belt and Road countries as well, facilitating also, for example, the procurement of visas and the fulfilment of other

110 | *Ibid.*

111 | Li, B., & Gou, Y. (2017, May 09). *What 'Belt and Road' Brings: Popular Chinese and Connected Hearts.* Retrieved November 06, 2017, from http://english. hanban.org/article/2017-05/09/content_683672.htm.

112 | *Ibid.*

113 | Belt and Road Portal. (2017, October 12). *Education Action Plan for the Belt and Road Initiative.* Retrieved from https://eng.yidaiyilu.gov.cn/zchj/qwfb/30277.htm.

114 | Minsky, C. (2016, September 12). *Five reasons why you should study in China.* Retrieved November 05, 2017, from https://www.timeshighereducation.com/student/advice/five-reasons-why-you-should-study-china#survey-answer.

administrative requirements. Funding should also be provided to students who apply to the programme successfully.

In 2013, Hungary adopted a governmental decree on the *Stipendium Hungaricum,* which gives foreign students the opportunity to come to study in Hungary and aims to enhance the internationalisation of Hungarian higher education. The programme is based on bilateral educational cooperation agreements and is intended for students coming from outside the European Union[115]. Another example is the Gateway Chinese-Hungarian Student Exchange Programme, organised by the University of Szeged, Faculty of Law and Political Sciences, which provides Chinese students with the opportunity to come and study at the University of Szeged for a semester[116]. This kind of cooperation might be further reinforced in the framework of the BRI.

The Education Action Plan also supports research cooperation as part of independent or joint research projects. One example is the 'Young Belt and Road' series – within the framework of which this article on Hungary is written – which provides the possibility to young scholars to participate in an international project related to the BRI. Closer cooperation between researchers might be desired through the establishment of joint seminars, conferences or research scholarships.

The European Institute for One Belt One Road Economic and Cultural Cooperation and Development (EUOBOR) – established in 2015 and headquartered in Budapest – can be cited as an example. The aim of EUOBOR is to ensure smooth cooperation and exchange in order to execute the project and carry out research[117]. For example, a Forum was organised by, amongst others, EUOBOR in May 2017, entitled 'European Forum for Belt and Road Cooperation'.

In conclusion, Hungary has introduced its policy opening to the East, which could give rise to closer cooperation with regard to the BRI. The project offers several great opportunities for Hungary's economy and foreign economic policy, and could also contribute to the improvement of opportunities for young

115 | Governmental decree 285/2013. (VII. 26.) on the Stipendium Hungaricum.

116 | Varga, N. (2017). GATEWAY - Welcoming thoughts. *University of Szeged, Faculty of Law and Political Sciences website.* Retrieved from http://www.juris.u-szeged.hu/english/education-141120/kapu-program.

117 | EUROBOR. (n.d.). *About us.* Retrieved from: http://www.euobor.org/index.php?app=aboutus.

Freight trains in Hungary.

people in Hungary. The fields related to these improvements are investments, trade, connectivity and people-to-people exchange. Hungary's participation in the BRI would not eliminate the challenges presented in the paper, but it can contribute to solving them. Still, caution should be taken, as it should not be forgotten that the details of the project are still in development and that Hungary has to comply with EU law during its participation in the Initiative.

This paper studied Hungary's political environment and the main policies of the ruling government in general, with special regard to the situation of young people in Hungary. The relevant measures of the Belt and Road Initiative were discussed to examine what the opportunities and risks arising from participation in the Initiative are. However, there is still a place for research in the future, as the careful assessment of the details of participation still needs to be elaborated.

Panoramic view of the University of Latvia building and the National Library of Latvia. The Daugava River and Old Riga can be seen in the distance.

CHAPTER 9

by SINTIJA BĒRZIŅA
Riga Stradiņš University, MA
International Relations and Diplomacy
(Class of 2020).

LATVIA

ESTONIA

LITHUANIA

POLAND

CZECH
REPUBLIC

SLOVAKIA

HUNGARY

SLOVENIA

CROATIA

ROMANIA

BOSNIA &
HERZEGOVINA

SERBIA

MONTENEGRO

BULGARIA

NORTH
MACEDONIA

ALBANIA

1. Overview of the Political Structure of Latvia

Latvia's history, just like any other country's, is the foundation of its identity. Even though Latvians take pride in their history and its development through the years, it is a history filled with many strikingly different experiences – for example, periods filled with comforting peace and full of prosperity, as well as many dark times saturated with wars and conflicts. In those times, when conflict and war characterised Latvia, the country was devastated, and Latvians were on the edge of survival. Latvia's geopolitical situation, in combination with its being a crossroads for many different yet crucial trade routes, makes it a place of keen interest for other countries. Moreover, Latvia's political system has changed a great deal, influenced by not only European socio-economic processes but also the overall situation in international relations. Latvia underwent drastic intermittent changes during its independence, between 1918 and 1940, as well as at the end of the Soviet occupation and the renewal of her sovereignty in 1991.

At the beginning of Latvia's history, this territory became famous mostly because of its advantageous geographical location[1]. One of the most famous routes connected Latvia to Russia – at that time, the Byzantine Empire – via the river Daugava. Thus, ancient Balts were the ones who actively took part in building trading networks. Latvia's wide coastline was well known for its multi-functionality and as a place for obtaining amber. Around the early 10th Century, the Baltic people began to establish tribal realms, which led to the development of four Baltic tribal cultures – Curonians, Latgalians, Selonians, and Semigallians[2]. Due to its strategic significance, Latvian land was periodically invaded by neighbouring nations.

For example, the Germans founded Riga in 1201 and established it as the largest city on the eastern coast of the Baltic Sea[3]. In the early 13th Century, 'a confederation of feudal nations was developed under German rule and named Livonia'[4], whose territory is equivalent to today's Latvia and Estonia.

By 1282, Riga, and later other Latvian regions like Cēsis,

1 | Ministry of Foreign Affairs of the Republic of Latvia. (2005). *History of Latvia.* Retrieved from http://www.mfa.gov.lv/data/history_of_latvia_2005.pdf.

2 | Riga Stradiņš University. (2007). *History of Latvia, A Brief Survey.* Retrieved from: http://www.rsu.lv/eng/images/Documents/Publications/History_of_Latvia_brief_survey.pdf.

3 | *Ibid.*

4 | Ministry of Foreign Affairs of the Republic of Latvia. (2005). *History of Latvia.* Retrieved from http://www.mfa.gov.lv/data/history_of_latvia_2005.pdf.

Limbaži and Valmiera, were included in the Northern German Trading Organisation (Hansa)[5]. Eventually, the country became an essential centre for east-west trade relations as well as the wider eastern Baltic region, which formed close cultural connections with Western Europe until the 16th Century[6].

The 16th century can be described as a time of considerable change for the people of Latvia. This can be explained especially by the Reformation and the collapse of the Livonian nation. Around the 15th and 16th centuries, Livonia-Russia relations became tense. In 1501, Walter von Plettenberg, the Master of the Livonian Order, gathered Livonian forces against a Russian invasion[7]. At the same time, Plettenberg allied with the Lithuanian Grand Prince Alexander to resist the Russians[8].

After Martin Luther published his 'Theses' in 1517, his ideas spread across Livonia. By the early 1520s, Riga became an epicentre of reformist ideas. After the Livonian War (1558-1583), the territories of Latvia and Estonia fell under Polish-Lithuanian rule, bringing an official end to Livonia's existence. Meanwhile, reformist movements solidified the formation of the Latvian nation and the evolution of the Latvian language[9]. However, due to the Polish-Swedish war (1600-1629), the Latvian territory was once again separated, with Latgale remaining under Polish control, while Vidzeme and Rīga were overtaken by the Swedish.

An important cornerstone of Latvia's history is also the awareness of the Latvian identity, which began to evolve in the 17th century.[10] In the early 18th century, the Russian Empire's desire to expand its territory and obtain Latvia led to the Great Northern War[11]. One of the Empire's main priorities was to take hold of the prosperous city of Rīga. Consequently, the Russian Tsar, Peter I, conquered the whole of Vidzeme in 1710[12]. This was also a strategical move allowing Russia unobstructed access to Europe via the Baltic Sea. By 1772, the Russian Empire gained control over Latgale.

In the mid-19th century, as a result of national recognition throughout Europe, Latvians had a powerful 'rebirth' of national

5 | *Ibid.*

6 | Riga Stradiņš University. (2007). *History of Latvia, A Brief Survey.* Retrieved from: http://www.rsu.lv/eng/images/Documents/Publications/History_of_Latvia_brief_survey.pdf.

7 | *Ibid.*

8 | *Ibid.*

9 | *Ibid.*

10 | *Ibid.*

11 | *Ibid.*

12 | *Ibid.*

13 | Latvian Institute. (2015). *History of Latvia 1918-1940.* Retrieved from https://www.latvia.eu/history-latvia-1918-1940.

14 | Riga Stradiņš University. (2007). *History of Latvia, A Brief Survey.* Retrieved from: http://www.rsu.lv/eng/images/Documents/Publications/History_of_Latvia_brief_survey.pdf.

15 | *Ibid.*

16 | *Jaunlatvieši un latviešu valodas attīstiba [Young Latvians and the development of the Latvian language].* (n.d.). http://valoda.ailab.lv/latval/vidusskolai/literval/lit18.htm.

17 | Riga Stradiņš University. (2007). *History of Latvia, A Brief Survey.* Retrieved from: http://www.rsu.lv/eng/images/Documents/Publications/History_of_Latvia_brief_survey.pdf.

18 | LSDSP. (2017). *Latvijas Sociāldemokrātiskās strādnieku partijas (LSDSP) īsā programma.* Retrieved from: http://www.lsdsp.lv/images/Dokumenti/programma.pdf.

19 | Riga Stradiņš University. (2007). *History of Latvia, A Brief Survey.* Retrieved from: http://www.rsu.lv/eng/images/Documents/Publications/History_of_Latvia_brief_survey.pdf.

20 | *Ibid.*

21 | Latvian Institute. (2015). *History of Latvia 1918-1940.* Retrieved from https://www.latvia.eu/history-latvia-1918-1940.

22 | *Ibid.*

identity[13]. This led to the formation of the New-Latvians (*jaunlatvieši*), who were 'the most active members of Latvian social and cultural life'[14]. One of their main goals was to appeal for rights that had been established in other nations long ago. Further changes began to take place when Marxism spread into Latvia in the 1890s by the socialist movement 'New Current' (*Jaunā strāva*)[15,16]. The movement's main political activities manifested in protests against social issues like 'capitalistic exploitation'. The movement made demands concerning democratisation and 'the victory of the working class'[17], as they believed it would lead to prosperity for all people.

The first Latvian political party, the Latvian Social Democratic Workers Party (*LSDSP*), was founded in Rīga in 1904 and began to demand improvements in workers' conditions[18]. It primarily targeted the existing social order in the Russian-controlled Baltic provinces. On 13th January 1905, the LSDSP started a general strike, consisting of approximately 20,000 factory workers, in an act of civil disobedience to further their fight for better social conditions and expanded political rights.

To most Latvians, the 1905 Revolution was pre-dominantly aimed towards the Baltic German aristocracy. For the first time, several nationally-minded Latvian Social Democrats, notably Miķelis Valters, voiced the need and wish to unite the various inhabited regions into an autonomous state – Latvia[19].

However, this unification would not occur as quickly, or as seamlessly, as everyone had hoped for. The territory equivalent to present-day Latvia was split across several provinces under the Russian Empire[20]. To add to that, Latvian participation in the political process was, at that time, strongly restrained. However, World War I caused a compelling chain of events which had extensive consequences in the Baltic area[21]. The involvement of Latvian territory in WWI prompted a powerful pro-autonomy movement[22]. The Latvian military who fought on the Tsarist Russian side earned recognition across Europe. Thus, amongst the Latvians, it became a common assumption that the Latvian military was for the freedom of Latvians in a united Latvia.

In September 1917, the Democratic Bloc was formed in German-occupied Rīga, by a coalition of Latvian political parties[23]. Latvian organisations started working more actively and accomplished the formation of the representative Latvian Provisional National Council (LPNC).[24] At a session, which took place on 30th January 1918, the LPNC decided to establish a sovereign and democratic Latvia – one that would include all inhabited and previously divided regions.[25]

Next, the Latvian People's Council was formed on 17th November 1918[26].The following day, 18th November 1918, the independent Republic of Latvia was proclaimed at the National Theatre in Rīga[27]. Jānis Čakste was elected as the chairman of the Latvian People's Council, and Kārlis Ulmanis became the leader of government[28].

Unfortunately, the end of the war and the establishment of the independent Republic did not bring Latvia its long-awaited peace. By the end of 1918, the new Latvian government faced threats of attack from the Bolsheviks' Russian Red Army. After a short period of Bolshevik rule, on 22nd May 1919, the German-headed government recaptured Rīga from the Bolsheviks, which led to a more German-oriented political regime in Latvia.

On 11th August 1920, Latvia signed a peace treaty with Soviet Russia.[29] Accordingly, Russia acknowledged Latvia as a sovereign state and irrevocably gave up claims to Latvian territory[30]. Soviet Russia was the first country that recognised Latvia as an independent country. Despite this concession, history is witness to Soviet Russia's tendency to contradicting their promises.

After World War I, many countries began to officially recognise Latvia's independence, signalling other countries to acknowledge Latvia's sovereignty as well. 22nd September 1921 marked the significant acceptance of Latvia, and the other two Baltic countries (Estonia, Lithuania), at the League of Nations[31]. On 15th February 1922, the Constitution of the Republic of Latvia was adopted[32].The highest power in Latvia was given to the parliament, the Saeima. The Saeima was able to elect the president and approve the Cabinet of Ministers. Jānis Čakste

23 | Riga Stradiņš University. (2007). *History of Latvia, A Brief Survey.* Retrieved from: http://www.rsu.lv/eng/images/Documents/Publications/History_of_Latvia_brief_survey.pdf.

24 | Latvian Institute. (2015). *History of Latvia 1918-1940.* Retrieved from https://www.latvia.eu/history-latvia-1918-1940.

25 | Riga Stradiņš Universitry. (2007). *History of Latvia, A Brief Survey.* Retrieved from: http://www.rsu.lv/eng/images/Documents/Publications/History_of_Latvia_brief_survey.pdf.

26 | *Ibid.*

27 | Latvian Institute. (2015). *History of Latvia 1918-1940.* Retrieved from https://www.latvia.eu/history-latvia-1918-1940.

28 | *Ibid.*

29 | Riga Stradiņš Universitry. (2007). *History of Latvia, A Brief Survey.* Retrieved from: http://www.rsu.lv/eng/images/Documents/Publications/History_of_Latvia_brief_survey.pdf.

30 | Latvian Institute. (2015). *History of Latvia 1918-1940.* Retrieved from https://www.latvia.eu/history-latvia-1918-1940.

31 | Latvian Institute. (2015). *History of Latvia 1918-1940.* Retrieved from https://www.latvia.eu/history-latvia-1918-1940.

32 | *Ibid.*

became the first elected President of Latvia.[33] Even after the war, the LSDSP remained alongside the Latvian Farmers' Union (*Latviešu* Zemnieku savienība)[34], one of the most influential political parties in Latvia. Increasingly, national minorities in Latvia began to get heavily involved in the political process. By the 1920s, Latvia became known for its special care for the rights of such minorities[35].

The worldwide financial crisis of the early 1930s significantly impacted Latvia. It created not only economic but also political tensions in Latvian society. However, as more and more political actors started to appear, international tensions began to rise, and Latvia's national security began to decline. The League of Nations was inadequate in preventing international conflicts. This created a perfect environment for Germany to garner power[36]. The German leader, Adolf Hitler, planned to annex the Baltic territory[37]. At the same time, Stalin of the USSR was also working more actively towards re-instating Russian control over the Baltic countries. As such, the interests of Moscow and Berlin intersected, leading them to sign a treaty of non-aggression, which detailed a specially designed secret protocol surrounding the division of Eastern Europe into the two countries' spheres of influence. As a result, Latvia became part of the Soviet sphere of influence and, for the umpteenth time, fell under foreign rule. A new government was formed under Russian command[38]. Soviet authorities imposed a regime full of terror and repression, arresting and punishing hundreds of Latvian men.

This occupation, however, was interrupted by the German-Soviet War and, eventually, Latvia was conquered by the Nazi armed forces on 10th July 1941[39]. As always, Latvian control was passed around as a manifestation of the geopolitical struggles between powerful nations. Those who resisted the German regime, alongside those who co-operated with the Soviet Union, were killed or sent to concentration camps[40]. Overall, the population of Latvia decreased by around half a million (a drop of 25% from 1939) by the time the war had drawn to its conclusion[41].

33 | Diena. (2007). *Latvijas Valsts prezidenti: Jānis Čakste*. Retrieved from: https://www.diena.lv/raksts/pasaule/krievija/latvijas-valsts-prezidenti-janis-cakste-7279

34 | Latvian Institute. (2015). *History of Latvia 1918-1940*. Retrieved from https://www.latvia.eu/history-latvia-1918-1940.

35 | *Ibid.*

36 | Latvian Institute. (2015). *History of Latvia 1918-1940*. Retrieved from https://www.latvia.eu/history-latvia-1918-1940.

37 | *Ibid.*

38 | *Ibid.*

39 | *Ibid.*

40 | Latvian Institute. (2015). *History of Latvia 1918-1940*. Retrieved from https://www.latvia.eu/history-latvia-1918-1940.

41 | Riga Stradiņš Universitry. (2007). *History of Latvia, A Brief Survey*. Retrieved from: http://www.rsu.lv/eng/images/Documents/Publications/History_of_Latvia_brief_survey.pdf.

After World War II, Latvia was, once again, re-asserted as a Soviet territory and pre-war Soviet rule was re-established[42]. Immediately, Moscow tried everything in its power to establish and enforce Russian law by fully subjugating the population[43]. They tried to expel those who had co-operated with German Nazi forces, starting the *Russification* of Latvian society. The Communist Party of the Soviet Union, which included the Communist Party of Latvia, retained power[44]. The formation of other political parties or active movements was banned. On 25 March 1949, as part of Soviet repression, about 44,000 people were deported to Siberia[45].

The processes of *glasnost* (liberalisation) and *perestroika* (restructuring of the system) began in the Soviet Union under then-leader Mikhail Gorbachev[46]. Following suit, attempts at social change followed, culminating in a mass public demonstration on 7th October 1988, dedicated to Latvia's sovereignty and the formation of a judicial order. On 8th and 9th October, the Latvian People's Front held its first congress session[47]. This organisation united about 200,000 members and was led by Dainis Īvāns, a renowned journalist who was one of the main driving forces working towards the restoration of Latvia's independence.

However, more dramatic measures were required, which would attract global attention to the issue. Hence, the three Baltic countries held a huge political demonstration, called 'The Baltic Way'[48], in which individuals constructed a human chain spanning 600 km, starting in Tallinn (in Estonia), cutting across Rīga and ending in Vilnius (in Lithuania). This immense collaboration was an extremely successful expression of affection and the need to be united, which led to the victory of independence supporters in the new elections of the Latvian Supreme Soviet that were held on 18th March 1990.

A Declaration of Independence was promptly drafted, aiming for the renewal of pre-war Latvia and its previous Constitution[49]. But before attempts at implementation could be made, Moscow and USSR military forces intervened. In January 1991, pro-Moscow and pro-communist forces launched

42 | *Ibid.*

43 | *Ibid.*

44 | *Ibid.*

45 | *Ibid.*

46 | *Ibid.*

47 | *Ibid.*

48 | Riga Stradiņš University. (2007). *History of Latvia, A Brief Survey.* Retrieved from: http://www.rsu.lv/eng/images/Documents/Publications/History_of_Latvia_brief_survey.pdf.

49 | *Ibid.*

a brutal attack in a bid to overthrow the elected government[50]. However, their assault was negated by the Latvian People's Resistance – the exchange coming to be known as the 'Day of the Barricades'. Following the event, the Supreme Soviet of the Latvian Republic officially declared that end of the near one-year conversion period on 21st August 1991[51]. Latvia once again proclaimed itself a fully independent nation.

Upon finally re-gaining independence, Latvia promptly returned to the international arena. On 17th September 1991, Latvia was accepted into the United Nations and became a member of the Organisation for Security and Co-operation in Europe (OSCE)[52]. On February 1995, Latvia become a member of the European Council[53]. After Guntis Ulmanis was elected president in 1993, a system of political parties was then able to take shape[54].

Latvia's independence defined two of its foreign policy goals – membership in the European Union and in NATO. In order to fulfil the requirements for admittance, Latvia began implementing a series of social, economic and judicial changes in the 1990s. Latvia's geographical location enabled it to become a regional centre. For instance, Latvia houses the headquarters of the Baltic Battalion (BALTBAT), the united land forces of the Baltic States.

In 2006, Latvia hosted its first NATO summit in Rīga[55]. By then, Latvia had become one of the fastest-growing economies in Europe. Nowadays Latvia is a democratic, parliamentary republic. All voting rights in the Republic of Latvia belong to the people of Latvia, who also have the power to elect 100 deputies to the Saeima, the parliament of the Republic. Saeima members nominate and elect a president, who in turn nominates the prime minister. Both the president and the prime minister may not be from the Saeima. The nominated prime minister prepares the Cabinet and tries to get the support of the deputies. If this succeeds, the Cabinet of Ministers is considered to be approved and may start working. The Cabinet of Ministers possess executive power, and in practice hold more influence than the Saeima itself. For this

50 | Ministry of Foreign Affairs of the Republic of Latvia. (2005). *History of Latvia.* Retrieved from http://www.mfa.gov.lv/data/history_of_latvia_2005.pdf.

51 | *Ibid.*

52 | Riga Stradiņš University. (2007). *History of Latvia, A Brief Survey.* Retrieved from: http://www.rsu.lv/eng/images/Documents/Publications/History_of_Latvia_brief_survey.pdf.

53 | *Ibid.*

54 | Ministry of Foreign Affairs of the Republic of Latvia. (2005). *History of Latvia.* Retrieved from http://www.mfa.gov.lv/data/history_of_latvia_2005.pdf.

55 | Riga Stradiņš University. (2007). *History of Latvia, A Brief Survey.* Retrieved from: http://www.rsu.lv/eng/images/Documents/Publications/History_of_Latvia_brief_survey.pdf.

reason, it is important to elect a prime minister who preserves the interest of the people.

Similar to Latvia's political history, the prime minister's position is not stable. The prime minister may lose his position if he loses the support of the Saeima members and their political parties. Saeima members may express their lack of confidence in the prime minister, vote on it and cause him to lose his post. Consequently, the president nominates a new prime minister, who creates another cabinet and seeks to receive parliamentary support.

2. The Political Direction of Latvia

Overall, the development of Latvia's domestic political system has been stable. However, in terms of foreign policy, Russia has remained an important neighbour and regional actor since Latvia's independence. While Russia is viewed as an important trading partner, the annexation of Crimea and conflict in Eastern Ukraine have raised concerns in Latvia over Russia's intentions and its influence among its neighbouring countries. A Russian presence in Latvia does exist and it may have an influence on numerous issues, including trade, investment, security, the Russian-speaking minority, etc.

Thus, after gaining independence, Latvia decisively set integration into NATO as one of its main foreign-policy priorities. Right from its inception, this aim was widely different from Russia's plans for the Commonwealth of Independent States and became a sore spot in Russian-Latvian relations.[56] The motivation behind such a decision stems from Latvian mistrust of Russia's security guarantees. Numerous other

56 | Bikovs, A., Bruge, I., & Spruds, A. (2014). *Russia's Influence and Presence in Latvia.* New Direction. p.14.

matters like minority issues, contrasting perspectives on history and border agreements have also had a deep impact on this relationship. The 1990s and the early 2000s saw the development of many lasting border disputes, citizenship issues, matters concerning the Russian language and also arguments surrounding the World War II commemoration events[57]. However, the most significant concern is the usage of the Russian language by the large proportion of Russian-speaking minorities. Russian is the second most-used language in Latvia – 37.2% in 2011[58]. Political initiatives have also been made to increase the usage of Russian. This issue brought about a constitutional referendum in 2012 for the purpose of instituting Russian as the second official language of the country in the constitution. The initiator of the campaign, Vladimir Linderman (*Vladimirs Lindermans* in Latvian), stated that 'Russian-speakers were not second-class people'[59]. Overall, almost three-quarters voted against the proposed constitutional amendments.

Nonetheless, Russian being taught in education still remains a sensitive societal and political subject. Heated debates spread across the country when Minister of Education and Science, Kārlis Šadurskis, stated that after three years, all general education subjects in secondary schools will be taught only in Latvian.[60] Minority students are expected to retain the ability to learn minority languages, literature as well as cultural and historical subjects in their mother tongue. The ministry is planning multiple initiatives to achieve state social integration and the usage of the state language by 2024.

Šadurskis emphasised that the education system is the basis of a society's integration and the role of the Latvian language in general education has to be strengthened. He stated that 'we have done all the necessary work and we must make responsible decisions.'[61] He also explained that in kindergartens, it will be necessary to ensure that children, regardless of the language spoken at home, at the end of pre-school education, are able to continue their education in primary school in Latvian.[62] In primary education, minority education programmes will teach bilingually, gradually increasing the proportion of Latvian, and after completing elementary schools,

57 | *Ibid.*, p. 14.

58 | Central Statistical Bureau of Latvia. (2013, September 26). *At Home Latvian is Spoken by 62% of Latvian Population; the Majority – in Vidzeme and Lubāna County.* Retrieved from https://www.csb.gov.lv/en/statistics/statistics-by-theme/population/census/search-in-theme/1442-home-latvian-spoken-62-latvian-population.

59 | Krejere, D. (2013, February 18). *Gads pēc valodu referendum [One year after the language referendum].* Retrieved from: http://www.lsm.lv/raksts/zinas/zinu-analize/gads-pec-valodu-referenduma.a51833/.

60 | Latvijas Radio Ziņu dienests [Latvian Radio News Service]. (2017, October 6). *IZM plāns: pēc 3 gadiem vidusskolās mācības tikai latviski [MES plan: after 3 years in secondary schools only in Latvian].* Retrieved from: http://www.lsm.lv/raksts/zinas/latvija/izm-plans-pec-3-gadiem-vidusskolas-macibas-tikai-latviski.a252761/.

61 | *Ibid.*

62 | Latvijas Radio Ziņu dienests [Latvian Radio News Service]. (2017, October 6). *IZM plāns: pēc 3 gadiem vidusskolās mācības tikai latviski [MES plan: after 3 years in secondary schools only in Latvian].* Retrieved from: http://www.lsm.lv/raksts/zinas/latvija/izm-plans-pec-3-gadiem-vidusskolas-macibas-tikai-latviski.a252761/.

they will learn about 80% of the subjects in Latvian. In the 7th-9th form, there will be a unified Latvian language standard for Latvian and minority education programmes. Currently, there are two standards – one for Latvians and the other for ethnic minorities.

However, the Russian-speaking minority in Latvia is fully against the proposed reforms. For example, on 23rd October 2017, 300-400 Russian school advocates gathered in protest at the Ministry of Education, in the Old Town of Riga. They protested against the idea that, after three years, general education in the upper secondary schools would be taught in Latvian only. Protestors consisted mostly of the elderly and children, who held posters with inscriptions like 'Minority schools are the wealth of Latvia' and 'Hands off the Russian schools!'[63]

A series of pro-Russian organisations addressed the then Prime Minister of Latvia, Māris Kučinskis, in an open letter, expressing dissatisfaction with his support for the idea of a transition to teaching in Latvian. Russian school advocates have chosen to justify their arguments by recalling the recent similar reform in Ukraine on the language of instruction. 'Only one European country is currently destroying minority education. It is violence-affected and economically-suffering Ukraine. There is no such crisis and opposition in the Latvian society,' the letter states[64].

The prime minister was asked to instruct the government to supplement the laws and regulations guaranteeing opportunities to use Russian and other minor languages when studying in general subjects.

Rossotrudnichestvo is one of the main components of Russia's compatriot policy in Latvia[65] that aims to 'implement the state policy of international humanitarian co-operation and promote the dissemination of objective representation about modern Russia'.[66] It mainly works as an influencer and provides a system of support for learning Russian abroad, e.g. in Latvia. Furthermore, it is connected with the implementation and promotion of Russian educational services. Additionally, the

63 | Čunka, J. (2017, October 23). *Protestā pret skolu piespiedu pārēju uz latviešu valodu vairāki simti cilvēku [Several hundred people protested against the forced conversion of the school to the Latvian language].* Retrieved from: http://www.lsm.lv/raksts/zinas/latvija/protesta-pret-skolu-piespiedu-pareju-uz-latviesu-valodu-vairaki-simti-cilveku.a254682/.

64 | LETA. (2017, October 20). Krievu skolu aizstāvji pie IZM protestēs pret ieceri pāriet uz izglītību latviešu valodā [Defenders of Russian schools will protest at the Ministry of Education and Science against the intention to switch to education in the Latvian language]. *DELFI.* Retrieved from https://www.delfi.lv/news/national/politics/krievu-skolu-aizstavji-pie-izm-protestes-pret-ieceri-pariet-uz-izglitibu-latviesu-valoda.d?id=49361861.

65 | Cepuritis, M. & Andis, K. (2017). *Latvijas reģioni un valsts ārpolitika: Krievijas faktora izpēte [Latvian regions and state foreign policy: research of the Russian factor]*, p. 45.

66 | Bikovs, A., Bruge, I., & Spruds, A. (2014). *Russia's Influence and Presence in Latvia.* New Direction. p.17.

Rossotrudnichestvo provides an assistance programme to voluntary compatriots for their re-settlement to the Russian Federation[67].

With the aid of short-term visits to Russia, young people are introduced to the social-political, socio-economic and cultural life of Russia. This allows Russia to build a more positive and constructive image for itself, deepening relations between both countries via those cultural ties. It can be concluded that the *Rossotrudnichestvo* is mainly responsible for implementing Russia's public diplomacy in Latvia[68]. Nonetheless, the expanding Russian reach in Latvia's education system, particularly via Latvia's ethnic minority schools, has added to the increased tension in Latvia.

The Russian factor is also present in politics and business, which hold the potential to change the country drastically. Even after the independence of Latvia, both official and unofficial links to Russia remain. Latvian politics is divided along ethnic lines (Latvian vs Russian-speaking) and foreign policy views (pro-West vs. pro-Russian)[69]. This is reflected in parliament where Russian-speaking ethnic minorities have been represented since the 1990s and is best exemplified by the rise of the social democratic party 'Harmony'[70]. Here, the important question is whether Russia is able to influence Latvia through these types of political parties. It would be naive to say the reach of pro-Russian agents is limited to solely politics in a country saturated with deep Russian influences. For example, many local businessmen also have links to Russia and have the potential to cause crucial impacts – both direct and indirect. Due to anti-corruption protests in 2011, the influence of oligarchs has declined, although it still continues to exist. For example, transit is one of the main spheres in which Latvia is highly dependent on Russian supplies[71].

Latvia's political direction has remained mostly steady since its independence, even as Russia continues to have the biggest influence on the country's political direction, through its influence and presence in various sectors. Russia has taken advantage of formal and informal means, which are remnants

67 | *Ibid.,* p. 18.

68 | *Ibid.,* p. 18.

69 | *Ibid.,* p. 26.

70 | Saskana. (n.d.). *Par Mums [About Us].* Retrieved from https://saskana.eu/par-mums/.'

71 | Bikovs, A., Bruge, I., & Spruds, A. (2014). *Russia's Influence and Presence in Latvia.* New Direction. p. 27.

from its era of Latvian colonisation[72]. Furthermore, the largest minority in Latvia is the Russian-speaking population, which is the source of rising tensions regarding the issue of language reform. The simplest prognosis is that Russian minorities will not settle for this legal reform and will try to campaign for their representation. In the meantime, Latvia's education system has become thoroughly disorganised. The reformation of the Latvian education system has many potential variables.

72 | Bikovs, A., Bruge, I., & Spruds, A. (2014). *Russia's Influence and Presence in Latvia.* New Direction. p.36.

3. Evaluating the Existing Opportunities and Challenges Faced by Young People in Latvia

73 | Centrālā Statistikas Pārvalde. (2017). *Nodarbinātība un bezdarbs – galvenie rādītāji.* Retrieved from: http://www.csb.gov.lv/statistikas-temas/nodarbinatiba-un-bezdarbs-galvenie-raditaji-30263.html

74 | *Ibid.*

75 | Centrālā Statistikas Pārvalde. (2017).*Nodarbinātiba un bezdarbs – galvenie rādītāji.* Retrieved from: http://www.csb.gov.lv/statistikas-temas/nodarbinatiba-un-bezdarbs-galvenie-raditaji-30263.html.

76 | Klāsons, G. (2016, February 19). *Country Sheet on Youth Policy in Latvia.* Youth Partnership, p. 4.

To completely understand the opportunities and challenges faced by young people in Latvia, one must examine the main employment indicators in the country. In the first quarter of 2017, the unemployment rate in Latvia was 9.4%, according to the results of the Labour Force Survey conducted by the Central Statistical Bureau[73]. Compared to the previous quarter, the unemployment rate had increased by 0.1%, but over the year, it decreased by 0.9% points. Thus, in the first quarter of 2017, 91,300 people aged 15-74 were unemployed – 10,300 down from the previous year, but up by 500 compared to the previous quarter[74]. Over the preceding five years, the Latvian unemployment rate exceeded the European Union (EU) average (except for the first quarter of 2015, when it was equal). In the fourth quarter of 2016, the unemployment rate in Latvia (9.3%) exceeded the EU average by 1.1% (8.2%)[75]. In the first quarter of 2017, Latvia held the highest unemployment rate in the Baltics – in Estonia the unemployment rate was 5.6%, while in Lithuania it was 8.1%. Despite showing slow positive employment growth, Latvia still lags behind its neighbours. Additionally, there are 276,817 young people (13-25 years) in Latvia, accounting for about 14% of the total population[76].

Issues regarding the everyday challenges these youth faces persist, leading some to leave the country in pursuit of a better career elsewhere.

From a historical vantage point, the Latvian youth policy, like most other policy areas, started to develop aggressively from the mid-1990s[77], despite being at a crossroads, both in institutional and substantive terms. For example, leisure time activities (hobbies and interest education) continued to exist, but international co-operation, as well as counselling, introduced many other new components to the Latvian context. Such components led to the establishment of new NGOs, encouraged autonomy and non-formal methods of education, in addition to the development of civil society by promoting participation[78].

Both national and international research show that Latvia has the highest proportion of politically alienated young people in Europe, which is − 39.2%[79]. One way to characterise this phenomenon is by defeatist indifference. Those that can afford it, choose to resolve it into 'civic privatism' and those few who can actually afford to change politics choose to pursue their private goals. Whereas, those who want to change things can't afford to do so and relate helplessly to the public issues in Latvia[80]. For example, about 11% of young people in Latvia take part in some type of youth organisation. Statistics show that 79% feel incapable of influencing events in the world and only 10% felt they were 'powerful enough to influence anything in Latvia'[81].

However, analysing the alternative positive initiatives available to Latvian youth, there exist other possibilities to pursue individual freedom of not only action but also thought, such as engaging in a variety of forms of voluntary work, helping fellow human beings through donating, or joining social networks to help the needy, etc. Thus, it is feasible to promote and achieve mutual respect and also resilience towards diversity. On top of that, young people in Latvia can also pursue the establishment and creation of innovative business forms and economic activities.[82] For example, the use of age-old Latvian symbols in the daily

77 | *Ibid.*, p. 2.

78 | *Ibid.*, p. 2.

79 | Bite, D., Kronberga, G., Kruzmetra, Z. (2017). *Youth Reflections on the Development of Society in Latvia*. Proceedings of the 2017 International Conference 'Economic Science for Rural Development' No 46 Jelgava, LLU ESAF, p. 32.

80 | *Ibid.*, p. 32.

81 | *Ibid.*, p. 33.

82 | *Ibid.*, p. 33.

production of goods, ecological products and many other kinds of local products.

However, the negative attitude of youth can be linked to their childhood development, with safety and well-being between children, and bad relationships with parents and technological dependency contributing to their warped attitude[83]. These issues interact and hold the possibility of deepening future social fragmentation – akin to how Latvians view each other as a 'grey mass.' People are seen as passive, intolerant and unwilling to initiate changes in society, with attitudes that inherently pose an obstacle to societal development. The cultivation of this attitude has become even more substantial due to the unsafe history of Latvia, the need for safety and the inflow of displaced immigrants.

A few concrete steps can be taken to alleviate the hurdles that younger members of the community face. For instance, concrete efforts must be made towards reducing public fragmentation, i.e. through stratifying society by income[84]. More to that point, it is crucial for society to pay more attention to issues of race, ethnicity and language. To put the above into perspective, one can examine the divisive effects of religion on society.

The expanding international displacement of immigrants automatically affects the situation in Latvia. As Muslims have entered society, there has been a growing lack of intolerance among people, including youth, towards people who are normatively different, especially when it comes to religion. This is closely related to familial attitude, which influences people to make certain assumptions, based on the behaviour patterns and opinions of their own parents[85]. These biases can only be overcome by socialisation and a more active change in people's attitudes and behaviour, which is connected to the accountability of an individual and not the opposite – that of a state, an organisation or even a government. The most important take-away is that changes are most effective if they begin at an individual level, being directed at actions, behaviour and family values. Yet, the role of pro-active education and

83 | *Ibid.*, p. 33.

84 | Bite, D., Kronberga, G., Kruzmetra, Z. (2017). *Youth Reflections on the Development of Society in Latvia.* Proceedings of the 2017 International Conference "Economic Science for Rural Development" No 46 Jelgava, LLU ESAF, p. 32.

85 | *Ibid.*, p. 35.

initiatives that target differences, with the ability to maintain unity, is paramount.

4. The Opportunities and Challenges Offered by the Belt and Road Initiative to Latvia

First of all, it needs to be stated that China`s new strategy and the '16+1' formal of co operation – one of its most important diplomatic achievements – have become particularly important. In principle, this approach can be defined as a regional approach that addresses relations with Central Europe alongside Eastern Europe. With this co-operation format, China has expanded its European policy and has become increasingly involved with these European countries.[86] The actual start of this co-operation was the 2012 visit to Poland by premier Wen Jiabao which laid the foundations for further interactions between China and Latvia.

Creating an inter-governmental network is one of the priorities of this collaborative format. Likewise, it serves as a platform for co-operation, not only at a multi-lateral level but also a bilateral one. Although China is much larger, in geographical size, population and its economy, than the countries in Central and Eastern Europe, with this cooperation format, China fosters an equal partnership between them.

The '16 + 1' co-operation format, in which all are equal partners, can serve as a platform for each country to realise its national interests.[87] Similarly, the structure of co-operation in this '16 + 1' format is based on a voluntary principle. When the Secretariat for Co-operation between China and Central and Eastern Europe at the Foreign Ministry was established, in order to ensure successful coordination, the 16 countries involved

86 | Kong, T. (2015). 16+1 Cooperation Framework: Genesis, Characteristics and Prospect. *Medjunarodni problemi 2015,* Volume 67, Issue 2-3, pp. 167-183.

87 | *Ibid.*

were extended a voluntary invitation and given the choice of designating an equivalent department and coordinator.[88] In addition, this co-operation is based on free institutionalisation and the structure of the '16 + 1' format has evolved primarily on this basis. Thus, no tension exists within the institutional arrangements reached with this structural co-operation as each state or entity has participated in the relevant co-operation mechanisms on a voluntary basis.

The Belt and Road Initiative, which is implemented through the '16 + 1' format, has a significant impact on Latvia. This is primarily due to Latvia becoming the first country in the Baltic Sea Region to sign a Memorandum of Understanding, thereby joining the 'Belt and Road Initiative', when Chinese Premier Li Keqiang conducted his first official visit to the country on 4th November 2016[89]. In addition, Li and the Prime Minister of Latvia, Maris Kučinskis, signed another five agreements in areas such as small and medium-sized enterprises (SMEs), transport and logistics. This event has been highly appreciated, because Latvia recognises the importance of this co-operation with China, via the '16 + 1' format. This co-operation substantially complements both the bilateral relations of Latvia with China and the joint strategic partnership with the Central and Eastern Europe Countries (CEEC).[90]

Similarly, Latvia is aware of the importance and potential of the Chinese market, which is an attractive destination for Latvian goods and services. Although all countries agree on the general guidelines laid out in the '16 + 1' format, they, including Latvia, still hold the ability to realise and update areas of interest to them on the basis of the principle of equivalence, which is particularly topical in this co-operation. For its three most important priorities, Latvia has indicated, 1) additional opportunities for promoting and developing economic relations, which would primarily extend partnership with China directly in the fields of transport, logistics, tourism, culture, education and science; 2) the promotion of state image in the international environment and; 3) the expansion of the international legal base[91]. Although Latvia's initial priorities are promising, it is just as important to analyse the real long-term

88 / *Ibid.*

89 | China Daily. (2016, November 5). *Latvia to Join China's Belt and Road Initiative. Cooperation between China and Central and Eastern European Countries.* Retrieved from http://www.china-ceec.org/eng/ldrhw_1/2016lj/hdtj4/t1413091.htm.

90 | Latvijas Republikas Ārlietu Ministrija. (2016). *Noslēgusies 16+1 valdību vadītāju sanāksme.* Retrieved from: http://www.mfa.gov.lv/arpolitika/daudzpusejas-attiecibas/centraleiropas-un-austrumeiropas-valstu-un-kinas-sadarbiba.

91 | Latvijas Republikas Ārlietu Ministrija. (2016). *Noslēgusies 16+1 valdību vadītāju sanāksme.* Retrieved from: http://www.mfa.gov.lv/arpolitika/daudzpusejas-attiecibas/centraleiropas-un-austrumeiropas-valstu-un-kinas-sadarbiba

benefits of this co-operation, especially for young people, as well as the impact(s) on Latvia.

Although Latvia's collaboration with China includes various activities beyond economic commitment, they still remain the top-most priority for not only Latvia but also most Central and Eastern Europe countries (CEECs)[92]. Already, the world is experiencing various global changes in the economy, which has a particular impact on the transport and logistics industry. China is partially responsible for this change. Mutual competition is increasing, and the demand for Latvian transit corridors specialising in the transit of energy resources is decreasing[93]. Thus, products exchanged between East Eurasia and the vast Chinese economy, as well as the global market, are a major factor contributing to the development of the Latvian transit sector.

The road through Latvia is the fastest and most efficient way to transport Chinese goods to Scandinavian countries. A real example of this is that with the help of Latvian ports, it would take only 48 hours to transport freight from China to anywhere in the Scandinavian region. Thus, it is precisely this co-operation that has offered Latvia the opportunity to increase economic activity, attracting Chinese cargo. As early as the summer of 2016, real benefits could be witnessed as Latvian Railways (*Latvijas Dzelzceļš*) signed a formal agreement with China Railways. The first trial cargo container, organised within this framework, arrived at the Riga Central Railway Station less than four months after the conclusion of this agreement[94]. Although this container merely serves as an example here, the volume of trade and investment in Latvia and China has increased since the establishment of this co-operation.

One of the potential factors that make this region a favourable destination for China's investment is the availability of skilled labour in the areas of its interest at a cost relatively lower than the EU average, which is crucially connected with the possibilities for Latvian youth[95]. However, when analysing investments and, in particular, China's ability to invest in Latvia, it is worth mentioning that they are primarily based on

92 | Andžāns, M. & Bērziņa-Čerenkova, U.A. (2017). *'16+1' and China in Latvia Foreign Policy: Between Values and Interests*, p. 169.

93 | Bērziņš, E. (2016). Foreword on the Behalf of Partners of the Forum. In M. Andžāns (Ed.) *Afterthoughts: Riga 2016 International Forum of China and Central and Eastern European Countries*. Latvian Institute of International Affairs, p.11.

94 | Latvijas Dzelzceļš. (2016, November 5). *First Pilot Container Train Arrived in Riga from Yiwu City in China*. Retrieved from https://www.ldz.lv/en/first-pilot-container-train-arrived-riga-yiwu-city-china#:~:text=The%20first%20pilot%20container%20train,and%20finally%20arrived%20in%20Riga.

95 | Elteto, A. & Agnes, S. (2016). Chinese Investment and Trade – Strengthening Ties With Central and Eastern Europe. *International Journal of Business & Management* Vol. IV No. 1, p.33.

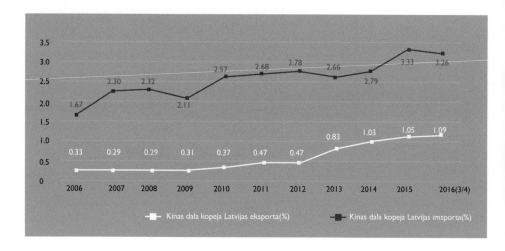

Figure 1: The blue line indicates the share of exports to China in Latvia's total exports. The red line indicates the share of imports from China in Latvia's total imports. The numbers on the lines are percentages.[96]

96 | Andžāns, M. & Bērziņa-Čerenkova, U.A. (2017). *"16+1' and China in Latvia Foreign Policy: Between Values and Interests",* p. 92.

97 | The Baltic Course. (2016, February 19). *China Interested in Investing in Rail Baltica and airBaltic.* Retrieved from http:// www.baltic-course.com/eng/transport/?doc=116991.

98 | Andžāns, M. & Bērziņa-Čerenkova, U.A. (2017). *"16+1' and China in Latvia Foreign Policy: Between Values and Interests",* p. 167.

99 | Latvijas Banka. (2017). 01 TI datu valstu dalījumā tabulas (atlikums perioda beigās). Retrieved from: https://statdb. bank.lv/lb/Data.aspx?id=128.

infrastructure projects, especially in the energy and transport sectors. An example is China's interest in investing in 'Rail Baltica' and 'AirBaltic'[97]. When analysing the possibility of investments, additional barriers arise, because in order to finance these infrastructure-related projects from Chinese loans, guarantees are needed from domestic institutions and ready funds are required from Chinese investors and sub-investors[98]. Thus, this situation most often does not coincide with the legislation adopted in the European Union and is better achieved in the partner countries of '16 + 1' that are not part of the EU. This, and other obstacles, severely impair China's ability to invest in Latvia, capping at only €74.6 million or 0.56% of total direct investment in Latvia[99].

Although import and export performances have both increased, it is still worthwhile to assess the impact they have had. Initially, it may seem that trade relations between Latvia and China are promising, they do not come without obstacles. One of which is that this co-operation is often reflected, by economic indicators, as imbalanced. The total Chinese share of Latvia's total exports is three times smaller than its share of total Latvian imports, as shown in Figure 1.[100] Although trade with China shows a general, gradual increase, its share will mostly remain limited when compared to overall Latvian foreign trade. This fact is particularly significant given that China's share in Latvia's total exports is just over 1%[101]. In comparison, Latvia's largest trade export partners are Lithuania (17.5%), Estonia (11.5%) and Russia (9%). China is a very small export partner.[102]

However, it is equally important to mention the benefits, alongside the various obstacles, when discussing Latvia's economic development. One of the most significant benefits is the popularisation of Latvia abroad. The 2016 summit on the '16 + 1' format in Riga included the presence of the Chinese Premier, Li Keqiang, and contributed to Latvia's international standing. Similarly, the 'Riga Guidelines' and the 'Riga Declaration' reflect the work done previously and will function as base documents during future cooperation[103].

In October 2016, a survey conducted on the perspectives of political science students revealed more than half of the respondents had never heard of the '16 + 1' co-operation format, while over two-thirds were unaware that this summit would take place in Riga. In contrast, according to a survey conducted in 2017, the situation is already quite different after the 2016 summit in Riga - 71.1% of the surveyed Latvian political students acknowledged that they knew the '16 + 1' co-operation summit had taken place in Riga, whereas 77.8% (almost two-thirds) had already heard about this collaborative format[104]. The knowledge of this collaborative format has grown significantly among young people and may also be related to the successful promotion of the Latvian public image within the framework of such a co-operation format, not only abroad, but also among young political students here in Latvia.

However, from the survey conducted, one has to conclude

100 | Andžāns, M. & Bērziņa-Čerenkova, U.A. (2017). *"16+1' and China in Latvia Foreign Policy: Between Values and Interests,"* p. 167.

101 | World Integrated Trade Solution. (n.d.). *Latvia - 2017 - Exports & Imports.* Retrieved from https://wits.worldbank.org/CountryProfile/en/Country/LVA/Year/2017/TradeFlow/EXPIMP.

102 | World Integrated Trade Solution. (n.d.). *Latvia - 2017 - Exports & Imports.* Retrieved from https://wits.worldbank.org/CountryProfile/en/Country/LVA/Year/2017/TradeFlow/EXPIMP.

103 | Andžāns, M. & Bērziņa-Čerenkova, U.A. (2017). *"16+1' and China in Latvia Foreign Policy: Between Values and Interests,"* p. 167.

104 | Based on the results of a survey conducted by the author in May 2017. The survey was titled "Assessment of Relation between the Latvia and the People's Republic of China – The Perspective of the Latvian Students of Political Science".

that the '16 + 1' co-operation format, which is closely linked to the Belt and Road Initiative, is faced with various difficulties. Amongst Latvian political science students, while knowledge about the '16 + 1' co-operation format is widespread, with 77.8% having had heard about it, many are still unaware of the Belt and Road Initiative (68.9% of all respondents). As well as analysing the impact of this co-operation on Latvia, students in political science are sceptical: since the 2016 summit, 62.2% have not experienced more active interchanges between Latvia and China. However, the main elements of this co-operation emphasise economic development and China's investment overall is rated neutral (37.8%) or positive (51.1%)[105].

105 | Authors survey, "Assessment of Relation between the Latvia and the People's Republic of China – The Perspective of the Latvian Students of Political Science"(May 2017).

5. Suggestions for the Belt and Road Initiative

More than two thousand years ago, China implemented the 'Silk Road', one of the more successful acts of mutual co-operation, which aimed at connecting Asia and Europe via not only land but also the sea. Thus, links, whether economic, political, cultural or ideological, began to develop between two different continents. The 'Silk Road Spirit' stood for 'peace and cooperation, openness, inclusiveness, mutual learning and value'. [106] Since then, China has undergone tremendous development through its history in order to achieve its present status and its readiness to define itself as one of the most important great powers of the modern world. As China's global influence has continued to proliferate, new initiatives have been launched.

106 | Xinhua. (2017, June 20). Full text of the Vision for Maritime Cooperation under the Belt and Road Initiative. *The State Council of the People's Republic of China*. Retrieved from http:// english.www.gov.cn/archive/ publications/2017/06/20/ content_281475691873460.htm.

In 2013, to promote development and trade both in China and other countries in the region, the '21st Century Sea Silk Road' and 'Silk Road Economic Belt' were introduced, jointly and colloquially coming to be known as the Belt and Road Initiative (BRI). In total, this initiative will include a total of 65 countries and 4.4 billion people, accounting for 63% of the world's population. One of the foundations of this initiative is the fight for equality, broad country-wide consultation, and joint contributions to the project, resulting in shared benefits and an *all-win* strategy.

The '16 + 1' co-operation has been one of the main highlights of Latvian foreign policy, especially in 2016, when the highest level '16 + 1' summit took place in Latvia. Although there are certain barriers to it, this co-operation has contributed to the economic development of Latvia, containing promising future prospects. As mentioned previously, Latvia aims to promote and develop economic relations, promote the image of the country in the international environment and expand the international legal base.

Economic development is considered to be the most important factor driving Latvia. It is closely linked with the involvement of young people, because economic growth is one of the main reasons that additional opportunities, for example, new working places or possibilities to get involved in mutual co-operation, are built. Each additional capital flow from China is a new opportunity to develop and move forward not only economically, but also politically.

To take advantage of mutual co-operation, transit corridors that specialise in the transportation of energy resources must be maintained and used more effectively to enable cooperation with the Eurasian East or China. Rail transportation has a shorter delivery time, showcasing a travelling time of merely 10-15 days between the Chinese and the European markets when compared to the 30-45 days that maritime shipping takes.[107] Thus, the shorter period of travelling would improve the efficient use of capital. What is more, improvement of the delivery time and effective use of Latvia's geographical situation would mean faster capital turnover and lower loss in interest rate payment for this capital[108].

107 | Gubins, S. (2017). *The New Silk Road: Latvian Branch.* Certus Think Tank, p. 5.

108 | *Ibid.*, p. 5.

China's influence in transit is also connected with the overall expenditure in the economy. The transport industry reaps the most benefits out of this cooperation but is still not fulfilling its promise and possible achievements, including the opportunities it holds for Latvian youth. For example, the Latvian transport and storage sector multiplier is 2.66, according to OECD data.[109] Consequently, €1 spent on the transport and transit industry equates to a €2.66 expenditure in Latvia's economy.

More attention should also be paid to Latvia's location, which can attract transit flows that head towards Northern and Northwestern Europe from Eastern and Northeastern Chinese provinces.[110] In this matter, the shortest geographical distance between the specified Chinese regions and Scandinavia and Northwestern Europe passes through the Baltic States; Latvia in particular. For example, Latvia is especially instrumental in the shipment of goods to Sweden and Norway, because Latvian seaports are well suited to handling containers headed there.

However, to exploit this maritime opportunity, an agreement must be signed, especially since these trade routes involve multiple countries and firms from China and Europe. Furthermore, the total price that customers face sometimes seems to be too extravagant, leading to low transit quantity and small profits for all involved parties.[111]

Additionally, China should seek to deepen its involvement in and co-operation with Latvia's high-speed project 'Rail Baltica'. This project aims to connect the three Baltic countries to Warsaw and plans to have a logistical terminal in Salaspils, which is near Riga.[112] Thus, China could seek additional opportunities on how to more effectively channel goods from the Baltics to China.

Among the obstacles that may deter investment opportunities are certain differences between EU and Chinese legislation. This collaborative format is more suitable for non-EU member states. However, since Latvia is a part of the EU and depends on its legislation, there should be additional differentiation in the implemented policies from China in Latvia that would

109 | *Ibid.*, p. 6.

110 | Gubins, S. (2017). *The New Silk Road: Latvian Branch.* Certus Think Tank, p.9.

111 | *Ibid.*, p. 13.

112 | The Baltic Course. (2016, February 19). *China Interested in Investing in Rail Baltica and airBaltic.* Retrieved from http://www.baltic-course.com/eng/transport/?doc=116991.

benefit all countries with no exceptions in the areas where regulations differ.

As analysed above, China's investment in Latvia is unbalanced. China's imports into Latvia have increased, especially since 2009, 40% of which are electrical and electronic equipment. By contrast, much more progress has been made in Latvia-to-China exports due to a special increase in 2013, or at a time when the '16 + 1' co-operation started to accelerate. Yet, the total share of Latvian exports to China is three times smaller than of Chinese imports to Latvia. To remedy this, there should be more active co-operation across international formats. This would benefit Latvia not only as a country but also a society, since, at the moment, only minimal attention has been given to building closer relations with China. Society seems distant, intolerant and not as active and participative as hoped.

Even though there have been many events like the '5th Meeting of the Heads of Government of Central and Eastern Europe Countries and China (16+1)' as well as some side events like a Business Forum, 'Symposium on Sinology Research and Chinese Language Pedagogy' and also a Think Tank Conference, there is still not enough information available to society.[113] The previously-mentioned survey conducted with political science students showed a lack of awareness regarding the Belt and Road Initiative, as well as scepticism toward Sino-Latvian co-operations and its impacts. However, it is doubtful whether the responses would be the same if the target group of the survey was to be expanded; in fact, the situation would predictably be even bleaker.

However, it still needs to be stated that the role of China itself keeps increasing and Latvia sees the possibilities, such as those presented by the '16 + 1' co-operation format and the BRI as good instruments for further enhancing co-operation between both countries – not only by a bilateral partnership but also through a wider, strategic EU-China Comprehensive Strategic Partnership. Given the continually growing size of its economy, China is an extremely attractive partner for Latvia and many other countries. Therefore, it is natural that countries like Latvia

113 | Bērziņš, E. (2016). Foreword on the Behalf of Partners of the Forum. In M. Andžāns (Ed.) *Afterthoughts: Riga 2016 International Forum of China and Central and Eastern European Countries.* Latvian Institute of International Affairs, p.7.

114 | Andžāns, M. & Bērziņa-Čerenkova, U.A. (2017). *'16+1' and China in Latvia Foreign Policy: Between Values and Interests.* p.165.

would aim to profit economically from this co-operation and further the possibilities afterwards. It is also to be noted that economics cannot be separated from politics completely, at least when it concerns big and small countries like China and Latvia respectively[114].

Wind turbines generating power in
Lithuania.

CHAPTER 10

by UGNĖ MIKALAJŪNAITĖ

Yenching Academy of Peking University,
MA in China Studies (Class of 2015).

LITHUANIA

ESTONIA

LATVIA

POLAND

CZECH
REPUBLIC

SLOVAKIA

HUNGARY

SLOVENIA

ROMANIA

CROATIA

BOSNIA &
HERZEGOVINA   SERBIA

MONTENEGRO

BULGARIA

NORTH
MACEDONIA

ALBANIA

LITHUANIA

Until recently, the Republic of Lithuania has not held much significance for the Chinese economists and policymakers. That is not surprising as the modern Lithuanian nation only formed in the late 1990s after the collapse of the Soviet Union (USSR). Previously, it had been a part of the Russian Empire. Despite having been one of the major nations of medieval Europe and an essential part of the ancient Amber Road, it was never significant enough (or perceived to be in the right geographic location) to become a part of China's trading destinations.[1] However, as China is turning its attention towards Central and Eastern Europe with the 16+1 Initiative; including the harbours of the Baltic Sea in the Belt and Road Initiative (BRI), the Sino-Lithuania relationship is bound to change.

Even though Lithuania is part of both the 16+1 Initiative and a country along the maritime route, it is still a minuscule player compared to the Central and Eastern Europe (CEE) giants such as Poland or Serbia, who have more impact on the region and can be more valuable partners for China. Nevertheless, there have already been tangible outcomes from the Lithuania-China ties. These advancements range from progress in economic trade to the growing interest in each other's culture through university majors and new institutes being opened up. As long as the Lithuanian policymakers and the public continue to take advantage of the prospects that these projects bring about, benefits will abound.

The cooperation between Central and Eastern Europe and China should bring a variety of opportunities and advantages to the Lithuanians. These range from new job openings in the areas of infrastructure, economics, high technology, and green technology, to possibilities of improving the wealth and national status of the citizens and enhancing Lithuania's international status. These initiatives are especially beneficial to Lithuanian youth. Never before have there been this many bilateral academic exchange programmes, language and culture courses, art and music events and career opportunities. As young Chinese learn about Lithuania through the visits of famous musicians and Lithuanian language programmes at University; Lithuanians are likewise gaining knowledge about China.

1 | The Amber Road was an ancient Europe-Asia trade route that transported amber from the Baltic Sea to the Mediterranean. Reference: Singer, G.G. (2008, December). Amber in the Ancient Near East. *Papyrus électronique des Ankou.*

This chapter shall discuss the ways the Belt and Road Initiative is affecting the Lithuanian nation and especially its youth. Part I will provide an overview of the current Lithuanian socio-political system and a preface on Lithuania's historical ties with China. Part 2 will discuss the future of Lithuanian politics, namely the upcoming elections and the government's strategies for the nation. Part 3 will overview the youth situation in Lithuania, including the challenges and opportunities that young Lithuanians are currently facing. Part 4 will delve in-depth about the Belt and Road Initiative, describing the specific challenges and opportunities it has within Central and Eastern Europe. Part 5 will analyse possible future outcomes for Lithuanian youth who are most likely to be impacted by BRI. Lastly, Part 6 will draw conclusions about the chapter and provide some future possibilities about the involvement of Lithuania in BRI projects.

1. The Domestic Politics and Foreign Affairs of Lithuania

Geographically, Lithuania (65,300 km²) is nearly the size of the Ningxia Hui Autonomous Region (66,400 km²), while Lithuania's population is almost eight times smaller than Beijing's – a mere 2.8 million people[2]. In medieval times, Lithuania used to be one of the largest European nations, yet numerous occupations challenged the sovereignty and diminished the physical size of the country. As a modern nation-state, Lithuania acquired its independence not once, but twice. The first time was on February 18th, 1918 right after World War I[3]. Unfortunately, the Soviet Union annexed Lithuania in the mid-1940s and maintained control of it for almost 50 years thereafter[4].

2 | World Bank. (2016). *Population, total*. Retrieved from World Bank: https://data.worldbank.org/indicator/SP.POP.TOTL?locations=LT.

3 | Eidintas, A., Bumblauskas, A., Kulakauskas, A., & Tamošaitis, M. (2013). *The History of Lithuania.* Vilnius: Eugrimas Publishing House, p. 152.

4 | *Ibid.,* p. 233.

The second time Lithuania regained its independence was on March 11th, 1990, after the fall of the Soviet Union when it immediately transitioned to a capitalist system[5]. Ever since its independence from the USSR, Lithuania has rid itself of virtually anything that is a reminder of the country's socialist past – be it Soviet-style city names or Lenin sculptures in the squares. Even though Lithuania has been a full-fledged part of the democratic and capitalist world for the past three decades, the state still shares numerous cultural and historical similarities with its Central and Eastern Europe neighbours to this day.

Furthermore, due to Lithuania's geopolitical position, it has become more comfortable for the nation-state itself and the European Union to utilise this region for managing relations with its Eastern frontier. Ever since the great European Union (EU) and North Atlantic Treaty Organisation (NATO) expansion in 2004[6], the Eastern Europe region has become a valuable crossing point between the democratic West and the more authoritarian East. However, even with membership in the EU and 30 years since its independence, Lithuania is still dwarfed economically. Lithuania's Gross Domestic Product (GDP) is a mere 0.3% of the EU's economy, amounting to only US$42.7 billion in 2016[7]. Lithuania, like the other Baltic States, did experience a significant GDP growth right after joining the EU. However, the economy began stagnating during the 2008 Economic Crisis. This crisis left many people jobless and hopeless and the official institutions became ridden by corruption consequences which last to this day.

Nevertheless, due to Lithuania's geographical position between Russia and Western Europe, the country remains an important strategic and economic point for both sides. Transfer roads filled with commercial trucks criss-cross the Lithuanian lands. Therefore, the cooperation between Central and Eastern Europe Countries (CEEC) and China in the 16+1 Framework could be of great significance to the Lithuanian economic and political future if efficiently utilised.

5 | *Ibid.*, p. 274.

6 | BBC News. (2017, May 18). *Lithuania Country Profile*. Retrieved from BBC News: http://www.bbc.com/news/world-europe-17536867.

7 | Eurostat. (2017, October 20). *GDP and main components*. Retrieved from Eurostat.

1.1. The Current Political System

Lithuania is an independent and democratic republic, run in a semi-presidential system with a multi-party parliament[8]. The first branch of the Lithuanian political system is the executive power, where the president is the chief of the state, working together with the prime minister, who is the head of the multi-partisan style Government (including the Council of Ministers who control the 13 ministries of Lithuania)[9]. The second branch is the legislative power, commanded by the unicameral Lithuanian Parliament (also known as the Seimas)[10]. In the last branch, judicial power is vested in a group of judges, who are appointed by the president. This branch is independent of the executive and the legislative powers and on the highest level consists of the Constitutional Court, the Supreme Court, and the Court of Appeal. As Lithuania functions on a multi-partisan system, not one but a few political parties govern the state - they must cooperate and form coalitions to reach effective results.

The 141 members of the Seimas and the president of Lithuania are elected every five and four years respectively. During the parliamentary elections, 71 members are elected by an absolute majority vote in single-member constituencies, while the other 70 members are chosen through a proportional representation system[11]. After the election, the Seimas appoints the Speaker, who is usually a representative of the majority party. The president, who is chosen by a plurality vote, later appoints the prime minister with the approval of the Seimas[12].

From the 2009 presidential election to the end of her second term in 2019, the President of the Republic of Lithuania was Dalia Grybauskaitė.[13] She was not only the first female president but also the first president to be re-elected for a second consecutive term. The president is an independent politician without any partisan affiliation. President Grybauskaitė had extensive experience in the economic and political affairs of Lithuania, having served as the Minister of Finance and a member of the European Commission responsible for finance and the budget. Therefore, a number of her initiatives relate to

8 | MoFA. (2015, May 09). *Government, politics*. Retrieved from Ministry of Foreign Affairs of the Republic of Lithuania: https://www.urm.lt/default/en/travel-and-residence/about-lithuania/government-politics.

9 | *Ibid.*

10 | *Ibid.*

11 | ElectionGuide. (n.d.). *Republic of Lithuania*. Retrieved from Election Guide: http://www.electionguide.org/countries/id/125/.

12 | *Ibid.*

13 | BBC News. (2017, May 18). *Lithuania Country Profile*. Retrieved from BBC News: http://www.bbc.com/news/world-europe-17536867.

domestic and economic matters – anti-corruption campaigns, protection of the people with lower incomes, and rescuing the nation from recession[14].

As for the Seimas, after the election of October 6th and 23rd in 2016, the Lithuanian Peasants and Greens Union became the leading party with 56 seats in the Seimas. Since the two parties gained the majority of the seats, they appointed the Seimas Speaker, Viktoras Pranckietis and the Prime Minister, Saulius Skvernelis.

After the election, the Peasants and Greens also created a coalition with the Social Democratic Party (who gained 19 seats), forming a majority government. However, in September 2017, the Social Democratic Party decided to separate from the Peasants and Greens Union, due to it being 'side-lined over policies ranging from alcohol age limits to subsidies on central heating'[15]. The remaining part of the Seimas as of March 2018 consisted of four political parties: the Homeland Union or the Lithuanian Christian Democrats (31 seats), Liberal Movement (14 seats), the Electoral Action of Poles (8 seats), and Order and Justice (7 seats).

In the 2017 Autumn Session, the central working themes of the Seimas were (1) decreasing social exclusion, (2) increasing the income of the population, (3) efficiently fighting corruption, and (4) pursuing continued education reform[16]. These initiatives are incidentally similar to the overall campaign focuses of the president of Lithuania and the Government of Lithuania (the Council of Ministers and their representative Ministries), who were partially monitoring the parliament due to the lack of productivity and efficient results[17].

1.2. Foreign Affairs of Lithuania and Relations with China

The Ministry of Foreign Affairs of the Republic of Lithuania, under the leadership of the Minister Linas Linkevičius, supervised the foreign relations matters of Lithuania. The Seimas outlined several significant themes for the foreign affairs of Lithuania as part of the Programme for the 17th Lithuanian

14 | CBC. (2009, May 18). *Lithuania president-elect vows to fight recession*. Retrieved from CBC News: http://www.cbc.ca/news/world/lithuania-president-elect-vows-to-fight-recession-1.798493.

15 | Sytas, A. (2017, September 23). *Lithuania gets minority government as junior partner leaves*. Retrieved from Reuters: https://www.reuters.com/article/us-lithuania-government/lithuania-gets-minority-government-as-junior-partner-leaves-idUSKCN1BY0GO.

16 | Baltic News Service. (2017, September 12). *Seimo rudens sesija: svarbiausi projektai [Autumn session of the Seimas: the most important projects]*. Retrieved from Delfi: https://www.delfi.lt/news/daily/lithuania/seimo-rudens-sesija-svarbiausi-projektai.d?id=75732173.

17 | Delfi. (2017, October 11). *A year after the Seimas election - hopes were left in the ballot boxes*. Retrieved from Delfi by The Lithuania Tribune: https://en.delfi.lt/corporate/a-year-after-the-seimas-election-hopes-were-left-in-the-ballot-boxes.d?id=76015005.

Government[18]. Some of the most important ones were first to take care of Lithuanian independence and democratic integrity, while adequately responding to regional and global threats. Lithuania's membership in NATO offers regional and national safety and Lithuania's membership and cooperation in the European Union enhance economic and trade freedom, in addition to managing issues such as terrorism, the massive influx of refugee and border safety. Lithuania focuses on strengthening relations not only with the neighbouring Central and Eastern Europe countries, and Eastern nations (such as Belarus, Ukraine, Russia, the Caucasus), but also the United States of America and East Asia.

With regards to East Asia, Lithuania cooperates with Japan and South Korea in the fields of biotechnology, informational technology, science, education, culture, and tourism. The goal is not only to create strong institutional links but also improve the visibility and appeal of Lithuania in the region. Lithuania has also continued its cooperation with China[19]. Lithuania signed the first trade agreement with China back in 1992, which started the bilateral economic relations between these nation-states. The official diplomatic ties between the countries began in 1991 after China recognised the Lithuanian nation-state.

Overall, the primary plans of the Government of Lithuania with regards to China involve stimulating political and economic relations. Lithuania wishes to make the best of its geopolitical position to facilitate China's access to the European market[20]. Such a goal can be achieved through a focus on bilateral trade cooperation and multilateral 16+1 or Belt and Road initiatives. Combined with the various domestic strategic plans for the nation's improvement, Lithuania should expect to not only advance its global status but also strengthen its relations with China.

18 | Seimas. (2016). *Programme of the 17th Lithuanian Government.* http://urm.lt/default/lt/uzsienio-politika/naujienos-kalbos-publikacijos/LR-vyriausybes-programa-UP-dalis.

19 | *Ibid.*

20 | *Ibid.*

2. The Future of Politics in Lithuania: Elections and National Projects

The Lithuanian government has begun outlining plans for the upcoming decades. Two of the most important initiatives that are worth mentioning are 'Lithuania 2030' and 'Create Lithuania'. The strategic projects for the next decades carry great importance for Lithuania's national well-being and economic status. And they might affect further cooperation with the neighbouring region and collaboration with China – both within the multilateral BRI and bilateral agreements.

2.1. Strategic and Political Initiatives

The Lithuanian government has been working on several significant initiatives aimed at increasing the influx of foreign direct investments (FDIs) to strengthen the national economy.

The first one is 'Lithuania 2030' (*Lietuva 2030*), which is a national strategy document outlining the vision of Lithuania's future. Although it was the government who proposed the initiative in 2012, the plan works on a bottom-up approach with the input provided by the whole Lithuanian society. Any community or organisation, including non-governmental ones and proactive citizens, can propose their ideas and projects that cohere with the main priorities of the initiative. These issues are 'smart society', 'smart economy', and 'smart governance'. The 'Lithuania 2030' project attempts to create a strong economy and society, allowing Lithuania to be open to the world, yet retain its strong cultural identity. Since its launch, the people have contributed more than a thousand ideas. However, propositions for cooperation with China or any projects relating to the BRI have yet to be made. The ideas that have been realised are allowing Lithuania to have a stronger economy to increase the country's significance in the European economy, attract more foreign talent and open the borders for further projects with its neighbouring countries.

Another important Lithuanian initiative for the future is 'Create Lithuania'. The initiative is a year-long professional development programme, which attracts Lithuanian youth with international experience to work for the domestic public sector and contribute their ideas. Some of the most important goals of the project are to improve the Lithuanian business and entrepreneurship spirit in addition to promoting FDI. As the participants tend to have extensive experience in studying and working abroad, they are able to work on a variety of national and regional issues with government officials as their mentors. Similar to the 'Lithuania 2030' project, 'Create Lithuania' focuses more on the improvement of the domestic situation, which aims to enhance Lithuania's competitiveness to create an environment that is more attractive and conducive for foreign investors.

Eventually, the training of young professionals should lead to more opportunities for regional development as well. With the growing interest in the Chinese market and the Lithuania government's willingness to invest in the country's youth as capital, the outcome may be a win-win situation with the progression of the 16+1 initiative. This is especially so when more students are gaining interest in green technologies, advanced technologies and infrastructure – some of the main focal points of the BRI. Nevertheless, Lithuanian youth still has to overcome some challenges that come with living in Lithuania.

3. Lithuanian Youth: Opportunities and Challenges

In Lithuania, any person aged 14 to 29 years is considered a 'youth'[21]. In the European Union, this definition tends to include the same age bracket (in other EU states, a 'youth' is someone aged 15 and above) and youths are divided into three age groups: 15-19, 20-24, and 25-29[22]. In Lithuania, there are around 600,000 young people, comprising 20% of the total population[23]. Teenagers and young adults in Lithuania face a variety of opportunities and challenges. One of the main issues cited in the media is the unsurprisingly high rate of youth emigration, an issue which will be discussed extensively in this chapter. The Lithuanian youth situation will be compared to both the Central and Eastern Europe and the Western

21 | The Department of Youth Affairs. (2016). *Overview of the Youth Situation in 2016.*

22 | European Commission. (2011). *Commission Staff Working Document on EU Indicators in the Field of Youth.* Brussels.

23 | The Department of Youth Affairs. (2016). *Overview of the Youth Situation in 2016.*

European youth statistics.

3.1. Contextual Information

It is important to provide an overview of the contextual and statistical information of the youth in Lithuania - namely their demographics, employment and education. As mentioned previously, around 600,000 young people live in Lithuania. Unfortunately, this number has decreased significantly. From 2011 to 2016, the youth population has fallen by 83,000[24]. High rates of emigration abroad are the main cause of this decrease in addition to lower birth rates. In 2015, around 20,000 youths left Lithuania for foreign countries[25]. Also, since regaining independence in 1990, the crude birth rate has dropped from 14.91 all the way to 10.56 (in 2015). This rate was even lower around the early 2000s – which was approximately 9.3 – the period during which the present-day 14-18-year-olds were born[26].

Another essential feature of the current youth situation in Lithuania is the state of the job market. Since 2011, the overall unemployment rate has dropped quite noticeably – from 13.2% to 7.6%[27]. The youth unemployment rate was considerably higher than the national rate, despite decreasing rapidly from 23.2% in 2011 to 12.7% in 2015[28]. Furthermore, Lithuania has a much lower youth unemployment rate (19.3%) compared to the EU average, which was 20.6% in 2015[29].

Lastly, Lithuania has had significant changes in the education sphere. Lithuania enjoys a higher secondary education completion rate of 91.4%, as compared to the 82.2% overall rate in the EU, and Lithuania also boasts less early leavers from education or training – 5.9% versus the EU's rate of 11.1%[30]. Despite these positive statistics when compared to the European Union, Lithuania saw an overall decrease in the number of its students by 120,000 from 2010 to 2016[31]. This trend is visible in all the stages of education, except for vocational training. The highest drops were in the secondary and tertiary education stages[32]. While these age ranges tend to be the ones falling in the same category of those most

24 | *Ibid.*

25 | *Ibid.*

26 | Knoema. (2015). *Lithuania - Fertility - Crude birth rate.* Retrieved from World Data Atlas: https://knoema.com/atlas/Lithuania/topics/Demographics/Fertility/Crude-birth-rate.

27 | The Department of Youth Affairs. (2016). *Overview of the Youth Situation in 2016.*

28 | *Ibid.*

29 | European Commission. (2015). *Draft 2015 Joint Report of the Council and the Commission on the implementation of the renewed framework for European cooperation in the youth field.* Brussels.

30 | *Ibid.*

31 | The Department of Youth Affairs. (2016). *Overview of the Youth Situation in 2016.*

32 | *Ibid.*

likely to migrate – secondary education students leave with their parents, whereas tertiary level students often leave for universities abroad – Lithuanian youth have experienced demographic and social changes that have improved their quality of life. Nonetheless, the Lithuanian Government, in cooperation with the European Union, is offering more initiatives to not only improve youth engagement but also entice highly qualified youth to stay and contribute to their home countries[33].

3.2. Opportunities for Lithuanian Youth

Lithuanian youth have been benefiting from more opportunities offered to them by the domestic government and the European-level administration. For example, Lithuania is one of the rising stars in the Information and Communications Technology (ICT) area with new tech companies starting up on almost a daily basis. Such technological growth is attributed to the incredibly high internet and internet technology (IT) services used by the youth – nearly every adolescent in Lithuania is computer-proficient and uses IT products every day[34]. Besides, Lithuania has some of the fastest internet speeds in Europe[35]. As Lithuania is located in a central geographic position on the European continent, crisscrossed by vital highways and railways, this strategic geopolitical advantage and its high-quality infrastructure make for easier transnational trade and provide more opportunities for both the young graduates and global companies.

In terms of empowering youths in Lithuania domestically, numerous initiatives have been launched in the past few years. For example, a critical player in promoting foreign investment in Lithuania and initiating projects for young professionals – 'Invest Lithuania' – consults global, export-oriented investors to expand their businesses to Lithuania. The 'Invest Lithuania team has started several campaigns with the intention to attract Lithuanian talents who have left Lithuania to return home so that the country can utilise their social capital to improve the nation's socio-political status and increase youth employment. As mentioned earlier, the 'Create Lithuania'

33 | European Commission. (2015). *Draft 2015 Joint Report of the Council and the Commission on the implementation of the renewed framework for European cooperation in the youth field.* Brussels.

34 | The Department of Youth Affairs. (2016). *Overview of the Youth Situation in 2016.*

35 | Invest Lithuania. (2015, March 3). *Lithuania Europe's No. 1 in fibre optic internet penetration.* Retrieved from Invest Lithuania: https://investlithuania.com/news/lithuania-europes-no-1-in-fibre-optic-internet-penetration/.

(*Kurk Lietuvai*) initiative vouches for the development and augmentation of youths' professional qualifications. This project attracts numerous young Lithuanians who have international experience and provides them with an opportunity to work in the public sector. The main aim of these national programmes is to gradually decrease the rate of emigration by creating attractive and unique opportunities domestically, especially for youth.

On the regional level, the European Union has always been a proponent of youth mobility and has been actively involved in social and political matters. Some noteworthy projects have been the European Community Action Scheme for the Mobility of University Students (Erasmus+) and European Employment Services (EURES). The Erasmus (EuRopean Community Action Scheme for the Mobility of University Students) Programme is primarily a student exchange initiative. It provides opportunities for European youths to study in other European countries in the spirit of promoting regional mobility and providing students with valuable skills, which they may apply when they return to their home communities. Erasmus+ helps improve the opportunities of European youths, including Lithuanians, who might not have been able to travel to other countries for exchange or volunteering positions. On the other hand, EURES functions as a job mobility platform which provides updates about various seasonal, part-time and full-time job opportunities. Such opportunities are imperative to help youths struggling to find summer work or training programmes in their city or region.

Therefore, while the domestic programmes aim to bring Lithuanians back home, the more influential and stronger European-level programmes may, in fact, contribute to more emigration from Eastern Europe to the Western regions. The variety of jobs and the availability of higher salaries in the more advanced European nations are especially attractive to Lithuanians, leading them to migrate, thus causing further issues for the domestic economy.

3.3. Challenges for the Lithuanian Youth

Despite the opportunities for Lithuanian youth from domestic and regional projects, they still face numerous challenges – from difficulties in the job market that are causing high migration rates to low education quality. Even though the overall rate of youth unemployment and the number of youths not in education, employment, or training (NEET) are decreasing, the emigration rate is still on the rise. From 2011 to 2014, this rate was steadily dropping; however, it increased once again in 2015 (by around 3,300)[36]. The majority of emigrants are in the age range of 20-24 and 25-29, which make up more than 80% of all emigrants[37]. Such a situation is troubling for the current Lithuanian economy and its future. As the nation's population is not only decreasing, it is also ageing, and the falling numbers of young workers are making it more difficult for the government to support senior citizens. As the pension ages are being delayed, youths' possibilities to join the workforce are also hindered. As observed, low salaries and the inability to find jobs are two of the top reasons for youth emigration[38].

Other reasons cited are the wish to study abroad and learn a foreign language – reasons which are more prevalent among young secondary education graduates, who leave for universities abroad[39]. For many youths, a degree from a well-known university in Scotland or Denmark implies more future opportunities in both Lithuania and other countries.

In the end, Lithuanian youth are experiencing a variety of opportunities and challenges in the face of the changing socio-economic situation. While there are benefits to be attained because of regional youth mobility provided by EU initiatives, the same initiatives also mean a significant brain drain from Lithuania. Nevertheless, cooperation with China and neighbouring Central and Eastern Europe nations may bring more benefits than problems; if utilised successfully. The main advantages and drawbacks will be discussed in further parts.

36 | The Department of Youth Affairs. (2016). *Overview of the Youth Situation in 2016.*

37 | *Ibid.*

38 | *Ibid.*

39 | *Ibid.*

4. The Belt and Road Initiative in Lithuania

Up to the early-mid 2000s, Chinese policymakers were not interested in cooperating with Central & Eastern Europe (CEE) nations. The Chinese focused on foreign direct investments (FDI) in other developing countries, such as the Indian subcontinent or the African continent. At the same time, the post-Soviet states were warming up to new geopolitical arrangements and environments such as NATO and the European Union. Thus, Lithuania's policy focus was on regional matters.

Lithuania's joining of these alliances was an essential step towards the close cooperation with China which Lithuania now has. The European Union has been a significant trading partner of China for a long time and the addition of a newly emerging group of countries from Eastern Europe and the Baltic States allowed for more opportunities. The significant improvements in the multilateral China-CEE cooperation came around 2013 when Xi Jinping became the President of the People's Republic of China. Since that year, Chinese foreign policy goals started shifting to include the CEE region, first as part of the 16+1 initiative and later on as a transfer route for the BRI. Lithuania became a vital part of both due to its geopolitical location, growing economy, and promising infrastructure.

4.1. China, Lithuania, and the BRI

Due to the multilateral initiatives of the Chinese government and its recognition which opportunities of the often neglected Central & Eastern Europe region can provide, Lithuania has benefited tremendously. From the opportunities in cultural and education cooperation to increased funding in technology and infrastructure, the benefits from the BRI have the potential to improve the economic competitiveness of Lithuania in the long term.

To begin with, the increasing interest of the Chinese government to cooperate economically and politically with the rapidly developing CEE region is an excellent opportunity for each nation involved[40]. Already before 2012, China noticed the economic potential of the area, especially the Baltics and the

40 | Martyn-Hemphill, R., & Morisseau, E. (2015, February 5). *Small Step for China - Giant Leap for the Baltics?* Retrieved from The Baltic Times.

emergence of the so-called Baltic Tigers in the mid-2000s as they represented the fastest-growing industrialising countries in Europe[41]. This rapid advancement only gave the Baltic States more credibility and potentiality for becoming essential members of the 16+1 initiative, it also positioned them as a transfer point on the Silk Road Economic Belt route.

To deepen its involvement with the region, China has made numerous developments in the CEE. One of the most important events that marked the growing relationship was the summit of China and 16 Central and Eastern Europe countries' leaders in Suzhou, Jiangsu province in late November 2015. Li Keqiang met with the prime ministers of the countries in the region and discussed China's policies and plans for the region in the future.

As one of the Chinese foreign policies is to enhance economic cooperation while maintaining a peaceful relationship with other countries, Li highlighted the government's plans to 'invest in infrastructure in the ports of the Baltic Sea, the Adriatic Sea, and the Black Sea to boost cooperation and connectivity'[42]. However, the projects cannot continue without stable political and diplomatic relationships. The current problem is that quite a number of nationals in CEE still oppose socialist dogmas due to historical reasons. Despite China's activities in the CEE being driven by capitalistic forces, among many post-Soviet countries, China is still viewed as a socialist nation. Furthermore, there is much influence from European Union authorities in Brussels. China has to seriously discuss its plans regarding the BRI with the EU if it wishes to continue its policies based on cooperation in the region and prove that it will indeed benefit everyone[43]. Nonetheless, the BRI may bring numerous opportunities to the participating nations, including Lithuania.

4.2. Opportunities of the BRI

The involvement of Lithuania in the bilateral cooperation with China and the multilateral initiatives of the BRI and 16+1 has been advantageous for the nation. Furthermore, these collaborations shall continue bringing more opportunities in the

41 | The Economist. (2003, July 17). *Baltic Tiger*. Retrieved from The Economist.

42 | Li, R. (2015, November 25). *China Boosts Ties with Central, Eastern European Nations*. Retrieved from Global Times.

43 | *Ibid.*

future. As of now, some of the most important advances for Lithuania fall within the areas of education and infrastructure.

First of all, more and more Lithuanian youth decide to study Chinese and plan their future in conjunction with China's increasing influence. A similar trend is also noticeable in China, as Lithuanian language courses are opening up and people are becoming more familiar with this Baltic country. For instance, in 2010, the Beijing Foreign Studies University began offering Lithuanian language courses[44], which is now one of the 60+ language options at that institution. Some of the students in the Lithuanian language course in China even attended the Lithuanian Language and Culture Summer Courses in the following years[45]. As for Chinese studies in Lithuania, the opening of the Confucius Institute in 2010 has also provided opportunities for Lithuanian youth. Furthermore, Chinese language schools are opening up in Lithuania each year, due to the rising importance of the Chinese economy in the world and the overall popularity of East Asian cultures.

Furthermore, as Lithuania is one of the countries that lie within the area of The New Eurasia Land-Bridge, the government has obtained great opportunities for receiving funding for the improvement of domestic infrastructure (roads, railways and seaports).[46] According to the information provided by the Ministry of Transport and Communications, some of the leading projects in discussion are the development of Klaipėda harbour, a bilateral agreement in the field of air transport connectivity, and the trilateral Lithuania-Belarus-China cooperation for railroad development[47]. Also, the accessibility to The New Eurasia Land Bridge allows for freer and more comfortable transportation of cargo between CEE and China, due to its primary idea of 'one declaration, one inspection, one cargo release'[48].

Overall, Lithuania should benefit in the fields of transport, education, culture, and investments. Although the main areas of focus for these China-CEE initiatives are pre-set by China and may not overlap with everyone's national priorities[49], they are flexible enough to push through domestic agendas. Fortunately,

44 | The Lithuania Tribune. (2010, March 25). *China's rising interest in Lithuania*. Retrieved from Alfa: https://www.alfa.lt/straipsnis/10330525/china-s-rising-interest-in-lithuania.

45 | Study in Lithuania. (2016, July 29). *Culture Summer Courses Organised by Five Lithuanian Universities Attract Greater Numbers of Foreign Students Each Year*. Retrieved from Study in Lithuania.

46 | The New Eurasia Land Bridge is a road and rail network that connects Asia and Europe. Reference: Norton Rose Fulbright. (2016). The Belt & Road Initiative: A Modern Day Silk Road. In *Nexus 2016 A Global Infrastructure Resource*.

47 | Embassy of Lithuania to PRC. (2017, May 15). *Lietuvos ir Kinijos ministrai tarėsi dėl akyvesnio bendradarbiavimo transporto srityje [Lithuanian and Chinese ministers discussed more active co-operation in the field of transport]*. Retrieved from Embassy of the Republic of Lithuania to the People's Republic of China.

48 | EUOBOR. (n.d.). *One Belt One Road Initiative*. Retrieved from http://www.euobor.org/index.php?app=OBOR.

49 | Pavlicevic, D. (2016, June 8). 16+1 Platform: Extent and Limit of China's Global Influence. *IPP Review*. Retrieved from https://ippreview.com/index.php/Blog/single/id/162.html.

these are also fields of interest for Lithuania, as the country has been attempting to improve the comparative advantage of its education and infrastructure in the region, in addition to enhancement in advanced technology, which may be the next big step in FDI distribution.

4.3. Challenges of the BRI

Although the BRI has brought a variety of opportunities and advantages to Lithuania and the CEE region, there are, nonetheless, numerous drawbacks which need to be considered.

To begin with, one of the major fiscal foundations for the Belt and Road cooperation is the establishment of the Asian Infrastructure Investment Bank (AIIB). However, Lithuania, as well as some other CEE countries are neither current non-regional members nor prospective members. Such a situation creates particular difficulties in managing the areas falling under the AIIB umbrella – infrastructure, financing, and production[50]. While Lithuania is involved in infrastructure development projects for railroad and sea harbour connectivity, the country's isolation from the banks might hinder future discussions and investment possibilities.

Another challenge for Lithuania is its small economy, which, besides being dependent on the EU, is overshadowed by regional cooperation and discourse, thus countries such as Poland, Serbia, or Hungary tend to obtain more funding and collaboration opportunities from China[51]. On the other hand, Lithuania can become a transfer spot but not an end destination for the Chinese. That is mostly true due to the already-existing economic and historical ties between China and the major European economies, thus leading to the smaller role of CEE states functioning as a 'bridge to Europe'[52]. The lack of achievements and projects may also be because of the local Lithuanian legislators, who are not too interested in such an initiative. The BRI and 16+1 projects require great resources and the involvement of the national governments[53]. However, some of the national governments in the CEE region feel that

50 | Tan, K. G., Sim, R., & Ho, J. W. (2017, June 14). The Grand One Belt-One Road Initiative: Opportunities, challenges and implications. *Business Today*. Retrieved from https://www.businesstoday.in/opinion/columns/the-grand-one-belt-one-road-initiative-opportunities-challenges-and-implications/story/254355.html.

51 | Goralczyk, B. (2017). China's interests in Central and Eastern Europe: enter the dragon. *European View*, pp. 153-162.

52 | Pavlicevic, D. (2016, June 8). 16+1 Platform: Extent and Limit of China's Global Influence. *IPP Review*. Retrieved from https://ippreview.com/index.php/Blog/single/id/162.html.

53 | Tan, K. G., Sim, R., & Ho, J. W. (2017, June 14). The Grand One Belt-One Road Initiative: Opportunities, challenges and implications. *Business Today*. Retrieved from https://www.businesstoday.in/opinion/columns/the-grand-one-belt-one-road-initiative-opportunities-challenges-and-implications/story/254355.html.

the whole project has been lacking a common agenda or explicit goals, thus were not as quick to join the initiative[54].

Lastly, some significant challenges for the 16+1 are the vast political, economic, and cultural differences within the European region in-discussion[55]. Before the Chinese initiation of the 16+1, such areas as the Baltics or the Balkans had little to no common interests or regional cooperation. Also, there are numerous divergences among the 16 countries in regard to their memberships in other organisations, such as the EU or NATO[56]. While Lithuania is dependent on the European Union, such states as Albania or Serbia have different foreign policy approaches and economic interests. Such a mismatch of national contexts hinders the success of the initiative, as each country tends to fight for prospects and funding according to their national interests. Furthermore, these European nations may not see the purpose of cooperating with each other unless they are geographically close-by or similar culturally. Therefore, one might see Serbian-Romanian or Lithuanian-Latvian initiatives with China, but not Bulgarian-Estonian cooperation, due to the mentioned differences.

To sum up, the majority of challenges for the CEE-China cooperation are mainly regional mismatches and national interest divergences. For Lithuania, the small size of its economy and dependence on regional foreign policy, combined with the passivity of its local legislators inhibits the incredible opportunities this cooperation might bring.

Despite numerous myths about China's intentions and various challenges, this multilateral cooperation might still bring great opportunities for the nations involved. The openness of the Chinese government for international collaboration should bring win-win opportunities for both domestic and regional economic growth, as well as improved connectivity between the two regions.

54 | Goralczyk, B. (2017). China's interests in Central and Eastern Europe: enter the dragon. *European View*, pp. 153-162.

55 | Pavlicevic, D. (2016, June 8). 16+1 Platform: Extent and Limit of China's Global Influence. *IPP Review*. Retrieved from https://ippreview.com/index.php/Blog/single/id/162.html.

56 | *Ibid.*

5. Future of the Belt and Road Initiative in Lithuania: A Case for the Youth

The past few chapters have highlighted the variety of opportunities and challenges that the BRI will pose for Lithuania, in addition to how such prospects might influence Lithuanian youth. The multilateral cooperation among the CEE region has a wide array of prospects for all the involved youth. The main opportunities fall within the areas of infrastructure, engineering, technology, and education. Unsurprisingly, the youths of Lithuania among other Central and Eastern Europeans might expect the corresponding benefits of cooperation as well. This generation of high school and university students are shaping the face of the project and their futures. As long as there is enough motivation from the youth and the government, the Sino-Lithuanian relationship is expected to grow and prosper.

5.1. Education and the BRI

One of the significant advantages for youth in Lithuania has already been mentioned – education. The foundation for the cooperation in education between China and Lithuania is secure. Many Lithuanian and Chinese university students decide to continue their education full-time in each other's country. For example, the recently established international programmes in Beijing – Yenching Academy of Peking University and Schwarzman College at Tsinghua University are bringing in hundreds of bright students from all around the world. Each year, several students from the 16 Central and Eastern European countries enrol in such programmes. These students can continue building international connections with their Chinese counterparts; strengthening bilateral ties. For instance, the global-minded Lithuanian graduates will have the possibility of applying Chinese language skills back in their country to help advance the BRI by easing communication barriers. Vice versa, the Chinese students, who have learned Lithuanian, can help to liaise between Lithuanian and Chinese companies and assist trade visits from the ministries. These bilingual global citizens shall be a great asset to any company wishing to engage in international business. Hopefully, through higher investment in the education sphere and increasing exchange programme opportunities, these young people will take on roles which support and enhance decision-making in the BRI and push the initiative forward through mutual cultural understanding.

5.2. Investment and the BRI

Another valuable prospect for Lithuanian youth is the rising Chinese investment in Lithuania, which should also bring greater career prospects and enhance the various bilateral cooperation projects. Since the beginning of the 16+1 Initiative, Chinese FDI has rapidly increased and in 2016 totalled to more than €21 million[57]. Also, during the past six years, Lithuanian exports to China have increased fourfold. In 2017, 17 Lithuanian dairy factories enjoyed the right to distribute their goods in China[58]. This number was expected to rise as the demand for Lithuanian dairy and other agricultural products continue to grow. Growing trade and export-oriented production in Lithuania should help the national economy, especially if more young people choose to pursue careers in these industries. Once again, bilingual experts in trade, agriculture, and export business will invaluable for the BRI.

Furthermore, there is a growing need for talents in the areas of infrastructure, engineering and higher technologies. As China is expected to invest further in the Lithuanian high technologies, more vacancies will open up for the youth interested in technology, laser science and computer science. Moreover, with infrastructure improvement projects rising throughout the region, these projects will require specialists in urbanisation and environment, legal experts and other related positions[59].

5.3. Youth as Shapers for Cooperation

The Belt and Road Initiative (BRI), together with the 16+1 Initiative have already proven to be beneficial, as long as the countries play an active role in the projects themselves. The young generations of the 16 Central and Eastern European (CEE) nation-states have the power to shape the collaboration as they wish. The current governments are seemingly stagnant and unsure of how to approach the Belt and Road Initiative as the older generations of politicians are waiting for directions or an action plan from the Chinese government.

57 | ekspertai.eu. (2017, April 15). *Tiesioginės Kinijos investicijos Lietuvoje 2016 m. viršijo net 21 milijoną eurų [Chinese direct investment in Lithuania in 2016 exceeded as much as €21 million].* Retrieved from ekspertai. eu: http://www.ekspertai.eu/ tiesiogines-kinijos-investicijos-lietuvoje-2016-mvirsijo-net-21-milijona-euru/.

58 | *Ibid.*

59 | Pukenė, R. (2015, February 7). *Žvilgsnis į priekį: kokių specialistų labiausiai reikės po 10 metų [Looking Ahead: What Kind of Professionals Will You Need Most in 10 Years].* Retrieved from Delfi https://www.delfi.lt/ news/daily/education/zvilgsnis-i-prieki-kokiu-specialistu-labiausiai-reikes-po-10-metu. d?id=66628194.

Two key suggestions can be made. First, in the CEE nations, the involvement of youth should be enhanced. Sometimes the youth in these countries, including Lithuania, have not been taken seriously or have been completely neglected in the decision-making process. However, if the future of the BRI is the future of the young generation, youths – who are more flexible and open to novel ideas, risky projects, and unknown matters – will be vital. By allowing the youth to be proactive, this abstract and colossal initiative called the BRI can become more concrete and localised in its implementation and benefits.

Second, the Chinese government should be more open about their ideas and ambitions for the project. While the BRI sounds incredible in theory, there are many practical matters to be taken into account, such as national border disputes, cultural differences, political instabilities, and other issues which cause misunderstandings. Working with the European and Chinese representatives (including the youth) and drafting specific shared goals, budgets, and deadlines, would enhance cooperation between all parties in the BRI.

All in all, the increasing interest of Chinese policymakers in this region, and especially Lithuania, is opening many doors for youth. As long as the governments allow the youth to be vocal about their future goals and set concrete plans based on those ideals, the BRI should bring prosperity to the region. The ties between Lithuania and China shall continue to grow stronger thanks to the perseverance and ambition of the young people.

6. Conclusion

Lithuania was put on the global map in the 1990s after it regained its independence. From the beginning, this country showed a great deal of determination and ambition to make a name for itself. However, its late entrance into the international theatre and the country's general geopolitical weaknesses proved somewhat hindering. Nonetheless, Lithuania – once ignored by the major global markets, such as China – is now being given the attention it deserves. With the help of the China-led multiregional initiatives such as the BRI and 16+1 initiative, this country is looking to improve faster than ever.

This chapter aimed to give an overview of Lithuania's place on the BRI. The first few chapters engaged with contextual information about the socio-political situation in Lithuania and the status quo of the youth matters. The government has engaged in several initiatives to rectify the brain drain and to enhance FDI. These projects have already proven to be useful as more companies are opening their subsidiary offices in Lithuania (for example, Uber, Booking Holdings, or Barclays, just to name a few), due to the laxer business-opening terms, cheap land, and affordable but high-quality personnel. Some of these initiatives have been launched thanks to the BRI projects. Lithuania is growing its trading links with Chinese companies in the areas of education, agriculture, infrastructure, and high technologies. If these collaborations continue, the future of the Lithuanian youth appears bright.

The chapter highlighted the inactivity of Lithuanian policymakers with regard to their cooperation with China. Such a trend has been visible throughout the region, as many neighbouring nation-states seem to be confused or unsure about the regional plans of China. The initiatives have proposed many fantastic ideas and projects; however, they appeared either unimportant or unattainable for the locals. Nonetheless, as long as the future generations, who are gaining interest in China and its economy, decide to maintain the initiatives and become more ambitious and active than the older generations of politicians, the 16+1 project will flourish. Consequentially, this will be beneficial not only to China and its eastern and western regions, but also the Central and Eastern European nations.

Library Courtyard at the university in the old city centre in Vilnius, Lithuania.

The International Balkan University, in
Skopje, North Macedonia.

by   IVANA JORDANOVSKA
New York University

ANASTASIJA MILKOVSKA
Ss. Cyril and Methodius University in Skopje

(Ordered by last name)

CHAPTER 11

NORTH
MACEDONIA

ESTONIA

LATVIA

LITHUANIA

POLAND

CZECH
REPUBLIC

SLOVAKIA

HUNGARY

SLOVENIA

CROATIA

ROMANIA

BOSNIA &
HERZEGOVINA

SERBIA

MONTENEGRO

BULGARIA

ALBANIA

NORTH
MACEDONIA

1. Overview of North Macedonia's Political Situation

1.1. North Macedonia after Yugoslavia

The Republic of Macedonia proclaimed independence from Yugoslavia on September 8th, 1991, following the decision to hold the referendum for independence on January 25th, 1991, by the first multi-party parliament of Macedonia. The question posed at the referendum was: 'Are you in favour of an independent Macedonia with the right to enter a future union of sovereign states of Yugoslavia?'[1] An overwhelming majority of 95of those voters of the new Republic of Macedonia voted in favour of independence and Macedonia became the only former Yugoslav republic to gain independence in the 1990s through peaceful means.[2]

Since then, Greece has objected over Macedonia's name, claiming that it signified territorial aspirations because one of the northern regions of Greece bears the same name. It meant a difficult and troublesome start for the country in the international arena, and joining the UN under a provisional name, the 'Former Yugoslav Republic of Macedonia' (FYROM) in 1993. While FYROM was a name used in both bilateral and multilateral relations with countries that hadn't recognised Macedonia under its constitutional name – Republic of Macedonia – more than 120 countries, including China, Russia and the United States, did recognise the country as the 'Republic of Macedonia'. Macedonia and Greece formulated their bilateral relations with the UN-mediated Interim Accord, signed in 1995.[3]

The tumultuous period that followed tested Macedonia's governments. Through the proportional representation system, the Social Democratic Union of Macedonia (SDSM), with a clear majority up until 1998, formed governments with smaller parties such as PDP[4] and the Liberal Party, which started out as the Union of Reform Forces. These governments dealt with a number of issues inherent to the Balkans of the 1990s, such as refugees pouring into the country from the wars raging on the territory of today's Croatia, Bosnia & Herzegovina, and Kosovo; the trade embargo to the north (today's Serbia) and south with Greece, which was a hard hit for the minuscule economy[5];

1 | Fazlagic, A. (2017, September 8). Македонија: Од осамостојувањето до денес [Macedonia: From Independence Until Today]. *AA*. Retrieved from https://www.aa.com.tr/mk/%D0%B0%D0%BD%D0%B0%D0%BB%D0%B8%D0%B7%D0%B0-%D0%BD%D0%B0-%D0%B2%D0%B5%D1%81%D1%82%D0%B8/%D0%BC%D0%B0%D0%BA%D0%B5%D0%B4%D0%BE%D0%BD%D0%B8%98%D0%B0-%D0%BE%D0%B4-%D0%BE%D1%81%D0%B0%D0%BC%D0%BE%D1%81%D1%82%D0%BE%D1%98%D1%83%D0%B2%D0%B0%D1%9A%D0%B5%D1%82%D0%BE-%D0%B4%D0%BE-%D0%B4%D0%B5%D0%BD%D0%B5%D1%81/904277.

2 | *Ibid.*

3 | *Interim Accord.* (1995, September 13). *Peacemaker.* Retrieved October 20, 2017, from http://peacemaker.un.org/sites/peacemaker.un.org/files/MK_950913_Interim%20Accord%20between%20the%20Hellenic%20Republic%20and%20the%20FYROM.pdf ; Maleski, D. (2012). Македонија во меѓународните односи (1991-1993): Бебето од катран [Macedonia in the international relations (1991-1993): The tar baby]. Skopje: Kultura ; Marusic, S. J. (2011, July 30). *Macedonia-Greece Name Dispute: What's in a Name? Balkan Insight.* Retrieved October 18, 2017 from http://www.balkaninsight.com/en/article/background-what-s-in-a-name.

4 | Party for Democratic Prosperity, predominantly an ethnically Albanian political party.

and the difficult and often murky privatisation process causing unemployment to rise.[6]

In 2001, armed conflict erupted between paramilitary ethnic Albanian groups and the security forces of the Macedonian state. A large number of former fighters and arms from the war in Kosovo spilled over to Macedonia after 1999, which fuelled the legitimate claims of Macedonia's own ethnic Albanians for greater political rights. Afraid of the earlier horrors witnessed in other former Yugoslav republics, the international community acted swiftly and brokered a peace deal, known as the Ohrid Framework Agreement, in 2001 within several weeks of intense negotiations. The full-blown war was avoided.[7]

The Ohrid Framework Agreement foresaw a number of significant structural changes to the governing structure of the country, effective to this very day. A new law on municipal boundaries reduced the number of municipalities to 80 as a step towards their improved sustainability. With the law on local self-governance and on local finances, a process of decentralisation began, giving municipalities greater decision-making powers and promoting better representation of various ethnic communities. On the central level, the equal representation of minorities was ensured in public institutions, as well as state universities, police, and the army. The use of languages and symbols other than Macedonian were guaranteed in the Constitution and subsequent laws.

1.2. Towards the EU and NATO

Macedonia then saw intense reformation through the adoption of the EU acquis[8]. Though the Stabilisation and Association Agreement with the EU was signed in 2001 (Macedonia is the first country in the region to do so) the Agreement was only ratified in its entirety in 2004.[9] The Agreement itself defines the relations between the EU and Macedonia, especially in the areas of political dialogue and economic and regional cooperation.[10]

On March 22nd, 2004, Macedonia submitted its application

5 | United States Bureau of Citizenship and Immigration Services. (1995, February 01). *Macedonia. Human Rights since 1990. RefWorld.* Retrieved October 24, 2017, from http://www.refworld.org/docid/3ae6a60d0.html.

6 | Maleski, D. (2012). *Македонија во меѓународните односи (1991-1993): Бебето од катран [Macedonia in the international relations (1991-1993): The tar baby].* Skopje: Kultura.

7 | Wood, P. (2001, December 29). *Macedonia: Step back from the Abyss. BBC News World Edition.* Retrieved October 20, 2017, from http://news.bbc.co.uk/2/hi/in_depth/world/2001/review_of_2001/1703711.stm; European Commission. (2001, August 13). *Framework Agreement. European Commission.* Retrieved October 20, 2017, from https://ec.europa.eu/neighbourhood-enlargement/sites/near/files/pdf/the_former_yugoslav_republic_of_macedonia/framework_agreement_ohrid_130801_en.pdf.

8 | Accumulated body of EU law and obligations. See: European Commission. (2016, December 6). *European Neighborhood Policy and Enlargement Negotiations.* Retrieved from https://ec.europa.eu/neighbourhood-enlargement/policy/glossary/terms/acquis_en.

9 | Secretariat for European Affairs. (2014). *Ten Years of the Enactment of the Stabilisation and Association Agreement.*

10 | European Union. (2001, March 26). *SAA Agreement between the EU and the FYROM. European External Action Service.* Retrieved October 20, 2017, from https://eeas.europa.eu/sites/eeas/files/saa03_01_en.pdf.

11 | European Council. (n.d.). *The Republic of North Macedonia.* Retrieved from https://www.consilium.europa.eu/en/policies/enlargement/republic-north-macedonia/. As of late 2020 North Macedonia was still waiting for the EU to accept its accession as a member state.

12 | "VMRO" is a historic acronym of a former militant group striving for independence of Macedonia from the Ottoman Empire. "DPMNE" stands for "Democratic Party for Macedonian National Unity".

13 | Balkan Insight. (2008, May 11). Who's Who In the Macedonian Election. *Balkan Insight.* Retrieved October 19, 2017, from http://www.balkaninsight.com/en/article/who-s-who-in-the-macedonian-election.

14 | Maleski, D. (2012). *Македонија во меѓународните односи (1991-1993): Бебето од катран [Macedonia in the international relations (1991-1993): The tar baby].* Skopje: Kultura.

15 | Nikola Dimitrov is the Minister of Foreign Affairs as of 2017.

16 | Spiegel Online. (2008, April 3). Jilted Macedonia Walks Out of NATO Summit. *Spiegel Online.* Retrieved October 24, 2017, from http://www.spiegel.de/international/world/dejection-in-bucharest-jilted-macedonia-walks-out-of-nato-summit-a-545214.html.

for EU membership and has been a candidate country since December 17th, 2005.[11]

After parliamentary elections in 2006, a new government led by the centre-right VMRO DPMNE[12] was formed. The new Prime Minister Nikola Gruevski, at first seemingly reformist, was once Macedonia's most popular politician, promised an economic revival after the stumbling years of Macedonia's privatisation process. For many, Gruevski was also seen as a patriot, especially with his stance on Greece.[13]

In 2008, a mere two years after the election of the first government led by Gruevski, Macedonia hoped to become a full-fledged NATO member at the summit in Bucharest, Romania. For years, Macedonia has prepared for and participated in NATO-led missions, and NATO membership has always been seen as a guarantee for the stability and territorial integrity of the country.[14] However, Macedonia's entry was blocked by Greece, due to the unresolved name issue. The Macedonian delegation left the Summit, and Nikola Dimitrov[15], Macedonia's then high official at the Ministry of Foreign Affairs (MFA), stated: 'it sends the wrong message in terms of moderate politicians and moderate forces.'[16] Indeed, the rejection became known as the turning point for the country's stability, but not in relation to its territorial integrity.

1.3. The Turning Point

The early parliamentary elections in 2008 saw VMRO DPMNE consolidate its rule by winning 63 MP seats out of a total of 123. In the aftermath of the election, a more nationalistic rhetoric took over the public discourse, partly manifested in the Skopje 2014 project that was initially announced in 2010.[17]

While the public was bombarded by a glorification of the ancient Macedonian Kingdom manifested in a surplus of neo-Baroque facades and monuments, a pervasive state capture happened on all levels. The extent to which it happened became widely known only in 2015 when the oppositional SDSM blew the lid off the biggest scandal in Macedonia since its independence by publishing excerpts of phone conversations illegally wiretapped by the government, as reported in *Foreign Policy*:

> 'Gruevski held power from 2006 until January [2016], when he stepped down as part of an EU-mediated deal after it was alleged that he personally ordered the tapping of 20,000 phone numbers, including those of his close associates. Contents of the phone calls were leaked by the main opposition party, the Social Democrats, which also raised allegations that Gruevski and key allies had organised election fraud, attempted to cover up a police killing and sought to control the judiciary. An investigation by the European Commission found that Gruevski misused national security services – led during his premiership by his cousin Sašo Mijalkov – to 'control top officials in the public administration, prosecutors, judges, and political opponents.' [18]

A grassroots-led uproar against the pervasive control over all areas of government forced the international community to also get involved and take a more decisive stance. During the two years of protests and popular discontent (2015 to the end of 2016), an intense and difficult process of negotiation between various political actors took place and eventually paved the way for somewhat free and fair elections. The elections that took place in December 2016 resulted in almost equal numbers of MPs for VMRO-DPMNE and SDSM (51 to 49 out of 120).

According to the constitution, the president awards the mandate to form a government. President Gjorge Ivanov awarded the mandate to VMRO DPMNE, and after that party failed to form a government within the legal timeframe, it was expected that SDSM would be given the chance to attempt to do so. However, President Ivanov refused to do so, claiming that an SDSM-led government would jeopardise the sovereignty of the country. The standoff contributed to VMRO DPMNE's tension-rising rhetoric and culminated with 200 of their supporters storming the parliament and injuring a

17 | Marusic, S. (2015, August 05). Skopje 2014: The New Face of Macedonia, Updated. *Balkan Insight*. Retrieved October 24, 2017, from http://www.balkaninsight.com/en/gallery/skopje-2014.

18 | Hopkins, V. (2016, June 19). Let Them Eat Alexander the Great Statues. *Foreign Policy*. Retrieved October 25, 2017, from http://foreignpolicy.com/2016/06/19/let-them-eat-alexander-the-great-statues-skopje-2014-macedonia-colorful-revolution/.; *Aljazeera America*. (2015, May 05). Macedonia Opposition Leader: PM Tried to Cover Up a Police Killing. *Aljazeera America*. Retrieved October 26, 2017, from: http://america.aljazeera.com/articles/2015/5/5/macedonian-opposition-chief-pm-tried-to-cover-up-a-killing.html.

number of MPs and journalists.[19]

After the fear of further instability becoming a real possibility, and with further international pressure, President Ivanov finally granted the mandate and the new government, led by SDSM and smaller parties with an ethnic-Albanian denomination, was approved by the parliament on May 31st, 2017.[20]

The new government introduced the '3-6-9 plan', a reform roadmap for taking the country back on track for NATO and EU membership, with evaluations after the first three, six and nine months. The main reform priorities were the judicial sector, intelligence services, public administration reform and media law. The successful completion of the '3-6-9' was followed up by the '18+' plan which should programme the implementation of reforms until the end of the mandate of the government.

19 | Hopkins, V. (2017, April 28). What Happened in Macedonia, and Why. *The Atlantic*. Retrieved October 25, 2017, from https://www.theatlantic.com/international/archive/2017/04/macedonia-parliament-zaev-nationalist-violence-eu-europe/524733/.

20 | Marusic, S. J. (2017, May 31). Macedonia Parliament Approves New Gov't after Prolonged Stalemate. *Balkan Insight*. Retrieved October 26, 2017, from http://www.balkaninsight.com/en/article/macedonia-parliament-approves-zaev-s-new-govt-05-31-2017.

21 | International Republican Institute. (2017). *Macedonia - National Public Opinion Poll*. Skopje: Centre for Insights in Survey Research.

2. What's Next for North Macedonia?

Changes in the political landscape and the promises of the new government gave a glimmer of hope for serious, progressive change. The national consensus was that Macedonia should become a full-fledged member of NATO and the European Union, with 72% of citizens explicitly supporting these policies in March 2017.[21]

However, the biggest external obstacle to achieving these goals was the issue with Greece over its name. Although a bilateral issue in essence, Greece had blocked Macedonia's entry to NATO in 2008, prompting a suit and victory for Macedonia in front of the International Court of Justice. As a result, any future developments on the prospects for NATO and EU membership were likely to parallel the eventual resolution of the name issue.

Macedonia and Greece signed a name agreement on June 17th, 2018, in the border region of Prespes/Prespa.[22] The agreement foresaw a number of steps to be taken by both governments to ensure the successful implementation of the new name (Republic of North Macedonia) and the unblocking of EU and NATO accession.

For these reasons, it appeared logical to regard these three events in consecutive order: first, the agreement for resolving the name issue was signed, requiring a number of steps taken by both sides, which resulted in an invitation by NATO and pre-screening for accession negotiations in 2019, dependent upon successful completion of the conditions set out by the Council.[23]

2.1. What's in the Name?

As mentioned before, the name issue has existed since the independence of the country in 1991. The blocking of the country's membership to NATO in 2008 resulted in a subsequent suit in front of the International Court of Justice that same year.

Macedonia's claim was that Greece breached Article 11, paragraph 1 of the Interim Agreement signed in 1995, which stipulates:

> 'Upon entry into force of this Interim Accord, the Party of the First Part agrees not to object to the application by or the membership of the Party of the Second Part in international, multilateral and regional organisations and institutions of which the Party of the First Part is a member; however, the Party of the First Part reserves the right to object to any membership referred to above if and to the extent the Party of the Second Part is to be referred to in such organisation or institution differently than in paragraph 2 of United Nations Security Council resolution

22 | Pamuk, H. (2018, July 11). NATO Formally Invites Macedonia to Join Alliance. *Reuters*. Retrieved from https://www.reuters.com/article/us-nato-summit-declaration/nato-formally-invites-macedonia-to-join-alliance-idUSKBN1K12AR.

23 | North Macedonia became a member of NATO in March 2020.

817 (1993).'[24]

However, it is widely accepted that Greece was not allowing Macedonia's accession to NATO or the start of negotiations for membership with the EU. Because both NATO and the EU require consensus on admitting new members, it was widely believed that without resolving the name issue, it would be very difficult to become a full member of both organisations.

The name negotiations were ongoing since the signing of the Interim Agreement in 1995, with various levels of intensity. Periods of low interest in resolving the name issue have usually occurred during a major internal crisis in either country (such as the 2008 Greek economic crisis or the political crisis in Macedonia in 2015, etc.). The Prespes agreement foresaw the ratification of the agreement in the Macedonian Parliament, as a first step. The following step was holding a referendum on September 30th, 2018 in which the citizens voted in favour or against the agreement. The law requires a 50%+1 turnout rate, out of which a majority is required for the referendum to be deemed successful. Once the referendum was complete, the parliament needed to adopt the constitutional changes with a two-thirds majority. For this to happen, support from the opposition, or at least parts of it was required.

2.2. Becoming a NATO Member

Macedonia started its NATO accession path in 1999 by joining the Membership Action Plan (MAP). MAP is a set of advice, assistance, and support covering political, economic, defence, resource, security, and legal aspects.[25] At a NATO Summit meeting in Brussels in July 2018. Macedonia received an invitation conditional upon the successful implementation of the Prespes Agreement.[26] Although joining NATO is not a precondition for joining the European Union, it is widely accepted as a 'boost' towards EU membership. The criteria laid out in the NATO accession talks are a subset of the EU Accession criteria. For these reasons, NATO membership can be seen as the second step in Macedonia's future political developments.

24 | International Court of Justice. (n.d.). *Application of the Interim Accord of 13 September 1995 (the former Yugoslav Republic of Macedonia v. Greece)*. International Court of Justice. Retrieved from http://www.icj-cij.org/en/case/142.

25 | NATO. (1999, April 24). Washington Summit Communiqu. *NATO Press Release*. Retrieved from https://www.nato.int/docu/pr/1999/p99-064e.htm.

26 | Schultz, T. (2017, December 07). Aspiring NATO Member Macedonia Angles for Membership Invite in 2018. *DW*. Retrieved from http://www.dw.com/en/aspiring-nato-member-macedonia-angles-for-membership-invite-in-2018/a-41682975.

2.3. Becoming an EU Member

Macedonia became a candidate country for EU accession in 2005. Ever since, it has received positive recommendations for the start of negotiations from the European Commission. However, while there has been a recommendation for starting the negotiations, they have never been opened due to Greece's objections.

With the Prespes agreement signed, the Council of the European Union adopted a recommendation to start negotiations with Macedonia in June 2019 conditional upon the fulfilment of several conditions laid out in the document. The conditions were focused on continued judicial reforms, intelligence and security services reforms, and public administration reform.

The negotiations were organised in 35 chapters, each focusing on one or more areas in which the country's laws and regulations needed to be amended in order to be compatible with the existing EU legislation. Through this process, Macedonia would prepare for the 'free movement of goods, capital, and workers, economic, energy, transport, and regional policy, fundamental rights and more'.[27] The negotiations' dynamics depend on the speed of adoption of the EU's acquis communautaire. The last country to join the EU, Croatia, finished negotiations in 2011, 6 years after starting the negotiation in 2005.

2.4. The Next Five Years

Having resolved the name issue, the path was cleared towards full NATO and EU membership. However, given the fragile political situation, one has to be careful and anticipate unforeseen developments and missteps. The extent to which this might delay EU membership was difficult to predict.

27 | European Commission. (2015). *EU Enlargement Factsheet. European Commission.* Retrieved from https://ec.europa. eu/neighbourhood-enlargement/ sites/near/files/pdf/publication/ factsheet_en.pdf.

3. Opportunities for Youth in North Macedonia

The following parts will cover some of the most important topics relevant for North Macedonian youth, the progress or stagnation of North Macedonian society, and an approximation for the further development of the opportunities and challenges faced by North Macedonian youth.

It is common knowledge that young people play a crucial role in every society in the world. Youth is considered to be the pillar that holds the future of society stable and that creates an intangible bond that connects the present with the future. Being an overexploited term, youth has more than one definition among different circles, this text will be referring to youth in accordance with the definition given by the UN – persons between the ages of 15-24.[28]

Another crucial variable in this definition would be the state of North Macedonia, which will help explain some of the processes that are happening in it. According to WESP (World Economic Situation and Prospects), North Macedonia is ranked as an economy in transition.[29] The word transition itself signifies the process or period of changing from one state to another.[30] The first years after the fall of Yugoslavia and the proclaimed independence of North Macedonia are usually those that are considered as synonyms of transition in North Macedonia since they represent the economic and political transition between two systems of governance.

Looking back, the situation of young people previously was far worse than it is now. Some of the main challenges that young people faced in earlier times were the lack of education, the fallen economy which resulted in a high unemployment rate, especially of young people, and not having access to information and opportunities for further development. These challenges created a vicious cycle that maintained a constant level of unemployment, even after the stabilisation of the economic transition. In 2006, the unemployment rate of young people in this country was 59.8%.

This alarming number stimulated various government measures to improve the situation of North Macedonian youth. 'National

28 | United Nations Department of Economic and Social Affairs (UNDESA) (2013). *Definition of Youth. United Nations.* Retrieved October 20, 2017, from http://www.un.org/esa/socdev/documents/youth/fact-sheets/youth-definition.pdf.

29 | United Nations Department of Economic and Social Affairs (UNDESA). (2014). *Country Classification.* United Nations. Retrieved October 20, 2017, from http://www.un.org/en/development/desa/policy/wesp/wesp_current/2014wesp_country_classification.pdf.

30 | Transition. (2017). *Merriam-Webster.com.* Retrieved from: https://www.merriam-webster.com/dictionary/transition.

strategy for youth' was voted in 2005 and implemented in 2007. This document was followed by various other action plans and programs among which included the 'National employment strategy' (2010) and 'The employment action plan' (2006-2008). There were changes in the laws for volunteering and employment, compulsory high school, accreditation and opening of new universities, implementation of the 'state quota' as a type of subvention for schooling etc. After all these efforts, the youth unemployment rate in 2016 is 49.5%.[31]

If we compare the numbers a drop of 10% may seem like significant progress, however, the odds are unfortunately not in our favour. As globalisation started to emerge and people became more informed about the opportunities offered around the world, our society got diseased with the 21st century plague – brain drain, which according to the World Bank Migration and remittances Factbook (2016) 'exported' 30.2% of our populace.[32] This trend of emigration has been present throughout the whole existence of the independent country. Over the long run (1998-2011) around 10% of the North Macedonian populace have moved out of the country.[33]

One of the most commonly used tools for discovering the reasons behind migration is the push and pull theorem, and in the North Macedonian case, it seems like it works one-sidedly. Results from the study for youth show that almost 80% of the statistical sample answered that they are considering moving out of North Macedonia. Some of the main factors for this answer were: better living conditions abroad (60.40%), better work conditions (37.40%), bigger chances of employment (32.9%), more income (28.10%), and more chances to advance in one's career (13.5%). A very small number of the youth answered that they would leave the country because of the mindset of the populace (9.8%), international career (4.70%), better education (5%), and access to services (3%).[34]

Corruption and bribery are other aspects of the North Macedonian reality. The current corruption score of North Macedonia is 37.[35] Mostly bribes are taken to speed up processes in the public sector – our country's biggest employer.

31 | World Bank. (2016). *Migration and Remittances Factbook 2016, 3rd edition. World Bank.* Retrieved from: https://openknowledge. worldbank.org/bitstream/hand le/10986/23743/9781464803192. pdf?sequence=3.

32 | *Ibid.*

33 | *Ibid.*

34 | Topuzovska Latkovic, M., Borota Popovska, M., Serafimovska, E., & Cekic, A. (2013). *Студија за младите во Република [Study for the Youth in Republic of Macedonia 2013].* Friedrich Ebert Stiftung.

35 | A scale from 0 (highly corrupt) to 100 (very clean). Transparency International. (2017) *Macedonia [FYR].* Retrieved from: https://www. transparency.org/country/MKD.

Public institutions are thus most prone to this type of activity, especially when it involves job applications.

According to a report done by United Nations Office on Drugs and Crime in 2011, 69.9% of government job applicants who did not get a job offer attributed the cause of unsuccessful application to 'someone got the job because of nepotism'. In other words, many unsuccessful job applicants found they engaged in unfair competition in finding a job in the public sector because the jobs are usually filled by people with special political connections.

3.1. A New Hope

However, every dark night comes with the promise of dawn. There are numerous notable initiatives, petitions and projects, mostly from the NGO sector and the youth. These mainly resulted in raising awareness about the 'swept-under-the-carpet' issues of our society, and provided young people with practical experiences, equipping them with new skills.

Another hope is in the entrepreneurial spirit of young people. In the past few years, significant numbers of startups have emerged. Efforts are put into developing both traditional and social entrepreneurship. Different stimulus programmes, grants, and competitions are now accessible to North Macedonian youth, which motivate them to start considering the creation of their own career opportunities and exploring new ideas. Our entrepreneurial culture is still below the regional and world average, but it is showing steady growth. According to the GEDI (Global Entrepreneurship and Development Institute), as of 2018, we were ranked as 32nd in the region, and 66th worldwide.[36]

To sum up, transition supplemented with the political crisis that North Macedonia faced managed to create a very challenging environment, especially for young people, many of whom only see prospects abroad. However, fresh efforts and new strategies are starting to be implemented in order to boost the economic growth of the country and provide more and better opportunities for the youth and their careers. A change

36 | Global Entrepreneurship and Development Institute. (2017) *Macedonia*. Retrieved November 29, 2017, from https://thegedi. org/countries/macedonia.

of government does not mean the automatic resolution of all problems. It will be a long and arduous journey transforming North Macedonia's rudimentary economy and inspiring each young individual in the country.

4. 'I move, Therefore I am.' – Haruki Murakami, 1Q84

Movement is an inevitable part of human life and progress. Throughout history, mobility enabled the discovery of new lands, acquiring knowledge about the world, understanding of new cultures, people and languages, and establishing trade routes.

World trade, defined as the exchange of goods and services, has a more profound and broader meaning than the vastly accepted definition. Throughout history, people used trade as a means to exchange goods between one another. This initial exchange developed a broad network of routes, on land and sea, and revolutionised the whole economic system of that period, thus providing goods and services that weren't accessible before. One of the more ground-breaking routes established, that introduced new perspectives and cooperation among nations, was the Silk Road. Primarily it represented the flow of silk, from China to the rest of the Asian and European countries. Even so, this route had a broader influence than just providing Eurasian people with precious material, it initiated the exploration of many possibilities and the exchange of knowledge on these continents. The trade boom, in Europe, contributed to a new and more profound root meaning of the word 'trade', and it is coined as 'a way of life'.[37]

37 | Trade. (n.d.) *Online Etymology Dictionary*. Retrieved from: https://www.etymonline.com/word/trade.

Nowadays trade is a common occurrence in every society.

One of the biggest, most strategically important revitalisations of one of the ancient trade routes, is the Belt and Road Initiative. This initiative represents new opportunities for both China and the other countries included.

The paper looks at the opportunities and challenges brought to young people in North Macedonia by the participation in the BRI as part of the 16+1, emphasising the economic objectives of this initiative as well as how it will contribute to the development of a small post-socialistic country.

4.1. 'BRI-nging' the Change

The Belt and Road Initiative (BRI), commonly known as China's New Silk Road (NSR) represents a core aspect of China's geopolitical and economic development blueprints. The initiative itself consists of two main components: The Silk Road Economic Belt and the Maritime Silk Road for the 21st century, with prospects for further developing the still unofficial Air Silk Road.[38]

The initiative itself provides China with 'soft power' and introduces novel non-western developmental strategies and policies on the market. In previous decades, since the introduction of new market reforms in 1978, China has propelled itself to becoming one of the leading economies in the modern world by providing exceptional growth. Maintaining a sustainable GDP growth of around 10%, Chinese policies and development managed to lift more than 800 million people out of poverty and reaching 2015 millennium goals (sustainable development targets) by significantly contributing to the level of their implementation globally.[39] While there have been benefits, the rapid economic growth has also brought many disadvantages including a broader polarisation gap, concentration of economic power and innovation on the east and south coast, excessive foreign currency reserves, labour migration, rapid urbanisation, growth of ghost towns, and an overheated economy.

Even though by strength the economy has made significant

38 | Han, B. (2015, October 13). China's Silk Road Takes to the Air. *The Diplomat*. Retrieved from https://thediplomat.com/2015/10/chinas-silk-road-takes-to-the-air/.

39 | World Bank. (2017). *The World Bank in China*. Retrieved on November 3, 2017: http://www.worldbank.org/en/country/china/overview.

progress, it is still considered as a developing economy, to be more precise as one of the leading developing nations with several focal points among which are: greener growth, more inclusive development, and establishing relationships and cooperation with other parts of the world.[40]

The Belt and Road Initiative incorporates all of the abovementioned goals, introducing the aspect of 'striving for achievement'[41] that provides an alternative to the globalised and westernised 'new world order', promising benefits to both sides. With around 70 countries included and 34 signed agreements, China is slowly but certainly setting the pace for international cooperation and improved trans-national relations with this initiative.[42]

North Macedonia was one of the countries included in the 16+1 strategy. A small economy with high growth potential, It can reap many benefits from this initiative. Chinese investment in the region has significantly grown. Investment in infrastructure in the region, as an initial part of the BRI, can have a huge positive impact on North Macedonian trade, thereby improving our balance of trade (BOT) and opening our market both ways.

Infrastructural innovation, as one of the main benefits from the BRI, would bring both hard infrastructure - including roads, railways, ports, pipelines, power grids, and communications infrastructure - as well as soft infrastructure, which is composed of finance networks and new banking solutions. Direct investments in that sphere can additionally contribute to the current value of our financial infrastructure and further explore the potential of our rudimentary finance market.

Meetings and cooperation between high government officials from China and North Macedonia[43] show an apparent interest on both sides regarding the infrastructure question. The initial investment of €574 million in road infrastructure is the first seed of prospective cooperation. The Chinese IE (Import-Export) bank credited North Macedonia in support for the construction of two important routes. The first route provides

40 | *Ibid.*

41 | Yan, X. (2014). From keeping a Low Profile to Striving for Achievement, *The Chinese Journal of International Politics*, Volume 7 Issue 2, June 2014, pp. 153-184.

42 | Ernst and Young, China. (2016). *China Go Abroad (4th Issue) Key Connectivity Improvements Along the Belt and Road in Telecommunications & Aviation Sectors.* Retrieved November 3, 2017 from http://www.chinagoabroad.com/en/guide/china-go-abroad-4th-issue-key-connectivity-improvements-along-the-belt-and-road-in-telecommunications-amp-aviation-sectors.

43 | Alsat M TV. (2015, November 24). *Груевски: Македонија е отворена за кинески инвестиции [Gruevski:Macedonia is ready for Chinese investment].* Retrieved from: https://www.alsat-m.tv/mk/223100/.

extended access to the pan-European corridor 10, connecting the municipality of Shtip and providing better connections between North Macedonia, Serbia (north) and Greece (to the south). The other is the highway between Ohrid and Kichevo, as part of the still-unfinished Corridor 8 that connects Albania and Bulgaria through North Macedonian territory.[44]

The prospects are good for future investment in transport infrastructure, especially for railway construction. Lacking a proper railway infrastructure debilitates the trading ability both of North Macedonia and neighbouring countries. Extended investment, in addition to the construction of a high-speed railway that connects Belgrade and Budapest, represents an amazing opportunity for the '21st century silk road' and the developing economies of the region. Thanks to its geographical position, North Macedonia represents a crucial link from this route, therefore providing a direct connection between Athens (as part of the Maritime Silk Road) and the rest of Europe. This infrastructural link (Budapest-Belgrade-Skopje-Athens) would provide North Macedonia with a direct connection both with Europe, on land, and with the countries that represent the Maritime road through the ports in Athens. This connection will help North Macedonia establish better connections and trade with both the European Market and China, therefore making our economy more open towards the rest of the world.

If we take under consideration the psychological factor that young people are more prone to take risks in starting businesses, as well as time element it can be said that the odds are in their favour and that they would be the ones to benefit the most from connecting the markets. In addition, this connection might result in providing new career choices for our youth in sectors that are not that popular nowadays but might develop. Opening new capacities would further provide fresh employment opportunities for the youth, which could result in reducing the high youth unemployment rate. The sectors that are expected to grow are transport, trade and industrial production since new markets will be more accessible.

Another bright spot is the North Macedonian government's

44 | BIRN (2013, October 01). Macedonia Taps Chinese loan for Motorways. *Balkan Insight.* Retrieved November 3, 2017 from http://www.balkaninsight. com/en/article/macedonia-taps-chinese-road-loan.

policies for supporting small and medium-sized enterprises (SMEs) and Chinese investments in this sector. As a global startup and innovation hub, China can make significant contributions to our start-up culture and climate. The cooperation and investments would open up a space for new ways of thinking that will lead to the creation of new products and stimulate demand for something novel which would result in improved innovation and establishment of more challenging and competitive companies.

State subventions, especially in the field of IT infrastructure and innovation, would also play a big role in conquering the challenges that might contribute to the self-employment and employment of many young people. This will directly influence the high youth unemployment rate and help North Macedonia tackle the brain drain problem. Furthermore, the connectivity and the improved communications network will give young people access to previously unexplored aspects of everyday life, which will contribute to the overall wellbeing of North Macedonia and provide experiential learning and job prospects in unexpected places.

As a leading power in the sustainable energy sector, China can also contribute to the transition of the North Macedonian energy sector. Through investments in the area of renewable energy sources, North Macedonia can fare better in the carbon and fossil-fuels war that debilitates our ecosystem and health. Since we have many students majoring in sustainable energy, the opportunities of the BRI mean many of them can gain experience and advance their careers in this sector.

4.2. Education and the North Macedonian Youth

One of the main challenges for North Macedonian youth is gaining quality education and developing hard skills. Education and vocational training in China can help North Macedonian students and graduates gain more practical experience and implement a more breakthrough-oriented attitude in their work. This will result in more shared experiences, gained skills and applied know-how which will directly open new ways for further collaboration, thanks to the relationships established by the students. Regarding this, the Chinese government has started several projects in North Macedonia, which include Chinese language courses and scholarships for gifted students at prestigious universities in China.[45]

Cooperation of this kind would support the second pillar of the development strategy of the government: having better skills and more inclusion in the labour market in order to further decrease the unemployment rate of 22.4% (in 2017).[46] By opening the cooperation routes first in the education and training sectors, many North Macedonians will have better opportunities of being accepted into the labour market and also be more competitive abroad. Through investments in the corporate sector, new jobs would become available, and the openness of the economy, in general, would provide more opportunities and greater demand for a qualified workforce.

With all previously said, the benefits North Macedonian youth might get are more than sufficient. However, in the real world, the knife almost never has one sharp side. Even though the BRI at its core could be a positive impact on the medium and long run for North Macedonia's economy, in the short run some challenges and threats might emerge. They are mostly with regards to the capacities used in the construction phase. A question of utmost importance is posed here, about whether the source of growth would be home firms or foreign firms. If the construction of the numerous infrastructure objects is done by engaging North Macedonian firms and thus workers, this will help to decrease the unemployment rate and will open up new job posts for people with the right profiles.

Another tipping point crucial for a small economy is leveraging investments. Taking the GDP of North Macedonia into consideration, as well as the current state and number of big firms in the market, it would be quite easy for a foreign investor to monopolise certain sectors of the economy. There may also be some problems with certain regulations and taxes. However, with good regulation, proper legislation, and leveraging, this threat can be eliminated.

45 | Development Solutions. (2012, April 4). Конкурс за доделување на стипендии на Владата на Народна Република Кина за додипломски(редовни) студии, магистерски и специјализации. [Concourse for scolarships from the government of Republic China for bachelor, masters degrees and specialisations]. Retrieved from http://www.deso.mk/mk/Item/606.

46 | The World Bank. (n.d.). Unemployment, total (% of total labor force) (national estimate) - North Macedonia. Retrieved from https://data.worldbank.org/indicator/SL.UEM.TOTL.NE.ZS?locations=MK.

In sum, being part of the BRI opens up many possibilities for North Macedonian youth, now and in the future. It represents one of the biggest and most well thought out plans that can create an era of mutual cooperation, with stimulated growth and expanded horizons on both sides. This initiative can make Eurasia great again by contributing to the direct reduction of inequalities among certain nations, providing access to multiple markets and creating new capacities in the economy of all included countries, which can create a ripple effect, boosting the Eurasian economy as a whole.

The outcome from investments in transport infrastructure, energy sources, overall improved communication channels, innovation, education, cultural development, and breaking the invisible barriers that stand between countries can contribute to the mutual creation of a better tomorrow, a world where inequalities would be reduced, countries developed, challenges overcome, and future generations satisfied.

5. What Could the Belt and Road Initiative Focus on in the Future?

5.1. Alternative Sources of Energy

Innovative green technologies are representing a new era of the industrial revolution which introduces better care for the environment while maintaining growth. Globally, many countries are implementing measures and developing technologies that are both environmentally sustainable and act as catalysts for economic growth, thus opening new green job posts along the time axis. Some of the main areas open for investment in North Macedonia are renewable energies and green construction. Energy is the area that has an excellent long-term green potential that could create additional jobs both directly and indirectly. Some of the sectors where foreign direct investment (FDI) can create positive effects are solar energy, wind energy, and the construction industry.

Having around 300 sunny days a year, North Macedonia is a geographical heaven for photovoltaic cells. Improvements can be done not only in the creation of solar farms but also in the integration of new technologies in already established locations, creating self-sustainable objects, especially in the southern regions of the country. New capacities create a need for certain skill requirements and provide specific occupational profiles for a suitable job. China as a global leader in the area[47] can provide education and skills development training to equip prospective workers with suitable qualifications.

The government of North Macedonia has pledged to reform energy sector legislation to make it compatible with EU standards. Bringing it in line with the demands of the European Energy Community would mean that any energy or energy technology produced in North Macedonia would be acceptable for EU markets.

Wind power is another renewable energy that started to emerge as a field for investments in our country. Although North Macedonia has the velocity of winds needed to generate enough energy, conventional wind turbines are not the most suitable direct investment because they have negative effects on the flora since the areas where the wind farms are installed cannot be used for agriculture. However, there is huge potential in vertical axis wind turbines, which create less noise and can be installed on smaller surfaces. Creating manufacturing capacity for the making of wind turbines or their parts is also an option since North Macedonia has a favourable investment climate with incentives for foreign investors in the industrial zones. As a country with many rivers and large hydro energy potential, innovations in this field can also help for the future development of new job posts.

Another booming area in North Macedonia is the construction

47 | The Climate Reality Project. (2016, February 03). *Follow the Leader: How 11 Countries are Shifting to Renewable Energy.* Retrieved from: https://www. climaterealityproject.org/blog/ follow-leader-how-11-countries-are-shifting-renewable-energy.

industry. Buildings are popping up like mushrooms in the main urban centres, creating a loud call for alternatives to the conventional construction methods in our country. Green buildings are another area where cooperation can have more than a beneficial effect. This also implies a need for new technologies and the creation of so-called smart buildings, both in residential and corporate environments.

5.2. Investing in the Youth

These types of investment would create opportunities regarding the employment of young people in our country, whose current state is characterised by a high percentage of unemployment. This problem can be tackled from two sides thanks to the BRI. The first side would be the investment side, which would enable the creation of certain capacities and help in resources supply needed for the proper functioning of the capacities (stocks, materials, machinery etc.). On the other hand, by providing educational opportunities, job shadowing, training for qualification or requalification of employees, the current qualifications structure in North Macedonia can be vastly improved. Also, an interesting perspective would be the implementation of internships abroad as well as introducing CEEDerships (Cultural Envoy for educational development) which are a form of traineeships for skills in the workplace. This is done by a CEEDer, an expert in the field from abroad, who comes to share the knowledge and expertise in the certain area by working every day with the employees and providing them hands-on training to conduct the job better in order to overcome bottlenecks and ensure better productivity.

This would not only create manufacturing and maintenance jobs for the youth, but also supplement administration, consulting, and innovation capacities that can further help propel the local and regional economies. Some of the other areas besides energy that have excellent long-term green potential[48] are recycling, transportation, agriculture and forestry.

48 | Worldwatch Institute. (2008). *Green Jobs: Towards Decent Work in a Sustainable, Low-Carbon World.* Worldwatch Institute with technical assistance by the Cornell University Labor Institute, 301.

5.3. Smart Cities

Another aspect which could be introduced to the BRI project is investing in ideas that contribute to the concept of 'smart cities'. The idea is that urban areas haven't fully explored the potential of the IT industry and digitalisation, for the purpose of improving the standard of living of citizens. By introducing new services in cities and digitalising already existing ones, North Macedonia would be creating 'cities of the future'.

North Macedonia has a large number of IT graduates each year. By fostering their entrepreneurial spirit and investing in their ideas for digitalisation, innovative solutions for existing problems could be realised. These could be in the aforementioned areas, such as green energies and recycling, or in areas such as transport and infrastructure. Similar to the experience of the Baltic states in the early 2000s, North Macedonia has the educated workforce and can be an excellent 'innovation hub' for new IT solutions. Furthermore, the ideas that are developed in this area, with Chinese investment, could be exported and implemented elsewhere.

The government of North Macedonia has a Ministry of Public Administration and Information Society, with a specific role of overseeing digitalisation processes. In cooperation with them and the local authorities, significant progress can be made in digitalising many aspects of urban life. An investment in 'smart cities' would also be beneficial for Chinese tech companies such as Huawei, which have already shown interest in expanding their market share in the Balkans, including North Macedonia.

Investment streams of this sort, in general, would result in benefits on both sides, but taking the size of the country and its needs into consideration, North Macedonia will get more of the benefits. Investments of this type would open up more opportunities for the youth, create more job openings, expand the horizons for innovation, and bring an air of devotion to results and striving for achievement in our populace.

This would help improve the overall economy of the country,

thus providing better services and better incentives for the employed and assist the unemployed to get employed. Last but not least, a strengthened economy in North Macedonia would mean greater purchasing power. With China as one of the main suppliers of goods in the world, it would have a positive effect on the overall trade balance.

Bay of Kotor in Montenegro (photo provided by the author).

by SLOBODAN FRANETA
University of Donja Gorica

LUKA LAKOVIĆ
PhD Student at the University of Donja
Gorica (Faculty of Information Systems
and Technologies)

RENATA
LJULJDJUROVIC
University of Donja Gorica

IVAN PIPER
PhD Student at the University of
Donja Gorica (Faculty of International
Economics, Finance and Business)

(Ordered by last name)

CHAPTER 12

MONTENEGRO

1. The Political Structure of Montenegro

Montenegro is a small country located along the coast of the Adriatic Sea on the Balkan Peninsula with a landmass of 13,812 square km and a population of 680,000. It is surrounded by Croatia, Bosnia & Herzegovina, Serbia, Kosovo, Albania and Italy across the sea. Montenegro is a member of NATO and a candidate to become an official member of the European Union. Montenegro is widely known not only for its glorious and very dynamic history, but also for being an attractive destination for tourists and investors because of its geopolitical significance.

This part of the paper aims to present the most important events which have defined the very dynamic and evolutionary history of Montenegro: starting from the wars and the different polities that governed the country, to the renewal of its independence and therefore the creation of the modern country as it is today. Furthermore, it will present the formation of rules and powers within the current political system and the ideology upheld by the government.

1.1. A Brief History of Montenegro

Previously known as Duklja, and then Zeta, Montenegro got its current name in the 14th century.[1] It is distinguished by its dynamic and turbulent history, which has been filled with wars against invaders, fights for independence as well as multiple regime changes. During the several-centuries-long Ottoman Empire occupation of the Balkans, Montenegro remained the only unconquered country and also the most rebellious one because it was the main centre of resistance against the expansion of the Ottoman Empire.[2] In the first half of the 19th century under the reign of Petar II Petrović Njegoš and subsequently Prince Danilo Petrović Njegoš, Montenegro began numerous reforms which had led to the modernisation of the country. It was the period when some of the biggest fights against the Ottoman Empire occurred. At the time, Montenegro became a Principality and Prince Danilo separated the church from the state, which marked Montenegro's transition from a theocracy to a secular state.

1 | Duklja, Montenegro's predecessor, was formed as a country in the early Middle Ages when Slavs (ancestors of Montenegrins) inhabited it.

2 | Montenegro is the only Balkan country that rejected the supremacy of the Ottoman Empire. Because of its frequent calls for uprising and resistance, Montenegro developed a unique form of autonomy within the Ottoman Empire.

Furthermore, Prince Danilo forbade the vendetta and other uncivilised customs during that period. In 1855, he developed the General Law of the Land (Danilo's Code), which significantly contributed to the further modernisation of Montenegro. In 1876, under the reign of Prince Nikola Petrović, Montenegro declared war against the Ottoman Empire, which signified the beginning of Montenegro's most intense fights for freedom and independence. After several major victories in the battles against the Ottoman Empire, e.g. the Battle of Vučji Do and Fundina, Montenegro's independence was finally recognised by the Great Powers in the Treaty of Berlin (1878)[3]. Besides, its territory became two times larger than the previous one. In the following years, Montenegro undertook numerous reforms and development plans.

For example, the number of schools notably increased, and several factories were built (including a brewery in Nikšić) in order to develop the industry and improve the country's economic situation. In 1910, after roughly 50 years of Prince Nikola's reign, Montenegro formed a monarchy and Prince Nikola became the first and only king. From 1912 to 1913, Montenegro fought in the First Balkan War alongside Serbia, Greece and Bulgaria against the Ottoman Empire. They defeated the Ottoman Empire, which was forced to leave the Balkan Peninsula. As a result, the Balkan Peninsula was finally liberated from Ottoman occupation. Montenegro managed to expand its territory, but under the pressure from great powers, it was forced to give up Skadar and leave it to the newly founded Albania.

Shortly after, the Second Balkan War started. This time, Montenegro fought alongside Serbia, Greece and Romania against Bulgaria. The cause of the war was a dispute over the division of Macedonia's territories liberated after the First Balkan War. After the two Balkan wars, Montenegro expanded its territory to a greater extent, despite being forced to surrender Skadar to Albania. Nevertheless, the Second Balkan War indirectly triggered the beginning of World War I.

In World War I, Montenegro fought against the Central Powers.

3 | Technically, Montenegro was an independent country even before the Treaty of Berlin. Unlike its neighbours, Montenegro rejected the supremacy of Ottoman Empire and developed a unique form of autonomy within the Empire. Its independence was informally recognised by every great power except the Ottoman Empire and the United Kingdom up until the Treaty of Berlin, when its independence was officially recognised by the two countries.

Once again, Montenegro proved itself as a strong resistance centre, which this time was against the Austro-Hungarian occupation. However, Montenegro gave a helping hand to its neighbours in the fight against the invaders and suffered from heavy losses in the process. For example, Montenegro's covering of Serbia's retreat towards Albania had ultimately led to the salvation of the Serbian Army from the Austro-Hungarian aggressors and the surrender of Montenegro. After the war, the irregular Podgorica Assembly decided to merge Montenegro with Serbia and create the Kingdom of Serbs, Croats and Slovenes[4]. Subsequently, the Kingdom of Montenegro lost independence and *de facto* ceased to exist as a sovereign country[5].

In World War II, Montenegro, as part of the Kingdom of Yugoslavia, became one of the main strongholds against the fascist regime in Europe. Montenegro was the initiator of the first and biggest people's uprising in occupied Europe on July 13th, 1941. After World War II, Montenegro joined the Federal People's Republic of Yugoslavia (which would later become the Socialist Federal Republic of Yugoslavia) as one of the six republics, with its state and national identity being fully recognised by the other republics.

After the dissolution, in 1992, of the Socialist Federal Republic of Yugoslavia, (which had consisted of Montenegro, Serbia, Slovenia, Croatia, Bosnia & Herzegovina and Macedonia), Montenegro remained part of the Federal Republic of Yugoslavia along with Serbia, which in 2003 became Serbia and Montenegro, a more decentralised state union. At the time, the Federal Republic of Yugoslavia had to cope with many problems, including economic sanctions, hyperinflation and NATO bombardment, albeit Montenegro suffered less than Serbia. That was the period when Montenegro began the process of separation from Serbia and further reaffirmation of its independence, followed by a strong dedication to democratic reform, friendly diplomacy with neighbours, and European and Euro-Atlantic integration.

Finally, Montenegro declared independence after a referendum

4 | Decisions made at Podgorica Assembly were legally unfounded, mostly because the members of the Assembly who made those decisions were illegally chosen.

5 | Andrijašević, Ž. (2016). *The History of Montenegro: From Ancient Times to 2006*. Beograd: Vukotić Media, Atlas Foundation.

was held on May 21st, 2016 with an absolute majority of citizens voting for separating from Serbia.

As a sovereign and once again an independent country, Montenegro promptly became a member of the United Nations and Council of Europe with international recognition. In 2017, Montenegro became a member of NATO, and it is expected to become a member of the EU by 2025.

1.2. Montenegro's Constitutional and Political Ideology

Montenegro is governed by a parliamentary system in which the country is represented by the president of Montenegro and is operationally run by the Government. The political authority of the country is divided into three branches of power, namely the legislative, executive and judicial powers. The legislative power is exercised by the parliament, the executive power by the Government, and the judicial power by the courts.[6] Their powers are limited by the Constitution and relevant laws.

There are three different political blocks in Montenegro, namely pro-Montenegrin, pro-Serbian, and minority parties. The ruling political party for the last three decades is the pro-Montenegrin Democratic Party of Socialists of Montenegro (DPS)[7].

According to the Constitution of Montenegro, the economic system of the country shall be based on liberal values such as a 'free and open market, freedom of entrepreneurship and competition, independence of economic entities and their responsibility for the obligations accepted in the legal undertakings, and protection and equality of all forms of property'[8]. It is a libertarian economic model that was introduced to Montenegro at the end of the 20th century and has been actively implemented since its independence in 2006.

After the breakup of Yugoslavia in the second half of the 20th century, the Montenegrin government started a gradual transition from socialism to economic liberalism and capitalism.

6 | WIPO. (n.d.). *Constitution of Montenegro*. Retrieved from https://www.wipo.int/edocs/lexdocs/laws/en/me/me004en.pdf.

7 | DPS is the ruling party since the introduction of a multi-party system in 1990.

8 | WIPO. (n.d.). *Constitution of Montenegro*. Retrieved from https://www.wipo.int/edocs/lexdocs/laws/en/me/me004en.pdf.

Liberal currents emerged which advocated a free and open market, privatisation, free trade, and development of entrepreneurship. After independence, there were certain tendencies to reform the economic system of Montenegro by the Austrian School of Economics' rules and principles with a goal to eliminate bureaucracy within the government and encourage entrepreneurship[9].

The idea of such an economic model, which is authored by Professor Veselin Vukotić, was to create a microstate based on the presidential system instead of the parliamentary system[10]. However, this reform never happened because it did not receive a positive reception from the majority of the population since the people were still committed to socialism and preferred a more paternalistic rule. Because of that, the actual economic policy of Montenegro is a mixture of liberal and socialist ideals which is significantly different from the initially-planned reforms. According to the World Bank, in the 15 years up to 2018 the GDP of Montenegro increased nearly four times (to €3.5 billion)[11]. Simultaneously, the public spending and budget deficit also increased significantly mainly due to the country's wobbly and incomplete transition to economic liberalism and capitalism.

One of the main obstacles for a smooth transition is the high level of government administration, bureaucracy and intervention which is precluding a boom in entrepreneurship and economic freedom. As a matter of fact, Montenegro is currently ranked 83 in the Index of Economic Freedom[12].

Since Montenegro is a small country, its future heavily relies on a strong and efficient small government and its active connection with the external global market. As a result, remaining open and liberal in both its economic and political policies is the key to developing the country in the long run.

9 | The Austrian School of economics advocates limited government intervention and reduction of its paternalistic role to protect an individual's rights and private properties (the so-called minimal state). It strongly opposes any kind of state intervention and promotes a *laissez-faire* policy that encourages free trade, globalisation and free market based on Adam Smith's principle of the invisible hand.

10 | Vukotić, V. (2007). *The Unaccepted Agenda of Economic Reform and Development of Montenegro 2007-2011*. Podgorica: ISSP.

11 | World Bank. (2018, January 19). *GDP of Montenegro*. Retrieved from https://data.worldbank.org/indicator/NY.GDP.MKTP.CD?locations=ME.

12 | The Heritage Foundation. (2018, January 18). *Index of Economic Freedom*. Retrieved from https://www.heritage.org/index/ranking.

2. Political Direction of Montenegro[13]

The prediction of the future political trends and directions of the country is pertinent to a deeper historical analysis of the role Montenegro has played in both global and regional contexts. The responses from society towards the past challenges and the constant shifts in the country's governance philosophies have put Montenegro in some turbulent incidents throughout its long history. Those past events not only could point us to the potential answers regarding the future of Montenegro in both political and economic terms, they could also provide us with a deeper insight into understanding the Montenegrin culture and society in a way that can serve as a good proxy to predict the future political direction of the country.

From the beginning to modern times, Montenegro has gone through various political, governance and economic development stages that have shaped the government's ways of thinking and how it devises both domestic and foreign policy reforms. These stages happened because of several important characteristics that have defined today's Montenegro. Firstly, Montenegro is located along the coast of the Mediterranean Sea which brings geostrategic significance to the country in terms of easy and open access to ports and resources. The country has also been dependent on foreign donations and aid because of its lack of comparative advantages in trade and subsequent underdeveloped economy.

Secondly, the constant wars and turbulent history of the Balkans contribute to the country's slow progress and limited prosperity. Since it is located in the heart of the Balkans region, the impact of wars is inevitable. The repercussions of bad political moves and decisions made by the Great Powers throughout history have cost the country's sovereignty and independence for a long time. Many times Montenegro was dragged into regional integration projects that mostly did not recognise Montenegro as an independent and sovereign nation. The most-discussed example is when Montenegro was violently attached to the new-born Kingdom of Serbs, Croats and Slovenes in 1918. Given the strong political power held by the Kingdom of Serbs after the end of World War I, its war

13 | This part takes reference to Andrijašević, Ž. (2000). *Short History of Montenegro from 1496-1918*, Contecto, Bar; Andrijašević, Ž. M. (2004). *A Nation with a Fault:(Historical Essays)*, National Library of Montenegro *Đurđe Crnojević*, Cetinje; Andrijašević, Ž. (2016). *The History of Montenegro: From Ancient Times to 2006*, Vukotić Media, Atlas Foundation, Beograd.

allies rejected Montenegro's independence and sovereignty and supported the integration of Montenegro with the Kingdom. Nikola Pašić, the Prime Minister of the Serbian Government at that time, ordered the creation of a special military force in September 1918 in Montenegro to prepare for a potential coup d'état. Its main mission was to overthrow the Petrović-Njegoš Dynasty and annex Montenegro to become part of the Serbian Kingdom.

The turbulent history and the challenges encountered in the past have defined three decisive elements that contribute to the creation of Montenegro's political culture. The three elements are:

1. **Pride in the nation**: the long history of wars and battles has stimulated Montenegrins' pride and courage to defend the country's territory and freedom from outside threats.
2. **Desire for a charismatic leader**: the history and past sufferings have made the citizens demand a charismatic leader who is wise and capable of leading the Montenegrins.
3. **Need for political diversity**: diversity in political opinions and beliefs among Montenegrins is common. Understanding their differences is important in understanding Montenegro's political scene and society because it is common to see various interest groups who actively defend their interests or the group of power that they belong.

These three elements are reflected in both the past and present times of Montenegro, especially when it comes to politics and political orientation of the people. Today, Montenegrins' pride in a reborn nation, their strong desire to have a charismatic and even paternalistic leader who is people-oriented, and the need for division among different national and political orientations have created a unique political scene where people are recognised as a member of a specific group rather than a free-will individual.

The political culture forces citizens to identify themselves as a member of a group and to adopt ideas of that group blindly. However, the existence of such a culture is inhibiting a society from moving forward and adopting values that would make both the country and the individuals feel freer and more prosperous. For instance, such political culture as exhibited by the three decisive elements has created an environment where dependence on the state and its bureaucracy is highly preferred. In such a society, the state acts as both a 'nanny' and a strong 'superhero.' As a result, it is unsurprising to see limited innovation inside the government. To many of the people, keeping the status quo is an undisputable ideal.

2.1. Lessons from History

In modern world history, there was no parallel case like Montenegro where a small country could maintain good relationships with powerful nations and Empires despite it was appeared in the interests and agendas of those powerhouses[14]. In 1878, the great powers at the time recognised Montenegro as an independent and sovereign state in Europe. This status, along with the internal problems with Serbia, opened an opportunity for Montenegro to become a closer ally with many powerful nations and empires such as the Austro-Hungarian, Ottoman and Russian Empires. Its traditional partner for a long time was the Russian Empire. Through the mutual agreement between the Russian Empire and Montenegro, it was concluded that the Russian Empire would help Montenegro financially in exchange for the latter's military support and loyalty in general. Very poor living conditions and the lack of resources for economic development made Montenegro dependent upon the assistance of the Russian Empire and the general Eastern influence.

Diplomatically, Montenegro maintained good relationships with the great powers where it sent its first ambassadors to those countries. For example, Montenegrin ambassadors were sent to France, the Ottoman Empire, the Russian Empire, the Austro-Hungarian Empire, etc. to negotiate the best conditions to secure financial assistance from the great powers. Securing

14 | Andrijašević, Ž. (2016). *The History of Montenegro: From Ancient Times to 2006*. Beograd: Vukotić Media, Atlas Foundation.

financial assistance was always the focus of Montenegro's diplomatic mission because of its resource scarcity and the lack of innovative ideas on how to improve the national economy.

The beginning of the 20th century brought to Montenegro many new problems and conflicts that threatened to weaken and destabilise the country in many ways. At the time King Nicholas was the absolute ruler of Montenegro whose power was unlimited[15]. As the only absolute ruler in Europe, King Nicholas decided to present the first constitution to the people of Montenegro in 1905 which allowed them to choose the representatives who would lead the country with the King. It was the moment when parliamentarianism was introduced.

In the initial phase of parliamentarianism, the King had power over the parliament and closely monitored its activities. In this sense, Montenegro became a hereditary Constitutional Monarchy with national representatives chosen through parliamentarianism. However, such political arrangements meant that the King could abuse the power and replace the ministers who did not share his same political attitudes and orientations. This was the reason for many citizens joining opposition parties to object to the abuse of power by the pro-establishment party. Subsequently, the tension between the two ends of the political spectrum stirred up an intense political battle in which the King and the pro-establishment party exercised all their political power to secure a majority in the parliament until the 1914 election.

In 1914, more oppositional voices were found in the parliament and the political contest between the pro-establishment and anti-establishment parties grew stronger. Political pluralism was increasingly celebrated by the people alongside the growth of media channels which allowed more political voices to be heard in civil society.

For example, the monarchy-owned newspapers had to face growing competition with independent newspapers to impress the readers and increase the readership. Independent newspapers were particularly attractive to the Montenegrins

15 | Andrijašević, Ž. (2016). *The History of Montenegro: From Ancient Times to 2006*. Beograd: Vukotić Media, Atlas Foundation.

at that time because they focused on reporting on domestic politics critically and encouraged people to discuss the potential political direction of Montenegro. With more media freedom and consequently more political debates among the members in civil society, Montenegro was gradually transiting from monarchism to parliamentarianism and democracy.

This brief heyday ended in 1918 after the Balkans Wars and the two World Wars. Montenegro was occupied by the Austro-Hungarian Empire, and the King and his ministers were exiled in France. After the Austro-Hungarian Empire left Montenegro, it opened a power vacuum which Serbia filled by moving its army and pushing the idea of a united Kingdom among Serbs, Croats and Slovenes where Montenegro would be a part of the Kingdom with no sovereignty and independence. The idea divided the citizens in Montenegro and the impact of such division is still seen even after a century.

The Podgorica Assembly brought a political question which divided Montenegro, namely whether or not Montenegro should become a part of the Serbian Kingdom or remain independent. On one hand, there was a popular party called Whites which proposed the idea that Montenegro should become a part of the Kingdom. On the other hand, a party called Greens opposed the idea by gathering those who opposed such ideas and who supported Montenegro's right to remain as an independent and free country.[16] After many rounds of heated debate, the Assembly announced that Montenegro would unite with Serbia and join the Kingdom of Serbs, Croats, and Slovenes. The political development of Montenegro then retreated from parliamentarianism to monarchy.

World War II changed the political direction of not only Montenegro but of the whole Balkans region as well. At the beginning of the War, the King and the dynasty were overthrown mostly because of the decision to sign documents with Adolf Hitler and Germany which guaranteed the Kingdom's neutrality and that the Kingdom of Serbs, Croats, and Slovenes would not join any side but only serve as a transit

16 | The name Whites and Greens came from the colour of papers that were used for voting on that day. The white papers meant for the unification with Serbia, while the green papers meant against it and for free and independent Montenegro.

zone. That was a motive for the Communist Party organising protests across the country to advocate for the people not becoming the slaves of Germany. However, the protests resulted in the bombing of Belgrade by German forces in 1941 and then the occupation of Montenegro and its territories by Italian forces.[17] As Živko Andrijašević noted in his book, the defeat of Yugoslavia in World War II was inevitable. King Petar Karađorđević and his government left the country and the Yugoslav army surrendered after a few days.

After that, the political ideology of Montenegro was communism. Following World War II, the Communist Party of Yugoslavia presented the federative system under which Yugoslavia consisted of six countries and nationalities including Serbia, Croatia, Bosnia & Herzegovina, Slovenia, Macedonia, and Montenegro. After almost three decades, Montenegro gained independence again and enjoyed rights as other sovereign countries did.

As part of the Socialist Federal Republic of Yugoslavia (SFRY), Montenegro followed the political direction of the republic in both domestic and foreign affairs. SFRY was a founding member of the Non-Aligned Movement, which remained neutral throughout the Cold War.[18] The Movement allowed the region to enjoy a peaceful period after all the turbulent and violent times.

17 | In March 1941 the Kingdom of Yugoslavia (Serbs, Croats and Slovenes) signed an agreement with Hitler's Germany where they would accept to be a transit zone for German and Italian troops. But a few days later, members of the Communist Group started a coup d'état after which they cancelled the agreement with Germany. This led to the bombing of Belgrade in April 1941 when the Yugoslav army surrendered.

18 | Nuclear Threat Initiative. (2018, May 31). *Non-Aligned Movement (NAM)*. Retrieved from https://www.nti.org/learn/treaties-and-regimes/non-aligned-movement-nam/.

In his book *The History of Montenegro*, Živko Andrijašević pointed out that the post-war period was a golden era for Montenegro mostly because of the visible changes in the mindset of the people. According to him, the highest level of development and living standards were achieved in that era. However, it was not possible to argue that such results were solely dependent on the leadership of the Communist Party of Yugoslavia because it was also the time when post-war industrialisation and global economic expansion began, and globalisation was achieved in an unprecedented way. Nonetheless, the so-called golden era of Yugoslavia did not last more than four decades. After the death of Josip Broz Tito in 1980, the country encountered several rounds of unexpected and unpleasant political and economic reforms.

In 1989, under the leadership of Ante Marković, Yugoslavia attempted to become one of the most progressive and prosperous countries in the world by following the principle of neoliberalism. However, most of the citizens and politicians in the country rejected privatisation and capitalism and put the constituent republics into a fatal war that negatively changed the political direction of several republics.

From 1991 to 1995, the new war divided the country into several parts. Slovenia was the only country that left the federation relatively peacefully, while the others ended up in bloodshed. The most intensive fight during the war was among Serbia, Croatia and Bosnia & Herzegovina when thousands of innocent civilians were executed, and thousands of families were expelled from their homes. Montenegro's role in this war was mostly political and rhetorical, except one part when Montenegrin reservists led by the Yugoslav National Army attacked Dubrovnik in Croatia in 1991.

Montenegro served as a secure zone for many refugees who sought asylum from the war zones. Such inclusiveness and a tendency to celebrate multiculturalism and diversity have helped Montenegro create a strong foundation for its transition to a pro-democracy, pro-EU, and pro-NATO independent country.

2.2. Modern Politics in Montenegro

In 2006, Montenegro gained its independence from the alliance with Serbia. The new political ideology of Montenegro was then based on democratic liberalism and respect towards the Constitution. The Constitution of Montenegro is a progressive document that declares Montenegro as a country that is founded on fundamental values such as rule of law and democracy. The Constitution is not only novel for the people who had lived under a long period of communist rule, it was also pioneering since it explicitly protected the rights and free choices of individuals.

The stipulated principles and values have opened the doors for Montenegro's economic and political development. Since then, the economy of Montenegro has been open and competitive in the global market as supported by the World Bank's official data. From 2004 to 2008, the GDP of Montenegro doubled from US$2 billion to US$4.5 billion.

Being a free market economy (as reflected in the increasing level of free movement of capital and goods, low corporate taxes, and high economic and political freedom) has attracted a flow of foreign investments to Montenegro that brought significant improvements in various industries such as tourism. Montenegro is becoming one of the most popular travel destinations in the Mediterranean Sea region with the help of a low-tax policy in the tourism industry. The policy attracts foreign direct investments (FDIs) and encourages local entrepreneurial initiatives that support the country's economic development.

The parliament is composed of pro-Montenegro, pro-Serbia and minority parties. There is a conflict between the pro-Montenegro and pro-Serbia parties, and such conflict has slowed Montenegro's political transition and economic development in general. What is interesting about the pro-Serbia party is that it never receives any official support and recognition from the Serbian government. Yet, its mission statement and political actions are mostly radical to voice out their historical connections with Serbia.

In the parliament, the pro-Montenegro and minority parties have enjoyed majorities that helped facilitate Montenegro's economic development in accordance with the principle of neoliberalism. However, the lack of opposition and dialogue because of the parties' wish to secure social peace and majority in the parliament, usually results in short-sighted and single-minded economic policies that do not benefit Montenegro's growth in the long run. Conventionally, the responsibility of pushing Montenegro's economic development is usually fulfilled by private enterprises and entrepreneurs.

Another problem in politics is the perception that maintaining a large public administration and bureaucracy is a necessary condition to secure social peace. The actual number of employees working for the public administration is still a mystery, even though some estimations are that almost one-sixth of the Montenegrin population is working or is related to the public administration and receive a salary from the state.

Since there is a high level of dependency on the government and the state by the people, it is unsurprising that all the political parties in the parliament are declared as socialists or leftists. In this case, what is Montenegro's probable future political direction?

2.3. Future Political Direction

Because of its turbulent history, Montenegro has experienced several important changes in its political and economic ideology. While Montenegro was once dependent on foreign donations and aid given by the Empires, the country has become a successful player in the global market economy through constantly developing industries where it possesses comparative advantages.

There are two probable future political directions that Montenegro may follow. The first one could be that the country would orient towards a paternalistic governance philosophy by which the government and its branches are the solutions to all the problems in Montenegro. This is not a surprising direction. Bearing in mind those three elements mentioned above that could determine the political approach of Montenegro - pride in the nation, desire for a charismatic leader, and need for political diversity - it is evident that the state and the political parties will still have a strong influence on the citizens in the future.

The division between pro-Montenegro and pro-Serbia followers will sustain the status quo and people will need to declare their own political stance in order to enjoy the opportunities offered by each side. If academics, politicians, media, and citizens decide to follow this path, then the economy, political life and social organisation will probably face a crisis that could take Montenegro backwards. This is neither a path that modern Montenegrins should follow nor the right future for thousands of young Montenegrins who will be investing their time and efforts in personal development to survive in the intense global market.

On the other hand, the second path could see them shaking off the shackles of the prevalent paternalistic and traditional values that have dominated the mindset of the people and prevented the country from being innovative and competitive. It is the way to get rid of tribal politics and encourage people to start thinking about individualism and self-interests. The question is not how *WE* can contribute to the changes in society but how *I* can? This path will not be widely accepted in Montenegro in the first place because it demands a minimal presence of the state in many issues. It is the path in which each citizen's individual liberty is respected regardless of his/her personal background as long as they follow the laws.

In this scenario, private enterprise and entrepreneurship, and the growth of capital should be given the highest priority in order to create an environment for progress and prosperity. This approach is advocated by the minority who believe that Montenegro could potentially be a microstate that is able to generate enormous economic power. The microstate in that concept is not correlated to the geographical size of the country but to the fact that public administration and governmental influence should be minimal, and that private enterprise and entrepreneurship should be the generators of economic power and future political direction of Montenegro.

Montenegro is standing at the political crossroads where there are two main options: either continuing with the paternalistic practice that is pulling the country back or switching to a freer and prosperous path that rejects all paternalistic and traditional values that have prevented the country from achieving its potential. The decision requires smart and brave politicians who will go beyond the three abovementioned elements that have defined Montenegro's political approach. Only then could Montenegrin society enjoy all the benefits brought by full-fledged development and growth.

Finally, Montenegro's turbulent history should remind the people of all the unwise decisions that pulled the country away from the path of progress and prosperity. The history is there to teach the people many valuable lessons that could prevent them from committing the same mistakes that their ancestors made in the past.

3. Opportunities and Challenges Faced by Young People in Montenegro

3.1. Human Capital in Montenegro

Human capital is one of the most important resources for a country. It contributes to productivity and prosperity which are important for the overall status of the country. There is a need to invest in people through increasing their motivation and thus their efficiency and productivity. It is also desirable to promote specialisation of people, which can improve the quality of work.

By the middle of 2016, there were 622,303 inhabitants in Montenegro. The labour force participation rate in the second quarter of 2017 was 54.8%, the employment rate 46.5%, the unemployment rate 15.1% and the inactivity rate 45.2%. Regarding the total number of the active population, 55.6% were men and 44.4% were women, while among the total number of inactive people 59.4% were women and 40.6% were men. Regarding the total number of employees, 55.9% were men and 44.1% were women. In terms of the total number of unemployed persons, 54.0% were men and 46.0%

were women[19].

3.1.1. The ethnic structure of Montenegro is diverse

Based on the results obtained by the 2011 Population Census, it was concluded that there is a diverse ethnic structure in Montenegro. Montenegro is a multicultural and multi-ethnic state. The ethnic structure of Montenegro is shown in the following table.

Montenegro is constitutionally defined as a civil state in order to overcome all ethnic differences. Diversity is good in every aspect of life. Considering that the world's population is 7.5 billion, people are 'condemned' to live with each other. In one territory, members clearly distinguish themselves from other people according to certain characteristics based on racial, national, linguistic or religious aspects. Together, they create the history and future of one country. This leads to the fact that individuals are a group, so each individual is equally important.

Diversity contributes to the development of the nation. By increasing people's awareness about diversity, people adapt more easily to the changes that are constantly happening. The fact that Montenegrins live with various ethnic groups represents a great advantage because it will broaden their horizons and open the door to development in every area of their life. Also, diversity promotes tolerance and an open mind. The synergy of difference motivates the improvement of human creativity.

According to the statistical data, Montenegro is a multicultural nation that has opened its borders and thus contributes to the desire for even greater development. When there is diversity in the ethnic structure, one can immediately see a great number of advantages and opportunities that will result. First of all, there are opportunities for advancement in personal and professional life. Every individual is given the opportunity to prepare himself/herself for the great number of challenges in life. People get a chance to live in the same country and with the presence of each other, they enrich their knowledge and

19 | Radojevic, G. (2016). *Population estimates and basic demographic indicators.* Monstat. Retrieved from https://www.monstat.org/ userfiles/file/demografija/ procjene%20stanovnistva/2016/ Saopstenje%202016.pdf.

experience. Also, the expansive network and the great number of contacts given the country's ethnic structure allow people to get new and better business opportunities in Montenegro and abroad. By combining different knowledge, customs and habits, they create a uniqueness that increases the abilities of people and influences their creativity and innovation.

3.1.2. Government's proactive financial support for students to complete post-secondary education in public universities

Young people in Montenegro are pretty much oriented towards progress, which is quite logical considering the fact they are living in a developing country. Education plays one of the biggest roles in the life of young Montenegrins.

The data of the education structure shows that persons with secondary education have the largest share. There were 62.0% persons with secondary education, 26.8% persons with completed higher education and 11.3% of persons with completed primary education or lower education[20].

In June 2017, the members of the Assembly adopted amendments to the Law on Higher Education, which stipulates that undergraduate and master studies for all students of the University of Montenegro are free. This has generated great fanfare among Montenegrin students.

This decision is encouraging for students because they don't have to worry about paying for their studies and in this way, they can be more focused to strive for a better GPA (grade point average). Students are encouraged to concentrate on their academic studies, knowledge development and personal growth. In this sense, the graduates are likely to serve as the human capital of higher quality and productivity. However, such tuition-free decisions may pose a challenge to private universities which charge students for tuition and probably provide a better quality of education. Nevertheless, time is the best indicator of all decisions and it is expected that providing high-quality education is the key to promoting a country's growth in the long run.

20 | Radojević, G. (2017). *Labour Force Survey Second Quarter 2017*. Monstat. Retrieved from http://monstat.org/userfiles/file/ars/2017/2/ARS%20saopstenje_2017_Q2_en.pdf.

	TOTAL	ETHNICITY (1)					
		Montenegrins	Serbs	Bosniaks	Albanians	Muslims	Croats
MONTENEGRO	622029	278865	178110	53605	30439	20537	6021
ANDRIJEVICA	5071	1646	3137	0	1	7	2
BAR	42048	19553	10656	2153	2515	3236	254
BECANE	33979	8838	14592	6021	70	1957	42
BIJELO POLJE	46051	8808	16562	12592	57	5985	41
BUDVA	19218	9262	7247	82	100	113	167
CETINJE	16657	15082	727	4	38	12	42
DANILOVGRAD	18472	11857	5001	16	81	38	55
HERCEG NOVI	30684	10395	15090	74	41	160	662
KOLAŠIN	8380	4812	2996	0	0	18	7
KOTOR	22601	11047	6910	29	102	64	1553
MOJKOVAC	8622	5097	3058	8	0	4	2
NIKŠIČ	72443	46139	18334	194	73	421	149
PLAV	13108	822	2098	6803	2475	727	5
PLJEVLJA	30786	7494	17569	2128	17	1739	16
PLUŽINE	3246	902	2131	0	0	0	2
PODGORICA	185937	106642	43248	3687	9538	4122	664
ROŽAJE	22964	401	822	19269	1158	1044	6
ŠAVNIK	2070	1114	878	0	0	2	1
TIVAT	14031	4666	4435	96	97	114	2304
ULCINJ	19921	2478	1145	449	14076	770	45
ŽABLJAK	3569	1800	1474	0	0	4	2

	ETHNICITY (3)						
	Italians	Yugoslavs	Hungarians	Macedonians	Muslims-Bosniaks	Muslims-Montenegrins	Germans
MONTENEGRO	135	1154	337	900	183	257	131
ANDRIJEVICA	0	5	0	2	0	3	0
BAR	8	105	38	87	3	65	13
BECANE	0	27	1	26	11	27	4
BIJELO POLJE	1	27	3	14	87	21	9
BUDVA	16	74	39	69	1	2	20
CETINJE	6	17	12	20	0	0	1
DANILOVGRAD	3	30	11	32	0	0	2
HERCEG NOVI	30	157	48	95	0	1	12
KOLAŠIN	2	22	2	5	0	0	3
KOTOR	31	93	35	54	3	0	14
MOJKOVAC	0	20	0	3	0	0	1
NIKŠIČ	1	161	6	68	0	36	3
PLAV	0	2	0	0	3	1	1
PLJEVLJA	0	20	3	9	25	34	2
PLUŽINE	0	1	0	0	0	0	0
PODGORICA	24	310	81	350	23	61	30
ROŽAJE	1	0	1	1	14	1	0
ŠAVNIK	0	3	1	1	0	0	0
TIVAT	11	61	43	48	3	0	9
ULCINJ	1	14	13	16	0	5	6
ŽABLJAK	0	5	0	0	0	0	1

Figure 1 - Population by ethnicity per municipality[21]

ETHNICITY (2)					
Bosnians	Bosiaks-Muslims	Montenegrins-Muslims	Montenegrins-Serbs	Egyptians	Gorani
427	181	175	1833	2054	197
1	0	0	17	0	0
53	12	48	95	33	16
5	28	21	43	170	18
6	84	10	32	0	5
42	0	0	46	144	10
7	0	1	37	0	4
13	0	0	97	2	0
48	0	0	72	28	0
4	0	0	57	0	4
22	0	3	60	63	0
3	0	0	11	0	0
28	1	8	335	446	22
5	0	0	2	0	10
6	15	20	14	1	0
1	0	0	1	0	0
105	28	56	850	685	85
12	13	3	3	74	19
0	0	0	5	0	0
35	0	0	28	335	4
31	0	5	9	73	0
0	0	0	19	0	0

ETHNICITY (4)							
Roma	Russians	Slovenians	Serbs from Montengro	Turkish	Other	Regional Qualification	Does not want to declare
6251	946	354	2103	104	3358	1202	30170
0	2	1	39	0	8	5	195
203	241	21	144	36	318	45	2097
531	6	5	179	2	93	3	1250
334	13	1	136	18	239	4	952
33	210	14	105	7	193	72	1150
97	6	7	14	0	60	6	457
28	29	8	57	1	62	15	1034
258	118	56	98	0	146	367	2908
0	8	5	58	0	47	7	323
74	70	29	58	0	162	179	1946
16	3	2	27	0	36	1	330
483	29	20	356	0	240	34	4846
0	1	0	1	5	32	2	107
12	5	1	84	4	115	5	1448
0	0	1	2	0	1	3	201
3988	135	117	650	22	1222	322	8892
0	3	1	3	6	37	0	72
0	1	0	14	0	1	2	47
35	56	57	34	2	174	109	1275
159	10	4	10	1	158	18	425
0	0	4	28	0	14	3	215

21 | Monstat. (2011). *Census of Population, Households and Dwellings in Montenegro* (2011), p. 7.

Another policy supported by the Government is to offer all Montenegrin students the paid opportunity to work for nine months during their undergraduate studies in the field for which they have been educated in order to gain more practical experience and be prepared to encounter and solve different challenges in real-life contexts.

3.2. Industries in Montenegro

Industrial production in Montenegro increased by 11.4% in November 2017 as compared to the previous month.[22] Observed by sectors, compared to the same month of the previous year, production increased in the mining and quarrying sector by 147.3% and processing industry by 8.9%, while it decreased in the sectors of electricity, gas, steam and air conditioning supply by 23.5%.

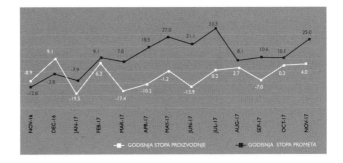

Figure 2 - Annual production and trade rates in industry in percentages[23]

Montenegro is a prime tourist destination in the Balkans

Montenegro is a thriving tourist destination. A defining feature is its diverse Mediterranian climate that offers a wide range of nature-related activities. There are also historical monuments as well as modern buildings with new architecture designs. In 2017 the tourism industry in Montenegro grew rapidly. For example, in November 2017, there were 35.2% more tourists, mostly from Serbia and Albania, compared to the year before.[24]

22 | Monstat. (2017). *Industrial production and turnover indices.* Retrieved from http://www.monstat.org/userfiles/file/industrija/promet%20u%20ind/2017/11/Indices%20of%20industrial%20production%20and%20turnover%20-november.pdf.

23 | Monstat. (2017). *Industrial production and turnover indices,* p. 1. The red line represents annual production rate; and the blue line represents annual turnover rate.

24 | Monstat. (2017). *Dolasci i noćenja turista u kolektivnom smještaju [Tourist arrivals and overnight stays in collective accommodation].* Retrieved from http://www.monstat.org/userfiles/file/turizam/dolasci%20i%20nocenja%202017/11/SAOPSTENJE%20novembar%202017.pdf.

Despite Montenegro being a very small country, it comprises many landscapes of true natural beauty. There are high mountains and deep canyons, wide plains, fast rivers and lakes. Vegetation is rich. It is a great advantage that tourists can climb the mountains and go to the seaside in just one day thanks to the short distance. Tourism is an important aspect of Montenegro's economic development. The industry is having an increasing impact on contributing to the country's GDP, national income and resources redistribution.

3.3. Thriving Civil Society in Montenegro

The number of youth organisations and initiatives are increasing. There are different NGOs for which the primary service target is young people. Today, almost all universities and faculties have active student parliaments and student organisations. Students have established national organisations such as AIESEC, MoMSIC, OMSA, Students for Liberty, Mladiinfo, ELSA, BEST, etc. Young Montenegrins are showing a growing interest in making progress in every area of their life, especially in education. There are organisations such as Student Business Centre, founded in 2008 by the students of the University of Donja Gorica, which promotes student entrepreneurship and innovation through communication with business and academic experts. The organisation is giving more opportunities for young people to work and encouraging them to start their own business.

Also, there is an event called Stock Market of Entrepreneurial Ideas, which is organised every year at the University of Donja Gorica 'to promote development of entrepreneurial and business ideas, especially among university students and high school students from schools across Montenegro'[25]. These events have provided valuable support opportunities for students to improve their business ideas continuously. In this way, young people can channel their new ideas and encourage innovation, entrepreneurship and self-employment.

Through exchange programmes such as Erasmus+ (EuRopean Community Action Scheme for the Mobility of University

25 | Idea Lab. (2016, April 19). *Call for Stock of Entrepreneurial Ideas at the University Donja Gorica, Podgorica.* http://www.idealab.uns.ac.rs/news/call-for-stock-of-entrepreneurial-ideas-at-the-university-donja-gorica-podgorica.html.

Students), Erasmus Mundus, and CEEPUS, they have had success in their purpose of establishing collaboration among universities, young professionals, and student organisations. Bodies such as the Organisation of Montenegrins Studying Abroad (OMSA) were established to promote Montenegro and its cultural and natural heritage, facilitate collaboration with organisations and communities abroad, collaborate with student organisations, NGOs, and public institutions in Montenegro, support Montenegrin high school students in their future study plans and use the established network of professionals to the end of supporting Montenegrin sustainable development, especially the development of science and research engagement of Montenegrin youths.

3.4. Gender Differences in Entrepreneurship

Throughout history, women in Montenegro were placed in a lower social status. For example, women were usually married at a young age following the arrangements designated by their parents in the past. Their wills were not respected, and their main obligations were to take care of the family and to manage the household. Despite a period of social and economic transformation, the traditional idea that women have to serve as a good mother and housewife still persists.

In Montenegro, Men are still predominately the breadwinners of a family. Regarding the total number of the employed working population, 55.9% are men and 44.1% are women. In terms of the total number of unemployed persons, 54% are men and 46% are women[26]. At first sight, the percentage difference between men and women is not very great. Nevertheless, while looking at the number of enterprises owned by women in Montenegro, only 9.6% of the companies (out of 21000 active business entities) are owned by women. This data depicts that entrepreneurship is a less developed concept among the female population and that men are usually the ones who own and work on private businesses.

26 | Radojević, G. (2017). *Labour Force Survey Second Quarter 2017*. Monstat. Retrieved from http://monstat.org/userfiles/file/ars/2017/2/ARS%20saopstenje_2017_Q2_en.pdf.

3.5. High Youth Unemployment Rate and Unpopular Entrepreneurship

This generation of young people will save future generations if they are awakened at the right time. In Montenegrin society, it is the belief that everything from abroad is better than the homegrown. This is the reason why entrepreneurship is not popular in Montenegro. Montenegrin society, especially among the young people, is uninformed about the great importance of entrepreneurship.

The unemployment rate represents one of the challenges for Montenegrin youth. Montenegro ranks 11th-highest in the world for youth unemployment rate (between 15-24 years) with a rate of 41.1 percent[27]. For this statistic, youth is defined as 15-to-24- years old. This is a quite concerning fact.

Entrepreneurship is key to solving the problem of youth unemployment. The development of human potential through education in the field of entrepreneurship is a key element of economic development throughout the world. Thanks to the knowledge of entrepreneurship, youths can think about the market and find new, better and more efficient ways to meet the needs of the market.

3.6. Brain Drain

Throughout Montenegro's history, there have been various reasons to emigrate, such as political, economic and educational causes. Some of the people left Montenegro to look for a better life in other countries which were more stable and economically strong. Some returned to the country after a while, while most of them stayed at the new place to start a new life. Those who stayed would acquire citizenship and settle down in the new country.

All down history, there were times when there were no opportunities for work and achieving a dignified life. Montenegrins have attempted to adopt a neoliberal market policy and follow the directions of economic development

27 | Central Intelligence Agency. (2011). Montenegro. In *The World Factbook*. Retrieved from https://www.cia.gov/library/publications/the-world-factbook/rankorder/2229rank.html#m.

of many developed economies such as Western Europe and North America. As a result, many Montenegrins chose to study abroad and learn the best practices from those Western liberal economies. More and more talented young people have moved their families and remain permanently abroad because there are better career opportunities and financial security there. Recently as a norm more and more talented young people move their families and remain permanently abroad because there are better career opportunities and financial security there:

> From a country of emigration, Montenegro has increasingly become a country of immigration which results in the need for adapted policies. The recent macro-economic stability achieved by Montenegro, and increased foreign direct investment has led to an increase in the number of labour migrants arriving in the country. 90% of the labour migrants are thought to come to Montenegro for economic reasons and most of them came from Serbia, Bosnia & Herzegovina, Kosovo and the Former Yugoslav Republic of Macedonia.[28]

Since there is a large wage gap between Montenegro and other countries, many intellectuals leave Montenegro because there are more prosperous opportunities abroad. It is obvious that measures are needed to be taken by the government in order to lessen the brain drain situation happening in Montenegro. Youth see working in foreign countries as a big and challenging opportunity.

Also, a large number of Montenegrin students, who graduated from American universities, decide to stay on after finishing their studies. Their decisions are made because of the more favourable living conditions and better earnings in the United States than in Montenegro. There are also negative long-term effects of such decisions. The emigration of highly qualified young people has imposed negative consequences on labour productivity, economic growth and the development of Montenegro. During an interview in October 2017, Dusko Markovic, Prime Minister of Montenegro, commented that the

28 | Chindea, A., Majkowska-Tomkin, M., Pastor, I. (2007). *The Republic Of Montenegro Migration Profile.* Retrieved from http://publications.iom.int/system/files/pdf/mp_montenegro_2007.pdf.

country faced a 'brain drain' of experts to Western Europe and emphasised the necessity of recognising talented young people in science, research and eductaion.[29]

The focus is on the future of young people. Information about the future. Taking responsibility for every action in life. What everybody is doing today, it is going to influence actions in the future. The mistakes committed by past generations still have repercussions for the current generation. When there is a willingness to work and reach full potential, it is not difficult to succeed. However, it is easier said than done. The main issue is that youth do not see the opportunity to achieve their goals in their home country. Youth have to fight for their rights. Entrepreneurship should be celebrated among the young people in Montenegro. The future is dependent on the liberation from limits and economic barriers.

29 | Markovic, D. (2017, October 6). *Dusko Markovic, Prime Minister of Montenegro, Interview for Global Citizen.* Retrieved from http://www. poreskauprava.gov.me/ pretraga/177492/Predsjednik-Markovic-u-intervjuu-za-Global-Citizen-Vracamo-Crnoj-Gori-staru-slavu-i-oplemenjujemo-je-novim-sadrzajima.html.

4. Opportunities and Challenges Offered by the Belt and Road Initiative

This Part discusses the importance of young people's role in the successful implementation of the Belt and Road Initiative (BRI) through an overview of opportunities and challenges offered by the Initiative. It attempts to recognise fields in which the BRI can positively affect the youth in Montenegro.

After Montenegro and China signed a Memorandum of Understanding in May 2017, all the relevant stakeholders in Montenegro have become aware of the Initiative's presence and the potential opportunities it could bring to Montenegro. A higher level of understanding, cooperation and exchange between Montenegro and China was presented after signing the agreement. The BRI, a highlight of the agreement, is especially impactful to young people in Montenegro given the incalculable potential it might bring to benefit this group of future leaders.

In Montenegro, 21.4% of the population is aged between 15 and 29[30]. Among this population, many are currently under the heavy influence of Western culture. Given such context, the promulgation of the BRI might serve as a peculiar way to introduce non-Western, i.e. Eastern, culture to young Montenegrins. Since young people are the future leaders of Montenegro, the Balkans and the whole world, it is believed that young people's innovative spirit could play a crucial role in determining the success of such a large-scale, multilateral and global cooperative project. This chapter attempts to outline how young Montenegrins can utilise their strengths and contribute to this project by first analysing how they have responded to the existing opportunities, then exploring future cooperation possibilities.

4.1. Existing Opportunities Offered by the BRI

The BRI has offered two levels of opportunities to the young people in Montenegro, i.e. general public level and university level.

At a general public level, the Confucius Institute was established in the University of Montenegro in 2015. The Institute aims at promoting Mandarin Chinese in Montenegro through offering Chinese classes at primary, secondary and university levels. Learners can be certified by passing the Chinese Proficiency Test (HSK) in Montenegro. Besides offering language classes, the Institute organises workshops and seminars on Chinese traditions and cultures, calligraphy, martial arts, Chinese food, etc. to promote Chinese culture to a wider global audience and even encourage cultural exchanges and understanding between China and Montenegro.

At a university level, the University of Donja Gorica (UDG), a private university in Montenegro, has been participating in the opportunities offered by the Initiative proactively. For example,

30 | Monstat. (2017). *Census of Population, Households and Dwellings in Montenegro*. Retrieved from http://www.monstat.org/userfiles/file/popis2011/saopstenje/saopstenje%20starost%204%2009%202011%20prevod.pdf.

the president of UDG attended 'The Belt and Road University Presidents Forum on Innovation and Entrepreneurship Education' in Xi'an, China in September 2017.[31] UDG is one of the 86 universities along the Belt and Road which was invited to the event. During the Forum, 'Xi'an New Silk Road University Students Alliance' was established with the goals of promoting joint university activities, summer schools, student exchange, etc. among students in the Alliance.[32] UDG is one of the founders of this alliance and currently the only university in the Balkans which participated in this initiative. Additionally, UDG has established partnerships with numerous universities in China which contribute to more frequent student and staff exchanges between the two countries.

Besides, the University of Donja Gorica co-organised the Miločer Development Forum with the Association of Economists and Managers of Montenegro in 2017.[33] The Forum featured the theme 'Globalisation and the Belt and Road Initiative,' by which the audience could learn about this subject from more than 30 presenters from all around the world in the two-day Forum.

Seven panels were held during the Forum and discussed a wide range of issues from economics and politics, to education. For instance, some speakers discussed the BRI from an economic perspective and the potential of Chinese investments in bringing business opportunities to Montenegro and the Balkans. The participants could learn about what Montenegrins could expect from Chinese investors and what would be the available possible ways of financing different investments within the framework of the BRI.

On top of that, the Forum hosted a sharing session about Chinese education and culture given by both Chinese professors and UDG students who had spent one semester studying in China. All in all, these panel sessions provided Montenegrins with an invaluable time to probe into the potential opportunities and challenges brought by the BRI and Chinese investments. As Ambassador Cui Zhiwei[34] said in the Forum, 'the Silk Road will connect and bring together the two

31 | Baltic Management Development Association. (n.d.). *Rector of UDG Participated in International Conference in China.* Retrieved from http://www.bmda. net/BMDA/members-area/news-from-bmda-members/rector-of-udg-professor-veselin-vukotic-participated-in-international-conference-in-china.

32 | *Ibid.*

33 | Miločer Development Forum. (n.d.). *Miločer Development Forum 2017.* Retrieved from http://www.demcg.me/files/download/1512987548_9795.pdf.

34 | Cui was the ambassador when this part was written. His successor was Liu Jin.

countries and two nations, making them true friends and good partners!'[35]. It is anticipated that there will be more frequent business, political and cultural exchanges between the two countries in the future.

4.2. Future Possibilities Brought by the BRI

The BRI could potentially bring in opportunities from various aspects such as scientific innovation, business entrepreneurship, political partnership, education, etc. As a result, the Initiative may improve the quality of life of young people from over 60 countries along the Belt and Road in terms of elimination of poverty, improvement in the quality of education, well-being promotion, etc.

Regarding the case of Montenegro, the positive impact of BRI is most significant in areas such as promoting entrepreneurship, infrastructure development, and tourism.

First, the BRI could encourage entrepreneurship in Montenegro. A survey conducted by China Youth Daily Social Survey Centre depicted that the BRI would encourage entrepreneurial potential significantly[36] Such potential is especially important in the case of Montenegro because it can enhance the country's global competitiveness and improve the economy in the long run.

When 99.8% of Montenegrin companies are small and medium-sized enterprises (SMEs), it is probably easier for these companies to adopt a flexible business strategy and exercise innovative entrepreneurial skill.[37] Given that China has a comparative advantage in internet technology and the BRI strongly supports knowledge sharing and technology transfer, it is reasonable to say that China could grasp the opportunity and actively engage in knowledge transfer (especially on the grounds of eCommerce) with Montenegrin companies. Since Montenegro does not have its own eCommerce platform, the country can perhaps adopt relevant technologies operated by Chinese tech giants such as Alibaba (AliPay) and Tencent (WeChatPay). Implementing a mature eCommerce platform

35 | Cui, Z. W. (2017). *Introductory words: Initiative "The Belt and Road" and Montenegro*. Retrieved from http://demcg.me/files/download/1509437776_4165.pdf.

36 | Xie, P. (2017, June 27). *Fifty Percent of Young People Believe That One Belt Initiative Will Have A Positive Impact On Their Future. China Youth Daily*. Retrieved from http://www.chinanews.com/sh/2017/06-15/8251045.shtml.

37 | Microfinance Centre (2017, December). *Supporting 'Generation Start-up': Opportunities For Montenegro*. Retrieved from https://mfc.org.pl/wp-content/uploads/2017/11/Case-study-MN_final.pdf.

could encourage entrepreneurship because it largely lowers the entry cost of starting a business, allows efficient business transactions through simple electronic gadgets, and enables convenient business cost management.

Besides bringing in eCommerce technology, the BRI offers a global market for young Montenegrins to explore and develop their entrepreneurship by starting up new business ventures. The global market does not only serve as an outlet for exporting goods and connecting to the world of globalisation, but is also a platform for knowledge sharing and enhanced business and cultural exchanges. By sharing and working on good practices, the BRI potentially offers a better living environment for everyone.

Second, the BRI opens a window of opportunity for Montenegro's tourism industry. Montenegro has been a very popular tourist destination in the Balkans given the distinct geographical advantage it enjoys. According to data published by the Statistical Office of Montenegro, 1.6 million tourists visited Montenegro in 2016[38]. The number of tourists increases steadily every year thanks to the active support from the Montenegrin government. In addition, Chinese companies have started exploring the opportunity of developing business partnerships with the Montenegrin tourism industry. One of the largest tourist agencies in China visited the Embassy of the People's Republic of China in Montenegro to explore the possibilities of creating partnerships with local travel agencies. Such collaboration is likely to introduce more Chinese tourists to visit Montenegro as well as provide more job opportunities for the Montenegrin youth.

Apart from benefiting the service sector, the growth of tourism will also encourage infrastructural development in order to support the industry in the long run. As a matter of fact, Chinese companies have already been involved in important infrastructural projects such as developing and enhancing Montenegro's telecommunication and transport systems. The more intensive cooperation, as supported by the BRI, will present more invaluable opportunities for young people to

38 | World Data. (n.d.). *Tourism in Montenegro*. Retrieved from https://www.worlddata.info/europe/montenegro/tourism.php.

participate in key developmental projects and learn the skills and technology e.g. building highways and railways from China. The experience of working on bilateral projects like these can be a stepping stone for young people and shows their ability to work with international projects on a global level in the future.

4.3. What Should Young People Do to Grasp These Opportunities?

There are mainly three ways that every young person in Montenegro could become well-prepared and grasp these approaching opportunities.

The most important skill that young Montenegrins should acquire is language skill. By learning a new language, we learn to communicate with the outside world more efficiently and effectively. This skill could also facilitate more meaningful cultural exchanges. By introducing the world to Montenegro through effective communication, it may encourage more foreign direct business exchanges and opportunities.

Second, young Montenegrins should be proactive in participating in cultural exchanges and international networking events to widen their horizons and understand more about the outside world. Young people are advised to travel as much as possible, make friends from other nations and countries, and learn the different distinctive cultures from people in order to understand the globalised world better and be prepared to get involved in new opportunities globally. The global labour market does not demand bookworms. Rather, the market is yearning for enthusiastic talents who are constantly curious and innovative.

Third, young Montenegrins could actively consider studying abroad given the many benefits that the experience may bring. For example, through studying abroad in China, the student not only can learn the language (some Montenegrin exchange students can even write their dissertations in Chinese), they can also observe and learn the best practices and ideas of the country and apply them in their home country with local creative adjustments. As Xi Jinping, the President of the People's Republic of China, mentioned in the opening of the Belt and Road Forum the goal of the Belt and Road Initiative is to 'embark on a path leading to friendship, shared development, peace, harmony and a better future'.[39]. More university-level academic exchanges are likely to foster more Sino-Montenegrin friendships that would be beneficial to relations between the two countries in the future.

To conclude, a consistent dedication to personal ongoing development, better education, and enhanced quality of life by both the government and the young people is required to grasp all the opportunities offered by the Belt and Road Initiative. Young people should first develop good language skills, then be proactive to get involved in the various exchange and business opportunities offered by the Initiative.

39 | Xinhua. (2017, May 14). Full Text of President Xi's speech at Opening of Belt and Road Forum. *Xinhuanet*. Retrieved from http://www.xinhuanet. com//english/2017-05/14/ c_136282982.htm.

5. Suggestions for the Belt and Road Initiative in Relation to the Young People of Montenegro

5.1. Overview of Projects Similar to the BRI

In order to understand and consider the possible alternatives for the Belt and Road Initiative (BRI), it would be useful to look back at some similar projects that happened in the past and their effects on countries that were covered by those projects. After that, this part will analyse the impact of those projects on the population of those countries, with the main focus on young people. Based on those analyses, suggestions for alternatives for the BRI will be presented.

The BRI aims to upgrade and improve infrastructure, trade and fellow-feeling in more than 70 countries – hundreds of billions of dollars of Chinese money are to be invested abroad in railways, ports, power stations and other infrastructure that will help those countries to prosper[40].

The most similar project to BRI, in terms of mission and global impact, was the Marshall Plan. In 1948, America passed the European Recovery Programme, better known as the Marshall Plan, in order to revive Europe's war-ravaged economies. The United States gave about US$13 billion (which is equivalent to US$130 billion today) in economic assistance to help rebuild Western European economies after the end of World War II[41]. Goals of the Marshall Plan were to rebuild regions heavily impacted by the war, improve European prosperity, and curb the spread of Communism. At the time, Europe was on the brink of economic collapse.

While the funds helped Europe's recovery, the financial impact alone was limited. The true impact of the Marshall Plan was found in the market-friendly policies it promoted. To receive aid, European governments had to implement liberal capitalistic policies, remove trade barriers, price controls, and narrow their deficits. These reforms and indirect economic effects had tremendous benefits, and they played a huge role in the process of European recovery.

It is hard to compare the Marshall Plan with the Belt and Road Initiative because no one knows how big the BRI will be. However, many expect that the BRI will be a far bigger project than the Marshall Plan, with bigger effects on the global economy. China's direct investment in BRI countries amounted to US$56 billion from 2014 to 2017. Predictions are that China would invest up to US$150 billion over the next five years.[42] The total amount needed for the whole project is expected to be about US$1 trillion. BRI investment already surpasses Marshall's billions, but the main difference is the source of that money.

Over 90% of the Marshall money was a handout (not a loan) and it came from the American government, while the BRI investments are from a variety of sources, including private entities[43]. There are differences between their goals too.

40 | The Economist. (2018). *Will China's Belt and Road Initiative outdo the Marshall Plan?*. Retrieved from https://www.economist.com/news/finance-and-economics/21738370-how-chinas-infrastructure-projects-around-world-stack-up-against-americas-plan.

41 | *Ibid.*

42 | *Ibid.*

43 | *Ibid.*

The Marshall Plan succeeded via granting markets a critical role in resource allocation. The BRI does not have any tendencies to export that principle abroad. The Marshall Plan represented the beginning of European integration, and it succeeded in its goal to stimulate the total political reconstruction of western Europe.

Now, new alternatives for the BRI are emerging – Australia, the United States, India and Japan are considering a joint regional infrastructure scheme as an alternative to China's BRI, primarily to counteract China's influence[44]. New projects should be an alternative to the BRI, rather than a rival. There should not be any bilateral challenges between these four countries and China or any kind of trade war. The global economy can prosper only if those countries cooperate and work together. If that is not the case, then the whole situation will be a zero-sum game – some will gain and prosper, and some will lose.

5.2. Economic and Political Implications of Similar Projects and Possible Alternatives for the Belt and Road Initiative

The Marshall Plan, which in some aspects reminds people of the Belt and Road Initiative, had a goal to reconstruct and develop the economy of Europe[45]. As such, billions of US dollars were invested in developing industries and infrastructure in Europe after World War II. France, the United Kingdom and West Germany were the countries that received the most cumulative aid through the Marshall Plan. Just these three countries received more than 50% of the total aid that was available between 1948 and 1951.[46] The idea of creating economies that are prosperous and progressive is presented in the Belt and Road Initiative as well.

Most of the countries that are covered under the BRI are not developed or their economies face stagnation. Montenegro, for example, is a country which enjoyed tremendous economic growth after it gained independence in 2006, but shortly afterwards it hit a stagnation or just small increase in production and GDP growth. In 2009, Montenegro's GDP was around €3 billion (approximately US$3.7 billion)[47], while seven

44 | Reuters. (2018). Australia, U.S., India and Japan in Talks to Establish Belt and Road Alternative. *Reuters*. Retrieved from https://www.reuters.com/article/us-china-beltandroad-quad/australia-u-s-india-and-japan-in-talks-to-establish-belt-and-road-alternative-report-idUSKCN1G20WG.

45 | Jaćimović, D. (2015). *Introduction to Economy of EU*. Podgorica: University of Montenegro, p. 72.

46 | Schain, M. A. (2001). *The Marshall Plan: Fifty Years After*. New York: Palgrave, p. 120.

47 | Djurovic, R. (2009). *Gross Domestic Product of Montenegro in 2009*. Podgorica: Statistical Office of Montenegro.

years later it was around €3.9 billion (approximately US$4.8 billion)[48]. For a country that is developing and going through various transitions, it is very slow progress.

The tremendous decline has been noticed in Russian GDP as well due to political instability and economic problems. According to World Bank Data from 2014 to 2016, the GDP of the Russian Federation fell from €2.06 trillion to €1.28 trillion (approximately from US$2.54 trillion to US$1.58 trillion). There are other examples as well, so it appears that the BRI could serve the same purpose as the Marshall Plan in reconstructing some economies.

A very important factor will be the distribution of resources that is initiated by the BRI. It is neither strange nor rare that economic policies and politics are oriented towards strengthening the bureaucracy, and not towards creating a strong and globally competitive market. It is an opportunity to create an environment in which young people will have all favourable conditions for creating new businesses and new values without barriers.

In the age of digitisation and technology progress, it is required for the BRI to stimulate countries to become hubs for start-ups and entrepreneurs who can use innovative ways to reach a global audience, and also to make their business models attractive to customers in different countries. It is important that the BRI stimulate the birth of economic alliance between members who will cooperate in the process of economic development, as was the case with the Marshall Plan and Organisation for European Economic Cooperation[49]. Cooperation among countries should be based on full commitment to the economic and political development of the People's Republic of China and all country members should join the effort to reduce trade barriers among all countries. Following the results of the Marshall Plan, it is worth noting that 'the first effects of liberalisation were not only the fast growth of many industries through trade activities among country members and increase of the income in Europe during the 50s, but also higher volume of export and import had strengthened

48 | Radivojevic, G. (2016). *Gross Domestic Product of Montenegro in 2016.* Podgorica: Statistical Office of Montenegro.

49 | Jaćimović, D. (2015). *Introduction to Economy of EU.* Podgorica: University of Montenegro, p. 72.

relationships between countries in Europe'.[50]

Such effects of the BRI could motivate the young generation to establish private enterprises and to trade with people from other countries by developing stronger relationships and cooperation. It could not only bring prosperity and progress, but also peace and stability to countries that would be part of the new Silk Road.

Just following the timeline after the Marshall Plan, one of the events that occurred was the Élysée Treaty signed by German Chancellor Konrad Adenauer and French President Charles de Gaulle. That meant that France and Germany made an agreement to exchange students and to develop their skills, so they could bring prosperity to both countries. Today, as a result of this cooperation, over 2,737 French companies are present in Germany employing around 360,000 people and there are around 3,200 German companies in France that employ around 310,000 people and generate a profit of €141 billion (approximately US$175 billion)[51]. It is a classic example of how cultural, educational, economic and political cooperation and reducing barriers can create an atmosphere that fosters progress and prosperity. This example demonstrates that further development of the BRI could be targetted at promoting and encouraging youth policies with language exchange and learning.

The Belt and Road Initiative should be built on these principles. It should motivate young people to open private enterprises and to think outside of their boundaries, to learn and examine other cultures, and finally to create new values for this and future centuries. According to the research of Freedom House, a U.S. Government-funded non-governmental organisation (NGO) that conducts research and advocacy on democracy, political freedom, and human rights, there are 21 countries that are not free, and 28 of them that are partly free when regarding freedom of the internet[52]. Most of these countries come from the Eastern Hemisphere of the planet. Those limitations on freedom limit young people's opportunities, and prevent them from being creative and informed about

50 | *Ibid*.

51 | France Diplomatique, France and Germany (2018). *Economic and Trade Relations*. Retrieved from https://www.diplomatie. gouv.fr/en/country-files/ germany/france-and-germany/.

52 | Kelly, S., Truong, M., Shahbaz, A. & Earp, M. (2016). *Freedom on the Net*. Washington DC: Freedom House.

the latest global changes and opportunities. The BRI should use its power and size as a project to push these countries to overcome all obstacles and challenges and to promote the development of freedoms in each sphere. It is the only way to create a place for young people to interact and exchange ideas that would help them create innovations and new values.

Young people in Montenegro, especially due to its small domestic market, could access all the global opportunities and try to be globally competitive in the Belt and Road market. It would help people in Montenegro to boost their economy and quality of life. The fact that Montenegro has a small domestic market explains why its aspirations are mostly directed towards the European Union. There is a need to be present to an even larger market where young people can show all their talents and skills that can help them to be better and stronger than other young counterparts.

5.3. Conclusion

It is understandable that if the BRI aims to incorporate and increase the involvement of young people, it must be based on principles such as:

- **Universal Legal Framework**, which will promote equal rights and laws for all competitors within the BRI network;

- **Free and Deregulated Market**, which will stimulate an atmosphere friendly to economic and political cooperation between countries;

- **Entrepreneurship**, which will encourage and empower young people to create new values and to solve complex social problems that can maximise their profits;

- **Intercultural Dialogue**, which will motivate young people to exchange their ideas with young people from other countries, but also which will motivate them to explore different cultures, languages and traditions.

Based on these four pillars of development, the BRI could become the most powerful project in the history of humankind and it could change the way we experience the world around us. It could bring new ways of competitiveness and most importantly it could boost the economies of different countries as a domino effect. It means that economic progress and prosperity of one country could lead to the economic prosperity of another, and so on. Alternatively, if the BRI does not succeed, one of the ways to achieve international progress and prosperity is to establish an International Custom Union that promotes trade without barriers and low taxes among competitors, while simultaneously following the above-mentioned four pillars of development.

Montenegro needs to find its place in this global opportunity and encourage young people to explore all possibilities and opportunities that the BRI offers. For a country with less than a million inhabitants, it is a great opportunity to boost its economy and to create a stronger position in Europe, by promoting Montenegro as a major hub of innovation and business development.

But these measures need reforms and politicians with vision. With the growth of populism and all sorts of collectivism, it is very difficult to expect that Montenegro will recognise the importance of such an opportunity and this global project. As was argued in Part II of this paper, the only way to create a prosperous and progressive society in Montenegro is to break chains with paternalism and to encourage individuals to strive for an entry to global market and entrepreneurship.

The Belt and Road will evolve from year to year, and it is now the opportunity for young people to involve themselves and to start working on projects that will benefit the entire network. However, policymakers and officials have to be careful when they promote policies that are not in favour of global economic development, mainly globalisation and liberalism.

Skyline of Warsaw, Poland.

CHAPTER 13

by   **GRZEGORZ STEC**
University of Oxford

SEBASTIAN SULKOWSKI
Jagiellonian University

(Ordered by last name)

ESTONIA

LATVIA

LITHUANIA

CZECH
REPUBLIC

SLOVAKIA

HUNGARY

SLOVENIA

ROMANIA

CROATIA

BOSNIA &
HERZEGOVINA

SERBIA

MONTENEGRO

BULGARIA

NORTH
MACEDONIA

ALBANIA

POLAND

1. Political Overview of Poland

Poland is a large European country located in the Central Eastern Europe (CEE) region. It is a unitary state divided into 16 subdivisions (voivodeships) that covers an area of about 313,000 square km km² with a population of 38.5 million.

Poland has been a member of the North Atlantic Treaty Organisation (NATO) since 1999 and the European Union (EU) since 2004. Despite its present status as a fully sovereign state, Poland was subordinate to its powerful neighbours – mostly Germany and Russia – for most of the 20th century. On top of that, Poland was an independent nation for less than one quarter of the past 220 years: 20 years during the Interwar Period and the period since the fall of communism in Poland in 1989.

For the past three decades, Poland had been considered as one of the most successful post-soviet states. Since 1989, its gross domestic product (GDP) per capita has doubled and the nominal GDP has grown continuously. A few years after joining the EU, Poland became one of the most important countries in the community and an informal leader of the CEE countries in the EU forum.

1.1. A Brief History of Poland and Sino-Polish Relations[1]

The beginning of the Polish state dates back to the year 966 when the first historically verified leader of Poland, Prince Mieszko I, decided to baptise his pagan country and connect it culturally and politically with the rest of the Christian kingdoms in Europe.

The history of Poland, like the history of most countries, could easily be compared to a sine wave. Despite the turbulence in history, Poland had a successful period during the 16th century – which Polish historians describe as a 'golden age' ruled by the Jagiellonian Dynasty. At that time, Poland was a state to be reckoned with on a regional scale. It was so powerful that it held the crowns of a few other countries such as Lithuania, Hungary and Bohemia. With its strong military, the Dynasty facilitated the unification of Poland and Lithuania, a unified state

1 | This section heavily draws from Davies, N. (2005). *God's Playground.* Columbia University Press.

powerful enough to challenge Russia, the Ottoman Empire (today's Turkey), and Sweden (see Figure 1). The first historical document that recorded Sino-Polish relations was regarding Polish missionaries' activities in China and also dates back to the 16th Century.[2]

Since the first half of the 17th century, Poland's position as the dominant regional power had deteriorated due to frequent conflict with other countries as well as civil wars. But the growing number of conflicts was not the only reason behind its diminishing power status. The political system at that time was also weak because of the 'Polish noble democracy' system that was highly unproductive.

The most recognised example of its inefficiency, called 'liberum veto', enabled a singular representative to break the plenary session of the parliament and revoke all the bills that were voted on in that session. 'Liberum veto' and many other unproductive laws in the political system hampered all attempts to reform the state and resulted in the underdevelopment of Poland in comparison to its neighbours.[3] Besides domestic factors, the rising power of neighbouring countries, such as Prussia, Russia, Sweden and the Ottoman Empire, was another factor that weakened Poland's dominant status.

In the year 1795, Poland disappeared from the map of Europe for the next 123 years because it was divided among three of its neighbours, namely Tsarist Russia, Austria, and Prussia. During that period, Polish nationality started to broaden among the habitants of former Polish territories – many uprisings of Poles against the partitioning powers began, even though all of them failed. It was also a period when Polish national literature was fully established with authors like Adam Mickiewicz, Bolesław Prus, Henryk Sienkiewicz and Władysław Reymont, of which the latter two received the Nobel Prize in Literature.[4]

World War I started in 1914 having a major impact on international relations all around the world. The Treaty of Versailles resulted in the recreation of Poland, mostly in its previous borders. Thanks to Józef Piłsudski's wits and strategic

2 | Kalicki W. (2016, January). *Jak polski jezuita ratował chińskie cesarstwo. Nasi tam byli. [How a Polish Jesuit saved the Chinese Empire. Ours were there]*. Retrieved from http://wyborcza.pl/duzyformat/1,127290,19539032,jak-polski-jezuita-ratowal-chinskie-cesarstwo-nasi-tam-byli.html? disableRedirects =true.

3 | Filonik, J. (2015). *The Polish Nobility's "Golden Freedom": On the Ancient Roots of a Political Idea*. Retrieved from https://zenodo.org/record/896895.

4 | Snyder, T. (2003). *The Reconstruction of Nations: Poland, Ukraine, Lithuania, Belarus, 1569-1999*. Yale University Press.

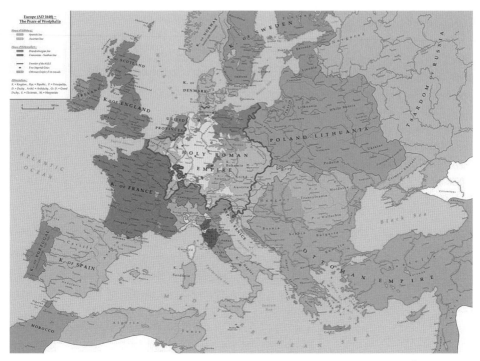

Figure 1: The area covered by the Poland-Lithuania union at the end of the 16th century is shown in the big mauve block, centre right.[5]

thinking, the words written in that treaty became a reality. It was one of the few moments in Polish history in which the political elite of the country was able to work together to achieve a common goal. The newly created Second Republic of Poland established its first diplomatic presences in China: in Shanghai and Harbin in the form of consulates.

5 | Retrieved from https:// i.pinimg.com/originals/6a/0b/8c /6a0b8c040d52ff43a0f6cc545e1 0ae14.jpg.

6 | Materski, W. & Szarota, T. (2009). *Polska 1939-1945. Straty osobowe i ofiary represji pod dwiema okupacjami [Poland 1939-1945. Personal losses and victims of repression under two occupations]*. Institute of National Remembrance.

Yet, the post-WWI independence lasted for only 21 years. The beginning of World War II, marked by the joint attack by the Soviet Union and the Third Reich, had put an end to Poland's independence for the next 60 years. During the six years of occupation by Nazi Germany and the Soviet Union, Poland lost almost 18% of its pre-war population. No other country suffered from such a large proportional death toll as did Poland.[6]

World War II resulted in Poland's submission to the Eastern Bloc and the Soviet Union. The government installed by Stalin in Poland was highly influenced by the Russian leaders. Despite the close linkage with the Soviet Union, the context offered a good occasion to develop closer political connections with the People's Republic of China (PRC) at that time. Poland was one of the first countries to recognise the PRC's government as the government of China in 1949.[7] Despite the Sino-Soviet split that had slightly frozen the development of Sino-Polish relations, Poland supported the candidacy of the PRC to the United Nations in 1971.

Meanwhile, in Poland, a significant part of the population rose against the government and created a political movement in the form of a labour revolt. Poor economic policies and continuous political repressions against the opposition by the ruling party resulted in massive protests from the workers all over the country. The uprisings subsequently ended the relatively peaceful communist rule that had determined the Polish political scene for over 40 years. The leader of the opposition, Lech Wałęsa, then became the president of the newly-founded Third Republic of Poland.

Poland experienced a massive wave of economic and political transformation during the 1990s after the collapse of the Soviet Union. For example, the authoritarian one-party system was replaced by a bi-partisan democratic system and the socialist economy was replaced by a free market notion advocated by Leszek Balcerowicz and his economic approach known as a 'shock therapy.' Balcerowicz's liberal reforms helped Poland become one of the fastest-developing countries in Europe for 25 years. Simultaneously, Poland intensively pursued its two main foreign policy objectives to join NATO and the EU. During that time, Sino-Polish relations were mainly founded on economic interests and exchanges. Trade between Poland and the PRC had grown rapidly during the 1990s, i.e. from US$144 million in 1991 to US$1.242 billion in 2001.[8]

7 | Xinhua. (2004, June 8). Major Events in Sino-Polish Relations. *China Daily*. Retrieved from http://www.chinadaily.com.cn/english/doc/2004-06/08/content_337598.htm.

8 | National People's Congress of the People's Republic of China. (2007, October 26). *China-Poland Bilateral Relations*. Retrieved from http://www.npc.gov.cn/zgrdw/englishnpc/Special/2007-10/26/content_1372503.htm.

1.2. The Political System and the Three Authorities in the Government: Legislative, Executive, and Judiciary

The political system of Poland is written into the Constitution that was voted by the citizens in the constitutional referendum of 1997. The Constitution is the most important act of law in Poland and it defines the competencies and relations among all the authorities in the country.

Poland is a democratic republic with a parliamentary-cabinet system. Therefore, no authority dominates over the other ones. Some of the central authorities, e.g. the president, Sejm[9], and Senat are chosen in general elections. Other positions such as the prime minister and his Government are selected by the Sejm itself. The local authorities of each voivodeship[10] are chosen in local elections held every four years. The Constitution places emphasis on the separation and independence of each of the authorities and attempts to create an efficient system of checks and balances among them. Thus, a significant aspect of the Constitution is the safeguard of the independence of the judicial authority. It is also The Constitution that defines the Polish voting system as a proportionally representative system.[11]

Sejm and Senat are the two houses of the Polish Parliament. There are 460 representatives in the Sejm and 100 senators in the Senat. The parliament as a whole is the place where drafts of the bills are further read, developed and voted. It also appoints the prime minister and the ministers.[12]

Direct democracy, in the form of referendums, is used for important matters concerning the whole country. There were two crucial referendums in Poland after 1989, namely the constitutional referendum mentioned earlier, and the referendum on the accession to the EU in 2003.

Despite some political turbulence in the initial period of the liberal reforms, the democratic political system promulgated in Poland has gradually stabilised in the later years. For example, the average term served by the elected prime ministers lasted

9 | The lower house of the Polish Parliament.

10 | An area administered by a voivode (governor).

11 | Sejm of Poland. (1997). *Constitution of Poland*. Retrieved from http://www.sejm.gov.pl/prawo/konst/polski/kon1.htm.

12 | *Ibid.*

for only 11 months between 1989 and 1997 (the initial phase). Nevertheless, as the political system matured, the prime ministers' average term of office between 1997 and 2017 was extended to over 25 months. Among all the Prime Ministers, Donald Tusk held the post for the longest period of almost seven years.

Formally, the president is the highest political office to be held in Poland and serves as the guardian of the Constitution and Commander-In-Chief of the Polish armed forces. On the other hand, the authority of the president of Poland is restrained by several factors. First, while the president has the right to legislate initiatives and veto bills, the veto can also be reserved by the Sejm with a three-fifths majority. Second, although the president also nominates the generals and ambassadors and is the highest representative of Poland abroad, his/her actions have to be coordinated with the government's foreign policy[13] Once the president is elected, he/she is required to leave his/her political party to avoid conflict of interests during legislations in the parliament, which limits his ability to create his own policies.

In practice, most of the political decisions are made and executed by the prime minister and the Government subordinate to the president. The Government is responsible for drafting the most important bills. The prime minister, often serving as the leader of the ruling party simultaneously, is responsible for gathering a majority vote for the bills. After the legislation, the governing administration is the most important organ to implement the laws. The ministers are usually respected public figures, experts and/or important figures in the ruling parties. Therefore, the Government meetings are also a regular form of meetings for the most important people in Polish politics.

13 | *Ibid.*

2. Modern Reforms and the Foreseeable Direction of Polish Politics

2.1. Reforms

A change of ruling party in 2015 has brought a visible shift in the directions of the country's policies in various areas. In this part, we focus on three important areas of internal state activities, namely economic, political, and constitutional reforms.

Law and Justice (Polish abbreviation: PiS) won the general election in 2015 mainly because of their economic propositions. They voted to reverse the liberal economic policy promoted by the previous government and implement some socialist programmes to improve the low standard of living of the poorest social groups in the country – the multiple-child families, pensioners, retired persons and rural population.[14] Many of the promises made in the election campaign were fulfilled. PiS revoked the decision made by the previous coalition to gradually raise the retirement age to 67 years old for both sexes and implemented a programme called '500+' which is the largest social programme since 1989. The cost of the programme is around 25 billion Polish złoty per year. Comparatively speaking, the budget of the Ministry of Defence is around 41 billion Polish złoty in 2018.[15]

Apart from the social reforms, the Government was implementing a strategic economic plan called 'Responsible Development Plan' that is promulgated by Mateusz Morawiecki, Poland's Minister of Development and also its Prime Minister of (as of November 2017). The Plan aims at supporting innovations in Polish companies, reindustrialisation of the country, attracting foreign investments, creating a 'Polish brand' in the international economy, and equalling opportunities for citizens.[16]

Political reforms are one of the most important areas of focus of the ruling party. The ideology behind the policies of the previous Government could be described in one word: decentralisation. However, the PiS reversed this ideology. It merged the functions of Minister of Justice and the Attorney General, required public television to openly support the ruling party's policies, and transferred some local authorities'

14 | PiS. (2014). *Program Prawo i Sprawiedliwość 2014 [Law and Justice Programme 2014]*. Retrieved from http://pis.org.pl/dokumenty.

15 | Dmitruk, T. (2017, November 5). *Projekt budżetu MON na 2018 rok [Draft budget of the Ministry of National Defense for 2018]*. Retrieved from http://dziennikzbrojny.pl/artykuly/art,2,4,10669,armie-swiata,wojsko-polskie,projekt-budzetu-mon-na-2018-rok.

16 | Ministry of Development. (2017). *Strategia na rzecz Odpowiedzialnego Rozwoju do roku 2020 (z perspektywą do 2030 r.) [Strategy for Responsible Development by 2020 (with a perspective until 2030)]*. Warsaw.

responsibilities to the central authorities. Controversies swirl around the reforms that have redefined the relations of the legislative and executive branches with the judiciary. The critics of the ruling party argue that most of these reforms are in a collision with the Constitution and will result in creating an authoritarian state. The supporters, however, claim that the Polish judiciary system was in such a bad shape that it needed swift and decisive actions to fix it.[17]

In the foreseeable future, PiS's most important potential areas of activity may be constitutional reforms. There have been many debates on whether the current political system of Poland is the most efficient one. And there have been a number of situations where defective laws stipulated in the Constitution led to competency disputes. Some of them had to be settled by the Constitutional Tribunal. Moreover, there are arguments that the parliamentary-cabinet system is unable to provide a firm and steady rule over the country. It has been suggested that the people decide between a chancellery system or a presidential one.

2.2. Important Political Movements

Analyses of the general elections in 2011 and 2015[18]

After comparing the results of the general elections in 2011 and 2015, one will conclude that the main shift was the switch of support from the centre and liberal parties to the conservative right-wing parties. The direct effect of this shift was the loss of power of the previous ruling coalition that consisted of The Civic Platform (PO) and Polish People's Party (PSL), and the victory of the right-wing populist Law and Justice (PiS).

The centrist parties (PO and PSL) received almost 50% of the votes in the general election in 2011 while the leftist parties could count on the support of about 18% of the voters. However, in 2015, the centrist and liberal parties altogether (PO, PSL, and newly founded Modern Party) gathered about 37% of the votes, which marked a 13% decline in support. Meanwhile,

17 | Cienski, J. & De la Baume, M. (2017, July 19). *Brussels warns Poland over judicial reforms*. Retrieved from https://www.politico.eu/article/poland-rule-of-law-constitution-brussels-warns-over-judicial-reforms/.

18 | SEC. (2011, October). *Elections 2011 to the Sejm and Senate The Republic of Poland*. Retrieved from http://wybory2011.pkw.gov.pl/wyn/en/000000.html; SEC. (2015). *Wybory do Sejmu i Senatu Rzeczypospolitej Polskiej 2015 [2015 elections to the Sejm and Senate of the Republic of Poland]*. Retrieved from https://parlament2015.pkw.gov.pl/349_Wyniki_Sejm.html.

the left-wing parties recorded a drastic decline from 18% to 10% of voters' support and were excluded from the parliament because none of them were able to reach the minimum entry threshold.

On the other hand, the right-wing parties recorded a significant increase in support between 2011 and 2015. In 2011, the conservative PiS was the only right-wing party with notable support[19] circled around 29% of the votes. In 2015, the support for PiS had grown up to 37%. In sum, an aggregate of about 50% of the voters supported the right-wing parties. This turned out to help Law and Justice in securing the majority in the parliament by forming a coalition with other right-wing parties.

Analyses of the important polls conducted in 2017

Another hint to predict the future political directions of the country is the use of opinion polls. CBOS, one of the most significant public opinion poll centres in Poland, conducts annual surveys on people's level of support for the leftist, centre and right-wing views of politics. The poll from August 2017 confirms the thesis that Poles are currently shifting to a more right-wing-inclined society than before. The trend is also visible among the youngest group of interviewees i.e. young adults between 18 to 24 years old).[20]

There are also some risks concerning this right-wing shift. Right-wing parties in Poland tend to be supportive of nationalist movements and therefore there is a slight possibility of rising radicalism in the country. The latest survey conducted by CBOS on the admission of refugees shows that almost 70% of the interviewed Poles oppose the idea of accepting refugees from the Arab states to their country.[21]

Despite rising nationalism as suggested by the election results and public opinion polls, another poll on people's support of Poland's EU membership showed that almost 83% of the Poles support Poland joining the EU.[22] As a matter of fact, Poland receives the highest level of support from its people regarding

19 | Other right-wing parties gathered less than 4% of the votes altogether and none of them was able to enter the Sejm.

20 | CBOS. (2017, August). *Czy młodzi Polacy są prawicowi? [Are young Poles right-wing?]*. Retrieved from http://www.cbos.pl/SPISKOM.POL/2017/K_102_17.PDF.

21 | Rosiejka, R. (2017, May 31). *Najnowszy sondaż CBOS: Większość Polaków przeciw przyjmowaniu uchodźców [The latest CBOS poll: Most Poles are against accepting refugees]*. Retrieved from https://wiadomosci.wp.pl/najnowszy-sondaz-cbos-wiekszosc-polakow-przeciw-przyjmowaniu-uchodzcow-6128546572019329a.

22 | Fakty TVN (2017, December). "Czy Polska powinna zostać w Unii Europejskiej?" Sondaż dla "Faktów" TVN i TVN24 ["Should Poland stay in the European Union?" Poll for "Fakty" TVN and TVN24]. Retrieved from https://fakty.tvn24.pl/sondaze-dla-faktow-tvn-i-tvn24,106/czy-polska-powinna-zostac-w-unii-europejskiej-sondaz-dla-faktow,800835.html.

joining the membership of the EU compared to all the other EU countries.

3. Current Opportunities and Challenges for Polish Youth

Poland still lacks sufficient funds to provide its young population with significant support for improving their education and career prospects. However, the need for state support for the youth has been increasingly acknowledged by politicians. New programmes targetting young people were introduced, especially in regard to professional and entrepreneurship support. There has been a clear shift in the state's approach to the issue since youth are increasingly treated as a separate social group whose problems have to be addressed with specially dedicated schemes and policies.

A significant obstacle to shaping policies that support the development of the youth is the substantial social inequalities that haunt the country. Divisions include an urban-rural gap and geographical discrepancies (i.e. the more prosperous West vs. the more economically challenged East as shown in Figure 2). The latter is a direct result of the partitions between the Austrian and Prussian part, and the Soviet part of the country. Consequently, a key challenge to the Polish authorities in the upcoming years is to provide equal access of opportunity to the rural youth, especially the ones coming from Eastern Poland.

The population's life satisfaction index corresponds with the West-East, urban-rural divisions. A 2017 poll by Ariadna Panel's research – where 60% of respondents were aged between 15 and 34 – found that the two most important factors defining

Figure 2: GDP per capita in voivodeships (in €).[23]

life satisfaction index are an individual's economic circumstances and whether or not he/she supports the ruling party[24]. In other words, compared to people living in the East, richer Poles who reside in Western Poland where there is a higher level of urbanisation have a higher level of life satisfaction.

The research also shows a correlation between politics and the citizens' quality of life. For instance, multiple 'youth schemes' were put forward by the coalition of Civic Platform (PO) and Polish People's Party (PSL) to improve young people's quality of education, support their career development and help establish a family. And a scheme called 'Start Support' (*Wsparcie w starcie*) was implemented to support young entrepreneurs to participate in business ventures by offering attractive credits to their start-ups.[25]

Apart from lacking opportunities, Polish youth face challenges that discourage them from climbing the social ladder. One of them is the high housing price. It bars many young families from purchasing a home and forces them to face high rents. Subsequently, a significant portion of the family income would be spent on housing instead of being invested in other areas such as education and business innovation. In face of the

23 | Retrieved from https://financialobserver.eu/wp-content/uploads/sites/2/2017/02/PIELACH-GDP-per-capita-jamnik.jpg.

24 | Ogólnopolski panel badawczy Ariadna (2017, June). *Zadowolenie z życia Polaków*. Retrieved from: http://ciekaweliczby.pl/wp-content/uploads/2017/06/Zadowolenie-z-%C5%BCycia-Polak%C3%B3w_Prezentacja.pdf.

25 | See http://wsparciewstarcie.bgk.pl/ for more information regarding the schemes that support the youths.

challenge, the PO-PSL coalition implemented the 'House for Youth' scheme (*Mieszkanie dla Młodych*) to provide the youth and young families with preferential credits to purchase an apartment with the state subsidising 10-20% of the costs. On top of that, PiS is currently implementing the 'Housing +' (*Mieszkanie+*) scheme to construct state-owned affordable housing. With the support of the scheme, the price of rent is expected to be three to four times lower than the market price. These apartments will be primarily offered to young households, particularly the ones having two or more children.[26] Additionally, this type of families is the primary beneficiaries of the '500+' scheme mentioned previously, which provides these households with self-managed financial support.

Besides the policies mentioned above, much has been done to include young people in the political discourse. For example, Youth City Councils are established to serve as advisory boards to the City Councils. In addition, a Sejm of Children and Youth is set up annually to provide the youths with a national platform to express their views on political, social and economic issues.

Another major challenge is the emigration of young Poles and the subsequent brain-drain effect. Many young Poles are moving to Western Europe with the hope of getting more job opportunities and better career prospects. There are around 2.8 million Poles who consider moving abroad to work.[27] However, it is believed that the problem of emigration can be controlled when the Polish economy starts developing and providing more jobs. Besides the Polish youths, it is worth mentioning the important role played by Ukrainian youths, who are moving to Poland to seek better education and job opportunities. Apart from the value that the Ukrainian youths might contribute to the Polish economy e.g. labour and skills, it is also crucial for the government to acknowledge their presence to formulate policies to facilitate their integration into the wider Polish society.

26 | For more information, see https://narodowyprogram.pl/mieszkanie-plus/mieszkanie-plus-biala-podlaska/?gclid=Cj0KCQi AsqLSBRCmARIsAL4Pa9Rgqc xssXvfi5rl2zyb3k4U1ixW3XAG X25vu71HmBXQYl7ScTwljyga AvGUEALw_wcB.

27 | Ratajczak, M. (2017, November 15). *Polacy wciąż chcą szukać szczęścia za granicą. 2,8 mln rodaków myśli o emigracji zarobkowej [Poles still want to look for happiness abroad 2.8 million compatriots think about economic migration].* Retrieved from https://www.money.pl/gospodarka/wiadomosci/artykul/emigracja-wyjazd-do-pracy,4,0,2389252.html.

3.1. Summary

Despite being placed in a complicated geopolitical competition between Germany and Russia, or widely speaking, Russia and the West, and being deprived of an independent status over a long period, Poland has demonstrated itself as an exemplary story of successful post-Soviet economic transformation. The three decades after the fall of the Berlin Wall comprise a period of continuous economic growth, the growing prosperity of society, maturing democratic state institutions, and strengthening status in the EU's multilateral regional structure.

The economic growth, ever-growing political influence, increasing military capabilities and its population clearly set Poland to be the leader of the CEE region. However, it is neither going to be easy to secure such a position nor to play such a leading role. CEE is still a developing region with quite often contradictory interests among the small countries located in the region. Without a doubt, Poland has an important role to play in both the BRI and the 16+1 framework. All the aforementioned factors, as well as the continuous positive approach to engaging with China, will potentially shape Poland to become a strategic partner of China in the CEE region.

4. Opportunities and Challenges Offered by the Belt and Road Initiative

4.1. The Role of Poland Within the Belt and Road Initiative

Poland is a key actor within the Central and Eastern Europe (CEE) arena and is related to a number of Belt and Road projects. Three qualities of Poland form its particular position, namely its market potential, geographic location, and the political role played in the region.

First of all, Poland is the largest economy in the CEE and has great potential to be the leader of the region to develop diplomatic relations with China within the 16+1 China-CEE Cooperation framework. Poland's domestic market is the largest among the 16 CEE countries engaged by China within the framework given that the Polish market on its own constitutes one-third of the entire CEE market (see Figure 3).

According to the data published by the Central Statistical Office of Poland in 2016, Poland is also China's largest trading partner in the region indicated by the €21.67 billion-worth imports and the €1.73 billion-worth exports.[28] However, there is a significant trade imbalance between Poland and China, as depicted by the fact that Chinese exports to Poland are over 12 times higher than Polish exports to China which focus on exporting manufactured goods (particularly copper products), machinery and transport equipment, and agriculture products.[29]

In terms of receiving foreign direct investments (FDI), Poland is among the top destinations within the CEE that attracts Chinese capital. As of 2014, Poland was one of the six CEE countries[30] which attracted 95% of the Chinese investments with an estimated value of €329.4 million.[31]

Since then, Poland has seen a considerable inflow of Chinese investments. Notably, the overall value of Chinese foreign direct investments (FDI) in Poland increased significantly throughout 2016. China Everbright International's acquisition of Novago, a Polish innovative ecological company, with €123 million in 2016 is a case in point.[32] The transaction was the biggest recorded Chinese acquisition in Poland and received approval and support from the Polish government. With many other deals taking place, the total value of Chinese FDI in Poland by the end of 2016 amounted to €757.6 million, according to the estimations shared with the authors by the Economic Section of the Embassy of the Republic of Poland in Beijing. That means Chinese investments in Poland nearly doubled over two years.

28 | Statistics Poland. (2017, October 26). *Yearbook of Foreign Trade Statistics 2017.* Retrieved from https://stat.gov.pl/en/topics/statistical-yearbooks/statistical-yearbooks/yearbook-of-foreign-trade-statistics-2017,9,11.html.

29 | *Ibid.*

30 | Other countries include Bulgaria, the Czech Republic, Hungary, Romania, and Slovakia.

31 | Stanzel, A. (2016, December 14). *China's Investment in Influence: The Future of 16+1 Cooperation.* Retrieved from http://www.ecfr.eu/publications/summary/chinas_investment_in_influence_the_future_of_161_cooperation7204.

32 | TVN24. (2016, August 29). *Gigantyczne przejęcie śmieci w Mławie [Giant Waste-sector Acquisition in Mława].* Retrieved from http://tvn24bis.pl/z-kraju,74/china-everbrightkupila-novago-z-mlawy-najwieksza-chinska-inwestycja,672100.html.

2016	MARKET SIZE (POPULATION, MN)	PURCHASING POWER (PER-CAPITA GDP, US$)
ALBANIA	2.9	4,200
BOSNIA AND HERZEGOVINA	3.9	4,300
BULGARIA	7.1	7,400
CROATIA	4.2	12,100
CZECH REPUBLIC	10.6	18,300
ESTONIA	1.3	17,600
HUNGARY	9.8	12,800
LATVIA	2.0	14,100
LITHUANIA	2.9	14,900
MACEDONIA	2.1	5,300
MONTENEGRO	0.6	6,600
POLAND	38.0	12,300
ROMANIA	19.8	9,500
SERBIA	7.0	5,400
SLOVAKIA	5.4	16,500
SLOVENIA	2.1	21,300

Source: IMF

Figure 3: Market Size of the CEE Countries as of 2016 with Visegrad Group Countries Highlighted [33]

Nevertheless, similar to other countries in the CEE, Chinese FDI forms less than 0.01% of the overall FDI received by Poland.[34]

Poland occupies a strategic position along the Eurasian Land Bridge, a transportation link established as part of the BRI.[35] Along the road, Poland is located on the Eastern border of the European Common Market, which makes it a natural transportation hub. The transported goods can reach any destination in the EU within three days when they are sent from a logistics headquarter in Poland (see Figures 4 and 5). Much attention is given to the city of Łódź (Lodz or 羅兹) located in Central Poland, where is expected to play the role of a major logistics hub within the EU.[36]

Another project relevant to Poland's strategic location is the Chinese plans of constructing the Land-Sea Express Route connecting the Greek port of Piraeus with Belgrade and Budapest.[37] Poland could then serve as a confluence of the Eurasian Land-Bridge and the Land-Sea Express Route.

33 | Graph retrieved from HKTDC Research. (2017, May 29). *The Visegrad Four (V4) Nations: Early Adopters of the Belt and Road Opportunity.* Available at https://hkmb.hktdc.com/en/1X0AA60I/hktdc-research/The-Visegrad-Four-V4-Nations-Early-Adopters-of-the-Belt-and-Road-Opportunity.

34 | Narodowy Bank Polski. (2017). *Foreign Direct investment in Poland and Polish direct investment abroad.* Retrived from http://www.nbp.pl/publikacje/pib/FDI_report_2015_en.pdf.

35 | Debreczeni, G. (2016). *The New Eurasian Land Bridge - Opportunities for China, Europe, and Central Asia.* Retrieved from http://publicspherejournal.com/wp-content/uploads/2016/02/02.eurasian_land_bridge.pdf.

36 | The cooperation between the city of Łódź and łódzkie voivodeship and Chinese partners (in particular Chengdu and Sichuan province) has been developing rapidly over the years. You can contact a university think tank which advises on the local authorities of the voivodeship. See http://osa.uni.lodz.pl for more information.

37 | Kynge, J., Beesley, A. & Byrne, A. (2017, February 20). *EU sets collision course with China over Silk Road rail project.* Retrieved from https://www.ft.com/content/003bad14-f52f-11e6-95ee-f14e55513608.

38 | Graphs retrieved from http://cctv.cn.

Figure 4: An illustration of the Łódź-centric transportation hub[38]

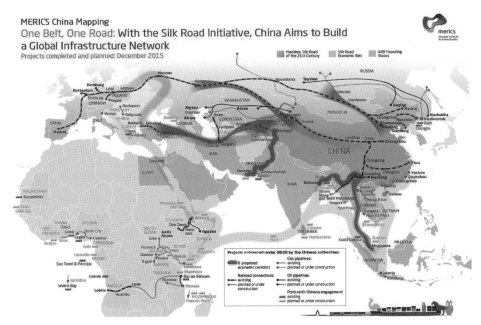

Figure 5: A map showing China's vision of building a global infrastructure network[39]

Finally, Poland possesses political significance given that it is a prominent member of the EU as well as a key actor in the BRI. Since it joined the EU in 2004, the position of Poland has grown significantly which makes it the key EU member state within the CEE region. As a matter of fact, Polish politicians have been the only ones from the CEE to take up top EU positions. For instance, former Polish Prime Minister Jerzy Buzek served as the President of the European Parliament between July 2009 and January 2012, whereas former Prime Minister Donald Tusk served as the second permanent President of the European Council (December 2014-December 2019).[40]

This example also demonstrates the diplomatic tension between Law and Justice's government and Brussels, which are related to political controversies over policies of the Polish government discussed in previous parts of this paper. In spite of the tension existing at the governmental level, 83% of the Poles

39 | Figure 5 was retrieved from http://www.merics.org.

40 | Herszenhorn, D. (2017, September 3). *EU leaders defy Warsaw to reappoint Donald Tusk*. Retrieved from https://www.politico.eu/article/eu-leaders-reappoint-donald-tusk-as-council-president/.

support the country's EU membership and its strengthening integration with the EU.[41] In the long run, Poland is expected to become one of the influential EU actors given the country is the largest economy in the EU's Eastern flank. Such political significance should be taken into account by China while formulating its BRI policies with Poland.

4.2. BRI as an Opportunity for Poland

Although the Belt and Road Initiative can offer significant opportunities to Poland, it is important to keep the plans realistic. The China-CEE cooperation can be seen as setting up the fundamentals for future development. However, applying overly ambitious rhetoric may backfire. For instance, a prolonged period of limited success would be seen as a failure if there was too much grandiose rhetoric. And in 2018 China itself was considering backtracking on the formula, as discussions were held to move China-CEE Summits from an annual mode to once every two years[42]. Irrespective of lack of spectacular success stories, the CEE countries remain earnestly interested in the economic dimensions of the BRI as explained in the following four points below.

4.2.1. Invigorating Polish exports to China

As described earlier, the trade exchange between China and Poland is highly imbalanced. With the establishment of the 16+1 framework in 2012 and the announcement of the BRI, there has been much hope for bolstering Polish exports. Particularly as the development of Chinese initiatives addressing the region overlapped with the 2014 Russian embargo on European food products, which remained in place as of 2017[43]. In 2013 Polish agri-food exports to Russia amounted to €1.25 billion from the €20.4 billion total and due to sanctions, the value dropped from €1.25 billion to just €398 million between 2013 and 2015[44]. This pushed Polish exporters to search for new markets. Naturally, given the establishment of 16+1 China-CEE cooperation and its message of increasing trade and agriculture cooperation, there were high hopes that the Polish food products could rapidly enter

41 | Fakty TVN (2017, December). *"Czy Polska powinna zostać w Unii Europejskiej?" Sondaż dla "Faktów" TVN i TVN24 (http:// www.tvn24.pl).* Retrieved from: https://fakty.tvn24.pl/sondaze-dla-faktow-tvn-i-tvn24,106/czy-polska-powinna-zostac-w-unii-europejskiej-sondaz-dla-faktow,800835.html.

42 | Barkin, N., Emmot, R., & Tsolova, T. (2018, March 12). *Exclusive: China may pare back 'divisive' Eastern Europe summits.* Retrieved from https://www.reuters.com/article/us-china-easteurope-exclusive/exclusive-china-may-pare-back-divisive-eastern-europe-summits-idUSKCN1GO1PI.

43 | Michalopoulus, S. (2016, June 30). *Russia extends embargo on EU food products.* Retrieved from https://www.euractiv.com/section/agriculture-food/news/russia-extends-embargo-on-eu-food-products/.

44 | Wyciszkiewicz, E. (2017, January). *The impact of Russian sanctions on the Polish agri-food sector.* Retrieved from http://cprdip.pl/en,projects,russias_influence_activities_in_cee,the_impact_of_russian_sanctions_on_the_polish_agri-food_sector.html.

the Chinese market. However, the trade cooperation turned out to be more complicated than expected and the procedure was only completed in 2016 after the state visit of President Xi Jinping following over four years-long application process, which some Polish exports suspected of being intentionally delayed by China to be used as a bargaining chip[45]. Nevertheless, as of 2014, the volume of Polish food products exported to China was the largest within the region.[46] Specifically, Polish apples and dairy products made up of over 30% of all the Polish food exported to China in 2015.[47, 48]

Responding to China's growing economy, rising quality of life and demand for a more stable supply of food, the PRC needs to increase its exports as well as its imports to better satisfy the needs of the Chinese people who have more purchasing power and demand for goods. This opens a window of opportunity for enhancing Sino-Polish trade relations through an increase in food imports.

However, the lack of branding of Polish products in the Chinese market constitutes a considerable challenge to Polish food exports. New opportunities may appear alongside the potential collaboration between the growing Chinese e-commerce market and the logistic hubs in Poland. Such collaboration could allow a more agreeable trade relationship and facilitate the promotion of Polish exports in the Chinese market.

4.2.2. Innovative business cooperation

Another potential area of cooperation is business innovation in terms of intellectual property (IP) development.

In recent years, Chinese authorities have put much emphasis on developing innovative technologies and attracting foreign start-ups to the country's 7500+ incubators.[49] But there is also much room for utilising Chinese FDI to cooperate with innovative Polish businesses through mergers and acquisitions.

A case in point is the aforementioned acquisition of Novago

45 | Jones, M. (2016, June 23). *China grants access to Polish apples.* Retrieved from http://www.fruitnet.com/asiafruit/article/169111/china-grants-access-to-polish-apples.

46 | Ministry of Treasury Republic of Poland. (2015). Poland as China's Key Partner in *CEE.* Retrieved from https://kujawskopomorskie.trade.gov.pl/en/news/171706,poland-as-chinas-key-partner-in-cee.html.pdf.

47 | Jones, M. (2016, June 23). *China grants access to Polish apples.* Retrieved from http://www.fruitnet.com/asiafruit/article/169111/china-grants-access-to-polish-apples.

48 | Xinhua. (2017, May 3). *Polish diary farmer turns east as China pushes Belt and Road Initiative.* Retrieved from http://news.xinhuanet.com/english/2017-05/03/c_136251859.htm.

49 | Xinhua. (2017, September 19). *China tops the world in incubators, makerspaces.* Retrieved from http://www.chinadaily.com.cn/business/2017-09/19/content_32203134.htm.

by China Everbright International. On one hand, the acquisition allows the innovative Polish company to gain access to capital and foreign markets. For the Chinese counterpart, the deal allows them to access innovative technology and acquire a company with high business potential. This acquisition constitutes a typical win-win situation for both Poland and China.

Poland can benefit from connecting its vibrant start-up environment with Chinese capital by setting up cooperation mechanisms among Chinese and Polish start-up incubators and establishing visa programmes that increase Chinese incubators' mobility in the EU (given the fact that Poland is located within the Schengen area where unrestricted travel within a territory of 26 European countries is guaranteed).[50] Similar programmes have already been implemented by countries such as Italy and Lithuania.[51,52] Such visa arrangement will be especially appealing to start-ups which require high geographical mobility to expand their markets and sources of capital.

However, such visa programmes may pose a legal challenge to the existing national and EU regulations regarding the protection of intellectual property as well as foreigners' mobility in the region. Since there is still insufficient trust and understanding of the Chinese market in relation to its intellectual property protection, the establishment of a reliable legal mechanism would be necessary to foster further cooperation between China and Poland.

In spite of the challenge, the Polish government practically supports this type of cooperation. As suggested by Radosław Domagalski-Łabędzki , the Deputy Minister of Economic Development in Poland, the Polish government will 'promote the projects related to high-technology transfer, which is why the investments in innovation, research and development and modern services sector enjoy support of the Polish government.'[53] Nevertheless, some Polish diplomats based in Beijing, who were contacted in the course of this study, emphasised the need for a reliable system to protect the intellectual property of Polish companies which accept Chinese

50 | Coleman, A. (2016, May 20). *Poland On Track To Becoming A Major European Tech Startup Hub.* Retrieved from https://www.forbes.com/sites/alisoncoleman/2016/05/20/poland-on-track-to-becoming-a-major-european-tech-startup-hub/.

51 | See http://italiastartupvisa.mise.gov.it/ for more information regarding Italy's visa programme for start-ups.

52 | See http://startupvisalithuania.com/ for more information regarding Lithuania's visa programme for start-ups.

53 | Polska Agencja Prasowa [Polish Press Agency]. (2016, August 30). MR: China Everbright International zainwestuje w mławską firmę Novago (komunikat) [Ministry of Development: China Everbright International to invest in the Mlawa company Novago (announcement)]. *Forsal.* Retrieved from https://forsal.pl/artykuly/971458,mr-china-everbright-international-zainwestuje-w-mlawska-firme-novago-komunikat.html.

capital.

4.2.3. Infrastructure investments

Infrastructure investments remain a complicated issue in Sino-Polish relations. Much of the offers put forward by China as part of the 16+1 framework are not attractive to the EU member states, which enjoy access to the EU's Cohesion Fund and follow the EU regulations (e.g. regulations related to the bidding system).

Although a failed public contract between Poland and Chinese company COVEC (Chinese Overseas Engineering Group) took place in 2009, there are still those hoping to see fruitful cooperation in the future.[54] COVEC's failure proved the need for a better understanding of one another, as Chinese partners did not comprehend Polish public investment's bidding system or the importance of EU regulations. And the two partners had a different understanding of the nature of a public contract, which cannot be renegotiated under Polish law. Furthermore, COVEC had to face unfavourable increases in material costs due to market change and pressure asserted on suppliers by the company's German competitor. As of today, two infrastructure investment projects stand out, namely the construction of Via Carpatia (a road link going across the Baltic area), and the Central Communication Port. However, the successful execution of the projects demands the Chinese government and companies' active involvement in tackling the legal constraints in order to keep the deals attractive for both sides.

4.2.4. Tourism

Finally, the growing number of Chinese tourists creates a big potential market for Poland's tourism industry to tap into. According to a report by Goldman Sachs, China's outbound tourism to Europe is expected to be doubled from 2015 to 2025.[55] CEE is a novel travel destination for Chinese tourists and the region is garnering more attention and attracting an increasing number of Chinese visitors. For example, Poland

54 | Kanarek, P. (2017, August 26). *Perspectives for development of China-EU relations in the infrastructure investment sector: a case study of COVEC's investment in Poland*. Retrieved from http://www.jpolrisk.com/category/europe/poland/.

55 | Sho, K. et al. (2015, November 20). *The Chinese Tourist Boom*. Retrieved from http://www.goldmansachs.com/our-thinking/pages/macroeconomic-insights-folder/chinese-tourist-boom/report.pdf.

itself attracted 50,000 Chinese tourists in 2014 and it was estimated that more than 130,000 Chinese tourists visited the country in 2017.[56] The number is expected to increase gradually in the upcoming years.

4.3. BRI as an Opportunity for Young Poles

The challenges mentioned above seem to have a common theme, i.e. there is a need to develop new human capital capable of working in a cross-culture environment and helping the countries bridge the cultural, legal, and intellectual gaps. Such gaps offer invaluable opportunities for young people to play a pioneering role in advancing Sino-Polish relations. And the young people of the CEE have a greater opportunity than their Western European peers, as the relations between the region and China have begun to gain impetus only recently and there is still a high need to develop new human capital to facilitate the growing cooperation[57]. Consequently, young Poles can be a catalyst in developing Sino-Polish relations and the BRI by participating in education schemes and setting up channels that foster closer communication between the two countries. In this way, young Poles can offer their human capital to facilitate Sino-Polish relations.

4.3.1. Education opportunities

Under the influence of the BRI, more education programmes are becoming available to young Poles, notably the language programmes offered mainly by the Confucius Institute. Educational and business exchange programmes are in high demand because they are likely to attract young Poles with diverse backgrounds to engage with China in a more effective manner. For instance, the 'Warsaw-Beijing Forum: Youth for Business' organised meetings and workshops for Chinese and Polish youths who are interested in business and economic cooperation.[58] In November 2017, a workshop was held in cooperation with the Polish Investment and Trade Agency and China University of Political Science and Law for students from the University of Warsaw to learn more about entering the Chinese market and protecting trademarks.[59] Another

56 | Raczynska, I. (2017, November 27). *Chinscy turysci wybieraja Polske [Chinese Tourists Choose Poland]*. Retrieved from https://turystyka.wp.pl/chinscy-turysci-wybieraja-polske-rekord-przyjazdow-pobity-6192257361311873a.

57 | Turcsányi, R., Qiaoan, R., & Kiriz Z. (2014). *Coming From Nowhere: the Chinese Perception of the Concept of Central and Eastern Europe*. Retrieved from http://dspace.uni.lodz.pl:8080/xmlui/bitstream/handle/11089/11318/11.155_171_kriz.pdf; sequence=1.

58 | See http://warsaw-beijing.pl/about for more information regarding the Forum.

59 | Polska Agencja Inwestycji Handlu [Polish Trade Investment Agency]. (2017, November 23). *Zagraniczne Biuro Handlowe w Szanghaju edukuje młodych prawników [The Foreign Trade Office in Shanghai educates young lawyers]*. Retrieved from https://www.paih.gov.pl/20171123/biuro_w_szanghaju_edukuje_mlodych_prawnikow.

example is that there are increasing postgraduate-level courses that target the BRI. The course 'Business in China – how to act efficiently in the era of the Belt and Road' organised by Kozminski University in Warsaw is a good case in point.[60] Given the expanding significance of the BRI, more programmes of this type are needed to facilitate exchanges between Chinese and Polish youths.

4.3.2. Work opportunities in China

With expanding investments abroad, Chinese companies start searching for employees who are capable of managing foreign assets. The demand for talent presents an opportunity for young Poles based in China to work for outgoing Chinese firms.

It is possible that there will be an increasing number of young Poles living in China who will join business consultancies to advise Polish companies that plan to enter the Chinese market. Since the digitisation of Chinese business and its growing e-commerce platforms, it is insufficient to be familiar with only the tech side of the business. It is also crucial to stay attuned to China's incredibly fast development and changing business environment. In this case, young Poles living in China can possess a competitive advantage over others who have never been to China.

4.3.3. Work opportunities in Poland

Developing Sino-Polish relations have a capacity to create new jobs in logistics or in representing Polish companies in dealing with Chinese partners. New positions can also be created when Chinese companies enter the Polish market.

4.3.4. Start-up funding

Chinese investment schemes established for Polish start-ups can target young people specifically because 'more than one in four managing directors at start-ups is under 30 years old.'[61] These young people are also more open to exploring alternative sources of funding and may be more responsive to

60 | For more information regarding the programme, see http://kozminski.edu.pl/pl/oferta-edukacyjna/studiapodyplomowe/kierunki/biznes-chinski-jak-dzialac-skutecznie-w-czasach-jedwabnego-szlaku/o-studiach/.

61 | Beauchamp, M., Kowalczyk, A., Skala, A. (2017). *Polish Startups Report 2017*. Retrieved from http://startuppoland.org/wp-content/uploads/2017/10/SP_raport17_ENG_singlepages_lr.pdf.

the investment programmes initiated by China.

When discussing these issues, it is crucial to remember that Sino-Polish relations are still developing. Therefore, the establishment of necessary fundamentals for cooperation is needed – be it human capital, trust, or institutional framework. The BRI offers stimulating and new opportunities for many young Poles, but many of the benefits are not immediate or still at an early stage of development. It is then important to be ambitious, but to keep the plans realistic so that the expectations would not be likely to backfire.

5. Suggestions for Improving Sino-Polish Cooperation Under the BRI

There are four primary barriers that limit the level to which young Poles could participate in the BRI-related projects, namely the lack of knowledge, insufficient institutions to build trust in the Chinese business environment, superficial engagement, and lack of professional opportunities provided for young people in the country. All the obstacles come with proposed solutions that are aimed at helping to create a fundamental basis for attracting new human capital to work on the BRI projects.

5.1. Obstacle 1: The Lack of Knowledge About the BRI Projects

Potential Solution 1: To establish a single coordinated 'China Opportunity' or 'BRI Platform' that shares updated information regarding the available Chinese education programmes and China-related events for young Poles

Such a platform, if properly promoted, could efficiently

disseminate updated information about new BRI-related opportunities such as exchange programmes, China-related events and language classes to young people. Similarly, actively promoting these opportunities on social media would help increase the exposure of the BRI and encourage more young people to learn about the Initiative and the multifaceted opportunities it may bring.

Potential Solution 2: To establish more Chinese language classes at primary schools and high schools

It is still uncommon to see schools incorporating Chinese language education into their conventional teaching curriculum. However, such incorporation is on the rise in language teaching at the high-school level. For example, a bilingual class (Polish and Mandarin Chinese) was created in Świętochłowice in 2016.[62] Besides, the Confucius Institute in Cracow is collaborating with Bartłomiej Nowodworski High School to arrange Mandarin Chinese classes for the high-school students.[63] Having students to learn about the Chinese language as well as providing them with a chance to visit China for a language exchange programme at the pre-university level of education could significantly benefit students who want to learn a foreign language , be trained as a bilingual speaker, and explore their interests in new areas of study such as China Studies.

5.2. Obstacle 2: Insufficient Institutions to Build Trust in the Chinese Business Environment

Potential Solution: To establish reliable and transparent mechanisms for protecting the intellectual property rights of Polish partners

In an interview conducted in the course of gathering data on the potential synergy between Chinese investors and the Polish start-up community, a couple of young start-up entrepreneurs mentioned to me about their reserved approach towards accepting Chinese funds and engaging with Chinese partners. Their main anxieties include a potential unauthorised

62 | Grychtol, W. (2016, September 1). *Rozpoczęcie roku w ZSO nr 1: klasa uniwersytecka z językiem chińskim, nowy dyrektor i nauczyciele [Year Opening at ZSO no. 1: University Class with Chinese Language, New Director and Teachers]*. Retrieved from http://swietochlowice.naszemiasto. pl/artykul/rozpoczecie-roku-w-zso-nr-1-klasa-uniwersytecka-z-jezykiem,3844178,artgal,t,id,tm. html.

63 | Dziennik Polski. (2013, November 4). *Nowodworek: tradycja i prestiż od ponad 400 lat [Nowodworek: Tradition and Prestige for Over 400 Years]*. Retrieved from http:// www.dziennikpolski24.pl/ artykul/3278258,nowodworek-tradycja-i-prestiz-od-ponad-400-lat,id,t.html.

technology transfer and the lack of knowledge about the Chinese business culture. Therefore, setting up necessary mechanisms in cooperation with Polish lawyers and organising informative events introducing Chinese business culture to start-up incubators in Poland would be mutually beneficial.

5.3. Obstacle 3 - Superficial Engagement

With a growing number of programmes and exchanges, it is important to not only improve their quantity, but also their quality. Numerous programmes do not allow international students to fully embrace Chinese society and learn about it. This in-depth level of understanding is essential to constructing new human capital. The organised education programmes and events have to go beyond transferring superficial knowledge and get into deeper interactions and understanding.

Potential Solution 1: To help achieve greater integration into Chinese society among participants in the education programmes and to move beyond mere language teaching

This can be achieved by preparing more home-stay programmes or ones that emphasize on cultural interactions between Chinese and international participants. It is also important to move training beyond the language dimension and increasingly focus on business or legal training.

Potential Solution 2: To Arrange More Joint Events with Organic Interactions

An issue with many programmes is their top-down nature, as they are built around high-level decisions rather than more organic lower-level cooperation (e.g. regional cooperation, sister cities programmes, or school exchanges). Relying only on a structure established during high-level political summits such as China-CEEC Young Political Leaders Forum is, of course, beneficial, but as a singular, political event may not be fully effective in truly deepening mutual understanding. In order to form new human capital among the youth of both countries, genuine interactions have to take place and human links have to be established. It is necessary to run more programmes in which both Chinese and Polish participants who share similar goals are placed in a similar position, e.g. shared dormitories, joint trips, etc. The natural interactions form stronger bonds and learning through informal interactions can be a significant component of the quality of exchange.

5.4. Obstacle 4: Young Talent not Being Given Enough Professional Opportunities

Potential Solution 1: To establish a top-10 under-30 award (with five Polish and five Chinese awardees)

The number of awardees is not the key part of this proposition, but establishing a cyclical system of recognising young talents who are active in developing Sino-Polish relations might potentially motivate more young talents to join this effort. It would also help companies and institutions to contact or make career offers to these young talents as well as create a community of alumni awardees.

Potential Solution 2: To include young people in top events under a young experts programme

Allowing recognised young talents to participate in important business or political events through a system of internships offered by both Chinese and Polish companies would allow grooming young experts for future cooperation.

5.5. Conclusion

Sino-Polish relations have been developing rapidly over the course of the last decade and it is important to sustain the momentum. Although many would claim that the economic outcome has not met expectations, given the starting point of the relations a decade ago the results can be characterised as positive. The relations still suffer from lack of sufficient mutual understanding and of clear vision, but China's Belt and Road Initiative and 16+1 China-CEE Cooperation provide multiple platforms and frameworks to improve Sino-Polish relations.

The next steps to develop cooperation are to focus on improving mutual understanding through enhancing the quality of people-to-people exchange. Afterwards, this enhanced understanding will allow the setting of ambitious, but realistic goals that can be achieved. That in turn will allow forming a clear vision and strategy for the future of Sino-Polish relations, which seem to be drifting rather than consciously developing in a specific direction.

New and old Cernavoda bridges span the Danube River, Romania. They connect the cities of Cernavoda and Fetesti.

CHAPTER 14

by **VLADIMIR VASILE**
New York University Abu Dhabi

BIANCA VLAD
The Paris Institute of Political Studies

(Ordered by last name)

ROMANIA

ESTONIA

LATVIA

LITHUANIA

POLAND

CZECH
REPUBLIC

SLOVAKIA

HUNGARY

SLOVENIA

CROATIA

BOSNIA &
HERZEGOVINA

SERBIA

MONTENEGRO

NORTH
MACEDONIA

BULGARIA

ALBANIA

ROMANIA

1. Overview of the Political Structure of Romania

1989 marked a year of turbulent political unrest, not just in Romania, but in the vast majority of Eastern Europe. The fall of Nicolae Ceausescu's rule[1] at the turn of the decade in December of 1989 posed many questions to Romania's volatile revolution.

Would the country remain under the influence of a, perhaps weaker, Soviet Union? Would NATO's borders expand into what was promised as 'neutral' territory to the Russians during the Soviet dissolution of 1991? Without a doubt, many Romanians felt a sense of hostility toward both superpowers in the Cold War – the Soviets, for permitting an autocratic regime (that led to a move away from the Russosphere), and the Americans, for turning a blind eye to Soviet aggression in Eastern Europe.

These attitudes enabled the then President Ion Iliescu to lead Romania for much of the 1990s and early 2000s. The now-retired Iliescu is a former communist, who, in the 1990 election, won a landslide 85% of the vote[2]. The hostility towards both the USA and USSR meant that Romanians were reluctant to change from the non-partisan and isolationist ways of Ceausescu, therefore electing Iliescu, who kept Romania away from international integration for much of the 1990s.

Iliescu's vast popularity at the 1990 elections reflected fear of an overly dramatic and rushed transition away from a centrally-planned economy, and his rule – marred by allegations of repression, a communist infiltration of administration, vast corruption and poverty – largely helped shape the nepotistic, somewhat kleptocratic and occasionally illiberal political system Romania is accused of having today.

A mix of both anti-communist sentiment and political corruption (in parties led by former communists), as well as the desire to retain a social safety net (a benefit arising from the communist era that Romanians most valued about the regime), have meant that Romania's political system today is one in which vital decisions are made behind closed doors, leaving a politically uninspired population who are often in the

1 | Ceausescu was the communist dictator who ruled over Romania from 1965 to 1989. Coming to power through the communist government installed by the Soviets post-World War II, he was considered one of the more liberal Warsaw Pact heads that, although advocating for socialist policies, was not so hostile to the West. This changed in the latter half of his reign – in a bid to pay off foreign debt, he implemented harsh austerity policies throughout the 1980s and sought to create a self-sufficient autarky, leading to worsening living standards and his eventual downfall.

2 | Nohlen, D., & Stöver, P. (2010). *Elections in Europe: a data handbook*. Baden-Baden, Germany: Nomos.

dark about the state of affairs in Bucharest.

1.1. Romania's Political Framework

Romania functions as a semi-presidential, representative democratic republic with two elected chambers. The president is chosen in elections that occur every five years – they control foreign affairs and the armed forces, symbolise an impartial force in politics that helps resolve political upheaval, and represent the interest of the people (explaining why Presidents must resign from their parties upon election to the post).

The prime minister, who has to be appointed by the president, is the head of the government and is typically the leader of the winning party at the legislative elections that occur every four years. Romania functions through a multi-party system. The proportional voting system in Romania ensures that a single party rarely receives a majority of votes or seats in either chamber of parliament, meaning Romanian parliamentary politics is regularly prone to uneasy, fragile coalitions (that largely exist out of parties' desire to be in government, as opposed to ideological similarities).

The third branch of government – the courts – manifests itself largely through the independent Constitutional Court, which establishes whether legislative acts are constitutional (these decisions cannot be overturned by a parliamentary majority). It is the sole body that has the power to overturn parliamentary decisions.

The role of citizens is limited in politics; the 2009 referendum on parliamentary reform (to reduce the number of parliamentarians) was approved by the electorate with a majority turnout, but so far, no legal change has occurred. Similarly, a 2012 referendum to impeach former President Traian Basescu failed. Despite support for his impeachment at 87.52%[3], the turnout of 46.24% failed to meet the 50%+1 required threshold of approving a referendum. In many ways, this characterises the foundations of Romania's political framework – a system with *de jure* democratic checks

3 | Biroul National Electoral. (2012). *Comunicat privind rezultatele referendumului national din data de 29 iulie 2012 [communicated on the results of the national referendum of July 29, 2012]*. Government of Romania.

and balances in place, but an often misled or uninterested electorate that allows politicians to bypass national laws.

1.2. Romania's Communist Legacy

In order to fully ascertain the present circumstances of Romania's political system, one must assess its history. Romania's revolution of 1989 largely occurred in three cities – Bucharest, Timisoara (the origin of protests) and Cluj. The revolution, spearheaded by intellectuals who had been repressed under the regime[4], sought to remove communist apparatchiks from all public offices and achieve the economic liberalisation of Romania, as mandated in their 'Proclamation of Timisoara'.

In reality, many communists clung to power; the NSF, or the National Salvation Front (composed of many former communists who had 'defected' in 1989 out of fear of being prosecuted), was a transition group meant to assist with the process of establishing free elections and went on to win the elections with a sweeping majority.

Subsequent protests arose in 1990 to support the so-called 'neoliberalism' the intellectuals had advocated for, weakening the NSF. At the time Ion Iliescu called upon the assistance of, and manipulated, the Jiu Valley miners to violently crush these protests in what came to be known as the Mineriads[5]. Somewhat prematurely, the anachronistic suppression of the public that was supposed to have ended along with Ceausescu's regime sealed Romania's political fate and ushered in an era of hybrid, illiberal, corrupt rule, with many illicit, secretive and counter-intuitive decisions being made.

The manipulation of the Jiu Valley miners and the indoctrination of the largely isolated, but electorally crucial rural population by Romania's Intelligence Service[6] allowed former communists to remain in the political spotlight to this day. Romania's communist legacy has ensured that the secretive ways of the 1990s are still problematic in today's political system, explaining the political apathy of an electorate that feels powerless –

4 | Kotkin, S. & Gross, J. T. (2010). *Uncivil society: 1989 and the implosion of the communist establishment*. New York: The Modern Library.

5 | Baleanu, V. (1995). *The Enemy within: The Romanian Intelligence Service in Transition*. UK Ministry of Defence.

6 | *Ibid.*

evidenced by a 39% turnout at the 2016 legislative elections[7].

1.3. Nepotism and Corruption

Corruption has proven to be a major hurdle to the effectiveness of Romania's political system and its provision of civil liberties and economic freedom. This has occurred on two levels. Firstly, corruption itself has undermined the democratic institutions of Romania, inhibiting peoples' freedoms. Secondly, corrupt political choices have often gone against public interest.

Regarded as the fourth most corrupt nation in the European Union[8], Romania has a corruption problem that cannot be taken lightly. While EU accession mandated increased transparency in government and business dealings, the slow implementation of these laws – evidenced by persistent shortcomings in scores for both perceptions and actual experiences of corruption[9] – has inhibited the development of public projects (e.g. reports of state funds for hospitals being siphoned into private hands).

This has caused significant tension in government[10] that limits societal progress and has led to mass protests. That was most blatant at the start of 2017 when Liviu Dragnea, the President of the Social Democratic Party (PSD) that won a landslide at the 2016 legislative elections, was unable to become prime minister due to his corruption charges of roughly RON105,300 (£20,000), where RON is the Romanian New Lei currency[11]. Attempting to bypass these charges, his crony government secretly agreed to pass a bill that decriminalised corruption worth less than RON200,070, (£38,000)[12].

The attempted passing of the bill led to the largest protests in Romania's history (surpassing the 1989 revolution). While the bill was repealed and Mihai Tudose was elected, he is still largely seen as a puppet figure for Dragnea, who rules from behind the scenes. This highlights how Romania, more than 30 years after the collapse of communism, is still plagued by the issues it swore to abandon in the aftermath of Ceausescu.

7 | Biroul Electoral Central. (2016). *PROCES-VERBAL privind rezultatele finale ale alegerilor pentru Senat [MINUTES regarding the final results of the elections for the Senate.].* Biroul Electoral National al Republicii Romane. Retrieved 03 October 2017 (Biroul Electoral National al Republicii Romane), from http://parlamentare2016.bec.ro/wp-content/uploads/2016/12/3_RF.pdf.

8 | 'Transparency International. (2017). *Corruption Perceptions Index 2016.* Retrieved from https://www.transparency.org/en/news/corruption-perceptions-index-2016#table.'

9 | European Commission. (2014). EU Anti-Corruption Report. European Commission. Retrieved from https://ec.europa.eu/home-affairs/sites/homeaffairs/files/e-library/documents/policies/organized-crime-and-human-trafficking/corruption/docs/acr_2014_en.pdf

10 | Romania's 2012 political crisis that was sparked by nefarious health reforms.

11 | Fishwick, C. (2017, February 06). '27 years of corruption is enough': Romanians on why they are protesting. *The Guardian.* Retrieved October 07, 2017, from https://www.theguardian.com/world/2017/feb/06/27-years-of-corruption-is-enough-romanians-on-why-theyre-protesting.

12 | *Ibid.*

1.4. Political Parties and Civic Engagement

The 2016 elections proved to be a significant shift in the direction of multi-party politics. The Colectiv nightclub fire in 2015[13], which caused over 60 deaths[14], prompted the resignation of Prime Minister Victor Ponta and his official government cabinet later in the year and led to growing concerns over the inefficiency that Romania's corruption was breeding.

In many ways, this serves as a lens through which one can analyse political trends in Romania. The Social Democratic Party (PSD) has held power in parliament for the majority of post-communist Romania and has traditionally advocated for a social safety net. It holds significant electoral success in the more agrarian and rural provinces of southern and eastern Romania. PSD has also been accused of harbouring former communists and allowing them to extend their grip on the country, giving rise to the 'neo-communist' accusations.

The main opposing party of the PSD is the National Liberal Party (PNL), which officially promotes both social and economic conservatism[15]. In response to growing allegations of corruption and communist-era complacency, the 'Save Romania Union' was established as a grassroots party that sought to separate itself from all traditional parties. Having initially started out as the Save Bucharest Union and receiving many votes in the Bucharest mayoral election of 2016, the party decided to participate in elections at a national level. Through its belief in the rule of law and justice, the Union successfully siphoned many votes off the PNL – the main opponents of the PSD – in the 2016 elections, unintentionally giving the PSD a stronger mandate by splitting the opposition vote. While this was not an intended consequence, as the Union believes the PSD is responsible for many of Romania's issues, the Union won mostly metropolitan votes that typically go to the PNL, thereby unwittingly strengthening the PSD.

Aside from these major parties, there are ethnic minority parties (e.g. UDMR for Romania's Hungarian minority) and

13 | The fire was sparked by the illegal use of outdoor pyrotechnics inside the club – many of the protesters who reacted to the Colectiv fire linked it to corruption and the negligent attitude of government officials in ensuring the rule of law.

14 | Ilie, L. (2016, October 30). Thousands march in Bucharest one year after nightclub fire killed 64. *Reuters*. Retrieved October 07, 2017, from http://www.reuters.com/article/us-romania-fire-march/thousands-march-in-bucharest-one-year-after-nightclub-fire-killed-64-idUSKBN12U0R7?il=0.

15 | National Liberal Party. (2014) *Origins of the National Liberal Party.* Visegrad+ Studies.

several parties associated with specific politicians (e.g. PMP with former President Traian Popescu, ALDE with former Prime Minister Calin Popescu-Tariceanu). Most crucially, one must remember that Romania's politics, despite official manifestoes from all parties, does not revolve around party lines or partisanship, but rather around individual leaders and compromise so as to help the government gain power. This lack of clear choice has demoralised voters, who see an identical result regardless of their vote. Misleading rulers and bureaucracy in politics have created widespread apathy towards the day-to-day impacts of politics and its workings.

The power of the people is only witnessed at elections and when major controversies occur[16], with mass, yet short-lived, protests in response to such actions being a crucial form of expression for Romanians. Part III will further elaborate upon the importance of youth participation in elections and protests. This wave of direct protests has marked a move, firstly, towards a more multi-party state with interest groups often turning into political parties, and secondly, has sparked a more direct, socially orientated form of politics.

1.5. Conclusion

Despite the numerous problems endemic to Romania's political system, there is potential in its future; as the country progresses into the 21st century, it is increasingly complying with EU laws and suggestions on democracy, civil liberties and civic engagement. This is gradually providing Romania with the strong, incorruptible, fundamental framework it needs to truly become a liberal democracy.

Moreover, big political scandals and heightened tensions within the EU have served as an impetus for the population to be involved in politics, resulting in a recent uptick in political interest (USR, for example, the third biggest party at the 2016 legislative elections, was only founded in 2015 by a university professor). Populist policies and politicians have drawn the interests of larger crowds, and the gradual realisation of the importance of civic duty in a liberal democracy is aiding

16 | For example, the Colectiv fire, the 2012 protests over health reform and the 2017 protests over an attempted change to graft law.

Romania's democratic process. Moreover, anti-corruption bodies and judicial courts have been granted increasing power and independence in making decisions, allowing for both a fairer and more just system of justice.

Romania's political structure is somewhat paradoxical – flaws within it are evident, and their solutions are relatively achievable. The main obstacle Romanians face in achieving these are Romanians themselves; only greater political awareness and information, along with increased participation and a heightened sense of civic duty, can truly ensure a wholly stable political system in Romania.

2. The Political Direction of Romania

Romania's political future is almost certainly settled – EU integration will remain a key objective of almost every influential political party, regardless of its standing on the left-right spectrum. Despite a slowly growing nationalist, Eurosceptic movement, all parties in parliament as of 2018 were pro-European and all individuals with political influence advocated for greater integration with Europe. This is indeed dictated by the sizeable Romanian diaspora, the vast majority of which – 3.5 million, as per conservative government estimates[17] – reside in EU nations and rely on the free trade and non-visa movement of the bloc to live.

While many Eastern Europe states have sought to increase trade with Russia and oppose the migrant quota centrally decided upon by the EU Parliament, Romania has accepted its migrant quota, has remained staunchly anti-Russian (allowing extensive NATO militarization of its geopolitically vital access

17 | Alexe, A. (2018, June 8). 3.5 Million Romanians are Currently Living Abroad, Says FLAME Foundation President. *Business Review*. Retrieved from https://business-review.eu/news/3-5-million-romanians-are-currently-living-abroad-says-flame-foundation-172573.

to the Black Sea) and has sided with Western powers on most international matters. While opposition to this passive political adherence towards the West has organised itself, the unsustainable rate of emigration from Romania to the EU means such opposition lacks credibility, outreach and support.

Romania's political future is shaped by a ruling elite that seeks to capitalise off emigration to the West, using it to bind Romania to the EU through bilateral agreements and relations. These long-lasting and politically significant agreements are likely to stipulate Romania's subservience to the EU for decades to come.

2.1. International Relations

The pro-EU, pro-Western movement in Romania was born out of disdain towards Russia for what Romanians saw as a Soviet occupation and aggression towards Romania post-1945. While numerous Slavic EU states, such as Bulgaria and Slovakia, have taken a much greater liking to Russia due to cultural, linguistic and ethnic affinities, Romania – an ethnically non-Slavic nation with a Romance language – does not share this trait, making the resentment towards Russia much more profound.

Successive governments in the aftermath of 1989 have been willing to sacrifice sovereignty and accept Western-style shock therapy in exchange for stronger political ties. This is evidenced by the CIA interrogation black sites that Romania permitted on its territory for more than a decade[18], which were originally approved by Ion Iliescu in a bid for NATO accession that came in 2004[19].

Just as these political incentives have driven Romania's willingness to abide by EU laws and respect Brussels, prospective political incentives on the horizon still dominate the government's policies. This is two-fold – firstly, there is a desire to join the Schengen Area, and secondly, a move towards joining the Euro Zone. Romania's foreign minister has already confirmed that Romania wishes to join the Euro Zone by 2022[20], and an influx of Euros in remittances has made it

18 | Cole, M. (2011, December 08). Another Secret CIA Prison Found. *ABC News*. Retrieved October 03, 2017 from http://abcnews.go.com/Blotter/secret-cia-prison-found/story?id=15112990.

19 | Marty, D. (2007, June 7). Secret detentions and illegal transfers of detainees involving Council of Europe member states: second report. *Parliamentary Assembly*. Retrieved October 03, 2017 from http://assembly.coe.int/CommitteeDocs/2007/EMarty_20070608_NoEmbargo.pdf.

20 | Stone, J. (2017, August 28). Romanian government says it will adopt the euro in 2022. *The Independent*. Retrieved October 03, 2017 from http://www.independent.co.uk/news/world/europe/romana-join-euro-economy-2022-five-years-foreign-minister-brexit-a7916856.html.

a crucial currency. Despite this, 2022 is not an assured date — the government had previously promised to adopt the Euro by 2015 (this failing as the EU nations voted to postpone admission), and Brussels has the power to delay the Euro Zone admission, using promises over the Euro as a means to ensure Romania's allegiance to the EU and to achieve Brussels' desired changes in the country.

EU control is seen with Schengen area integration as well. When Romania was given its migrant quota, it, along with Bulgaria, demanded Schengen integration in exchange for accepting migrants.[21] While the Romanian government subsequently abandoned this, it has continued pursuing Schengen membership despite consistent rejections by nations such as Germany and Austria.

Despite this, Romania's future within the EU faces numerous difficulties. Firstly, as mentioned in the previous section, the issue of corruption still plagues the nation, with frequent protests against the government, causing political instability. It is ranked the fifth most corrupt country in the EU[22], and 57th most corrupt globally in the *Transparency International Index*, and a significant proportion of EU funds are not adequately implemented, meaning citizens suffer as a result.

Secondly, rural areas remain largely under-developed and lack the benefits reaped from the EU that are seen in major cities. With more than half of all rural houses lacking access to running water[23], the EU needs to address the vast inequality existing in its bloc should it wish for Romania's firm support. Most provoking of all, numerous Romanians increasingly denounce the West for supposed abuses experienced by young Romanians working abroad in the EU (the primary destination of the diaspora)[24]. With abuses such as rape and physical assault being commonly over-sensationalised[25], there is growing resentment in Romania towards what many believe is a Western Europe that demeans Romanian immigrants, often sparking outcry and protests domestically. This manifested itself at the last legislative election in December of 2016, where the Eurosceptic United Romania Party won 207,608 votes[26].

21 | Zhelev, V. & Bird, M. (2015, September 11). Bulgaria, Romania Tie Migrant Quotas to Schengen. *EU Observer*. Retrieved from https://euobserver.com/beyond-brussels/130202.

22 | Transparency International. (2016). *Corruption Perceptions Index*. Retrieved 03 October, 2017, from https://www.transparency.org/news/feature/corruption_perceptions_index_2016#table.

23 | Marica, I. (2015, March 06). Rural Romania: surf the net but no running water. *Romania Insider*. Retrieved October 03, 2017, from https://www.romania-insider.com/survey-rural-romania-internet-running-water-modern-toilets/.

24 | The diaspora holds a very special place in Romanian politics – because it is mostly young workers who have parents in Romania that go abroad, there is an emotional aspect for many of the elder voters left at home when it comes to abuses against the diaspora; their children. Similarly, there are over 100,000 children in Romania with parents working abroad, with stories of suicide and depression among such children playing a deep emotional role amongst Romanian voters.

25 | Kelly, A., & Tondo, L. (2017, March 11). Raped, beaten, exploited: the 21st-century slavery propping up Sicilian farming. *The Guardian*. Retrieved October 03, 2017 from https://www.theguardian.com/global-development/2017/mar/12/slavery-sicily-farming-raped-beaten-exploited-romanian-women.

Although this represents a very small proportion of voters (2.95%), the virulent movement has the power to spread, as it has done in neighbouring countries.

Internationally, despite outcry from some citizens, Romania is almost certain to remain within the EU sphere of influence for two reasons. First and foremost, emigration towards Europe is beyond a point of return. The diaspora not only provides crucial remittances for the country that the economy needs – US$4 billion in 2013[27] – but also influences the political votes of the domestic population, with both the 2009 and 2014 presidential elections having been won by the diaspora vote[28]. Consequently, the second reason arises: governments will be forced to obey the power of the diaspora (which now has special seats in Parliament dedicated to it), and, because the diaspora is liberal-leaning and based in the EU, will look to EU integration in a bid to ensure easy lives for its diaspora. Similarly, the government also wants closer relations to the EU and NATO to ensure its military protection and benefit from the generous EU funds that come its way. For the foreseeable future and beyond, Romania will remain firmly within the EU's sphere of influence.

2.2. Romania-Moldova Relations

The issue of unification with Moldova, a former Soviet Socialist Republic, has only arisen since the collapse of the USSR in 1991. The ethnic background of Moldovans is largely disputed, but most historians tend to agree that they are a Romanian people who underwent Russification at the hands of the USSR. This was furthered by forced migration of ethnic Russians and Ukrainians to Moldova under Stalin's post-WWII reign, meaning only 75% of Moldova's population is ethnically Moldovan[29]. Despite this, Romanians have increasingly pressed for unification with what was previously Romanian territory annexed by the USSR during World War II.

A 2015 poll indicated that more than two-thirds of Romanians were in favour of unification[30]. The Romanian government has taken action towards this (especially under former President

26 | Biroul Electoral Central. (2016). PROCES-VERBAL privind rezultatele finale ale alegerilor pentru Senat [MINUTES on the final results of the elections for the Senate]. *Biroul Electoral National al Republicii Romane.* Retrieved 03 October, 2017 from http://parlamentare2016.bec.ro/wp-content/uploads/2016/12/3_RF.pdf.

27 | Transparency International. (2017). *Corruption Perceptions Index.* Retrieved 03 October, 2017, from https://www.transparency.org/news/feature/corruption_perceptions_index_2016#table.

28 | Clej, P. (2014, November 17). Romania's expatriate voters overshadow presidential poll. *BBC.* Retrieved October 03, 2017 from http://www.bbc.com/news/world-europe-30078385.

29 | National Bureau of Statistics of the Republic of Moldova. (2014). *Population and Housing Census in the Republic of Moldova, May 12-25, 2014.* Retrieved October 03, 2017 from http://www.statistica.md/pageview.php?l=en&idc=479.

30 | Mihalache, M. (2015, July 31). *INSCOP: Două treimi dintre români vor Marea Unire cu Republica Moldova până in 2018 [PURPOSE: Two thirds of Romanians want the Great Union with the Republic of Moldova by 2018].* Retrieved October 03, 2017, from http://adevarul.ro/news/politica/doua-treimi-romani-vor-marea-unire-republica-moldova-2018-1_55ba236df5eaafab2c1877c0/index.html.

Traian Basescu) by granting 323,049 Romanian passports to Moldovan citizens[31] and by opening the border. However, progress has been hindered due to a pro-Russian communist party governing Moldova from 2001-2009. Progress has also been limited since a minority pro-European government in Moldova, that has ruled from 2009 onwards, did little to strengthen relations with Romania. In the midst of a US$1 billion corruption scandal (colossal for Moldova whose GDP is US$9 billion), a pro-Russian, anti-unification president was elected in 2016.

Although unification remains high up on the Romanian agenda, the same cannot be said for Moldova, where a 2016 poll showed 63% of Moldovans to be opposed to unification[32]. Moreover, because many Moldovan workers reside in Russia and send back crucial remittances from there, Moldova is unlikely to break their allegiance to Russia by joining Romania (just as Romania won't break ties with the EU for the same reason, paradoxically). Furthermore, Russian geopolitical interests and a breakaway, pro-Russian state (Transnistria) make unification almost impossible. In Romania, the Moldova question remains a way through which politicians win the nationalist vote, but also remains an unanswerable question that does little to alter Romania's political direction.

2.3. Conclusion

Romania's political future seems stable to say the least, if not stagnant. The emigration circumstances, beyond the government's control, dictate a pro-EU approach in Bucharest, and exaggerated fears of both a Russian threat and an economic collapse without the EU ensure the people's support for the bloc. Very gradually, the political system is normalising, too – efforts are being made to root out corruption at all layers of society, and democratic institutions are increasingly protected by a now-respected rule of law. As this occurs, Romania will become an increasingly attractive country for foreign direct investment (FDI), and political goals in the future (Schengen, Euro Zone) will mean that Brussels and Washington will have significant influence over the political decision-making in Bucharest.

31 | Fundatia Soros Romania [Soros Foundation Romania]. (2013, April 3). *Communicate de presa: o politică ce capătă viziune - redobândirea cetățeniei române [Press releases: a policy that gains vision - regaining Romanian citizenship]*. Retrieved October 03, 2017 from https://web.archive.org/web/20131029202540/http://www.soros.ro/ro/comunicate_detaliu.php?comunicat=220.

32 | FOP Research. (2014). *Sondaj socio-politic realizat la soliticarea ziarului "Timpul" [Socio-political survey conducted at the request of the newspaper Timpul]*. FOP Research. Retrieved October 03, 2017 from https://www.scribd.com/doc/305812783/Sondaj-socio-politic-realizat-de-FOP-la-solicitarea-ziarului-Timpul-martie-2016#.

The main problems Romania will face in the long term will be the constrictions that are placed on it by an EU Commission that is gradually diminishing in influence, seeking to integrate Europe to an unprecedented level and wanting to minimise national sovereignty and populist movements. With the Romanian government's goals clearly stated, the nation's future political direction has already been determined.

3. Opportunities and Challenges Faced by the Romanian Youth

In Romania, the youth is often portrayed in contradictory ways: either as the generation holding the key to a better future, or the failure of a never-ending transition to democracy. Naturally, this demographic is of great importance for the future development of any country and should be subject to in-depth analysis and political projects aimed at maximising its full potential. There should be a clear picture of the opportunities and challenges affecting young Romanians. However, Romania has yet to achieve a clear vision on how to maximise existing opportunities for the youth, how to create new ones and how to equitably distribute them at a national level. Moreover, it is difficult to act against challenges that have not been clearly outlined in any coherent governmental report.

According to data from Eurostat (as of 1st January 2016), the young Romanian generation aged 15 to 29 and residing on the national territory represents 17.8% of the total population. They are confronted with various challenges, determined by socioeconomic factors, recent political issues (political instability, corruption), demographics (mass emigration and negative demographic growth), religion (particularly when intertwined with politics for agenda-setting purposes), or other factors. There are, however, three recurrent themes when it comes to the rhetoric of greatest problems.

3.1. Challenges

One of the main challenges for the Romanian youth is poverty, a fact both acknowledged by the public authorities and strongly perceived as such by 64.1% of young people in a study of Friedrich Ebert Stiftung Foundation[33]. A report of the Romanian National Statistics Institute[34] concluded that 1 in 3 people aged 18 to 24 live in poverty[35], which leads to an acute problem of social exclusion. Moreover, the level of income inequality is high, as shown by the Gini coefficient, which reached 34%. In total, around 5.3% of the youth aged 18-24 years were found at risk of poverty and social exclusion. This probably explains the strong desire of many young Romanians to emigrate in search of better living standards[36].

The issue of social exclusion is also linked to the second main challenge – poor results yielded by the educational system. Not only is the school dropout level worrying, but also those who finish the mandatory 10 classes are inadequately prepared for the labour market. The socio-economic context fuels the rate of early dropout from the educational system and other forms of training among young people aged 18-24: in 2014, it reached 18.1%, compared to the European average of 11.2%[37].

Furthermore, of those who pursue a high school education, only 79.1% pass the final national examination. In addition, those who do graduate high school lack the necessary competencies and skills required by the labour market, a fact proven by the very low scores obtained at national PISA (Programme for International Student Assessment) exams.

Education in Romania is still based on old-fashioned principles and old school manuals, and extracurricular activities are still not very popular. The consequences are profound on Romanian society and, in particular, on young people who have great difficulties in finding a job – 24.1% of the active population in the 15-24 age group is unemployed[38]. Consequently, most young people are dissatisfied with the Romanian educational system. Nonetheless, politicians and policymakers fail to address the root cause of its problems or to adopt reforms for

33 | Umbres, R., Sandu, D., & Augustin S., C. (2014). *Romanian Youth: concern, aspirations, attitudes and life style*. Bucharest: Center for Urban and Regional Sociology.

34 | Cuturela, A., Gheroghe, A., & Simion, C. (2016). *Dimensiuni ale incluziunii sociale in Romania [Dimensions of social inclusion in Romania]*. Bucharest: National Statistics Institute.

35 | Defined as 'the situation in which people find it almost impossible to achieve the standards of living considered 'acceptable' in their society'.

36 | In the study for FES Foundation, 40% of the participant youth declared their willingness to emigrate, even for shorter periods or for studies. Also, Romanians have the largest diaspora in Europe, residing in countries such as Italy, Spain, United Kingdom and France.

37 | Ministry of National Education. (2015). *Raport privind starea invatamantului preuniversitar in Romania [Report Regarding the Condition of Pre-University Education from Romania]*. Bucharest: Official site Ministry of National Education.

38 | *Ibid.*

modernisation of the school.

That leads us to the third great challenge of the Romanian youth, namely civic and political disengagement. Apart from poverty and unemployment, young Romanians also pointed to corruption as one of the major threats to the future of the country[39]. There is a widespread feeling of deception, caused by the actions of public authorities and top officials, which tend to discourage the youth's engagement in civic initiatives and political life. Regardless of their lack of participation in traditional forms of political activity, starting in 2013 there has been an upsurge in alternative forms of engagement, such as large protests directed against corrupt decisions or legislative projects, which have been powerful to the point of changing the ruling government in 2015 and gaining significant international coverage.[40] Despite the disappointment at the current state of the country, young Romanians have confidence in a better future, encouraged by the opportunities offered by the European Union[41] (education, funds, free movement).

3.2. Opportunities

When it comes to what Romanians usually perceive as one of the greatest opportunities for development, the attention immediately focuses on the ICT (Information and Communications Technology) industry. Historically, there has been a long tradition emphasising education in mathematics and technology, investments in the ICT field made during the communist regime, and an innovative way to make use of the internet offer in the 2000s (i.e. infrastructure based on personal servers created in residential buildings or neighbourhoods which were afterwards directly bought by the internet providers)[42]. All these factors created the foundations for the sector's growth.

Nowadays, with more than 14,000 IT companies, of which upwards of 50% are located in Bucharest and Cluj, and numerous global corporations present, the Romanian ICT industry is expected to generate around €4 billion in annual revenue by 2020[43]. Romania is famous for its Internet speed,

39 | Umbres, R., Sandu, D., & Augustin S., C. (2014). *Romanian Youth: concern, aspirations, attitudes and life style.* Bucharest: Center for Urban and Regional Sociology.

40 | More on the protests from 2017, see Tara, J. (2017, February 6). Everything to Know about Romania's Anti-Corruption Protests. *Time*. Retrieved from https://time.com/4660860/romania-protests-corruption-problem/.

41 | According to Eurobarometer 2016, around 52% of Romanians still trust the European Union, compared to the European average of 36%. See more at Ro Insider. (2017, March 3). *Eurobarometer: Over Half of the Romanians Still Trust the EU Compared to Only a Third of the Europeans.* Retrieved from https://www.romania-insider.com/eurobarometer-romanians-trust-eu-52-still-european-average/.

42 | Morin, G. (2016, March 6). How Romania got one of the fastest internet services in the world. *Medium*. Retrieved from https://medium.com/@gabriel_morin/10-years-later-diy-romanian-kids-are-today-s-network-expert-ccb25cd1967.

43 | ANIS. (2016). *Software and IT Services in Romania 2016*. Retrieved from Employers' Association of the Software and Services Industry: https://www.anis.ro/wp-content/uploads/2016/05/Software-and-IT-Services-in-Romania-2016_General-Considerations-Excerpt.pdf.

44 | TeamFound. (2016). *Quick Guide to Romania IT and Software Industry*. Retrieved from TeamFound: https://teamfound.com/blog/romania-it-industry-report/.

45 | According to a reform in 2017, for both employee and employer in the ICT sector are exempted from paying the national income tax (~16%), in an attempt to encourage entrepreneurship and attract investments in the Romanian ICT. See Iteanu, I. (2017, March 23). Income Tax Exemption for IT Employees in Romania. *Fred Payroll*. Retrieved from http://www.fredpayroll.com/income-tax-exemption-it-employees/.

46 | Romania organised for the first time the International Mathematics Olympiad. Apart from being one of the countries with the most frequent participations at the final competition, it also ranks among top 5 countries with most Gold medals ever obtained-75 (as of 2017). See IMO Official Website Results Page at https://www.imo-official.org/results_country.aspx.

47 | At the International Physics Olympiad 2017, high school Romanian students got four gold medals, placing Romania fifth in the final ranking of participant countries.

48 | Law 78/2014 established the legal framework of any volunteering activity, while recognising it as a professional experience (if completed in the field of graduated studies).

49 | FDSC. (n.d.). *Voluntariat*. Retrieved from Fundatia pentru Dezvoltarea Societatii Civile: http://www.fdsc.ro/voluntariat.

having nine out of the 15 top cities in the world with the highest download speed on a broadband Internet connection, and an outstanding 3G and 4G coverage. Each year, more than 7,000 students graduate in the field of ICT. However, annually there is a growing demand for people with ICT skills that still remains unsatisfied in the labour market[44].

Consequently, there are incentives for Romanian youth to specialise in this field, either through working for existing ICT companies to access higher salaries, or start their own business in the ICT industry, using not only the benefits of good Internet and telecom infrastructure, but also the state incentives are given to ICT companies – e.g. tax exemption[45].

Despite the insufficiencies of the Romanian educational system, it still represents an opportunity for the young generation. In Romania, education is free and mandatory up until 10th grade. Higher education is also available for everyone, offering thousands of free university places on the basis of academic results. Romania is part of the Bologna Agreement, which harmonised European higher education and enhanced the international reach of Romanian universities and encouraged partnerships across borders. This gives the opportunity of pursuing studies abroad through scholarships like Erasmus (EuRopean Community Action Scheme for the Mobility of University Students) or other agreements. Numerous Romanian students obtain top results and medals each year in the most intensive international and global competitions, especially in mathematics[46], physics[47] and the social sciences.

Apart from the chance to achieve excellent academic results and access the most prestigious universities around the world, young Romanians have also started to dedicate more of their time to volunteering activities. Since 2001, when the Law for Volunteering Activity[48] was implemented, more than 20% of all Romanians have become involved in volunteering experiences, which is nonetheless a small percentage compared to the European average of over 30%.[49] The FES study found that young Romanians were especially involved in organising cultural events, community service, or helping those in need. Alongside this, there are also high-stakes opportunities, like becoming the UN Youth Delegate and representing the voice of young Romanian people for a one-year mandate[50] or doing a summer internship for the Romanian Government or Parliament. For high school students, they can get involved in the National Students' Council, which has gained attention and influence in previous years when it successfully fought to obtain an official Statute of Students that guarantees the students' rights[51].

3.3. Conclusion

The landscape of challenges and opportunities for young Romanians is complex. While the official data and the perception of the youth do not place Romania on a very promising development course, there remain factors that have the potential to bring about a positive turn.

For example, an improvement in, and stabilisation of, the political sphere would give the possibility to better counteract the challenges affecting youth, especially the inadequate educational system and the striking levels of poverty. In addition, the entrepreneurial spirit should be encouraged, so that the state efforts to diminish socio-economic issues are met by private initiatives.

It is important to maximise the potential of Romanian youth, especially in the ICT area, where Romania can become a hub of technology not only in the region, but also in the whole of Europe. Even though Romania mainly exports average-level services for the time being, more should be invested in Research and Development centres that could make use of the available highly skilled personnel in the industry and invest in future generations.

Finally, young people must individually look for opportunities and maximise them in their best interests. The international projects involving Romania ought to be regarded as opportunities *per se,* and the Romanian youth need to find ways to employ them for the benefit of both Romania and its partners.

50 | More on this can be read at Delegat de Tineret al Romaniei la ONU at http://unyouthdelegate.ro/despre-program/ and Youth Delegate Programme at https://www.un.org/development/desa/youth/what-we-do/youth-delegate-programme.html.

51 | Students struggled for a legal recognition of their rights within school for several years, during which they drafted the statute and advocated for their cause through the National Students Council. In 2016, for the first time, the Ministry of National Education adopted a degree recognising the Statute of Students aimed at clarifying the rights and obligations of students in pre-university schooling, useful, among many other things, in helping to fight abuses committed by professors (for example, discrimination or giving bad grades as forms of punishment). See Oana, D. (2017, September 12). RAPORTUL privind Statutul Elevului: 67% dintre elevi primesc note in catalog fără ca ei să fie informați. Ce probleme au mai fost reclamate [REPORT on Student Status: 67% of Students Receive Notes in the Catalog without being Informed. What Problems Have Been Reported]. *Mediafax.* Retrieved from https://www.mediafax.ro/social/raportul-privind-statutul-elevului-67-dintre-elevi-primesc-note-in-catalog-fara-ca-ei-sa-fie-informati-ce-probleme-au-mai-fost-reclamate-16726280.

4. Evaluation of the Challenges and Opportunities Offered by the Belt and Road Initiative in Romania

4.1. Context

Romania is located in a key geopolitical area. Placed at a crossroads between East and West[52], marking the edge of the Balkans and open to the Black Sea, Romania is connected to Europe by the Danube River. The various geographical reliefs provide natural wealth (significant natural resources, touristic attractions, etc.), while at the same time allowing connectivity inside and outside of the country, both on land and water.

Politically speaking, two decades after the fall of the communist regime, Romania continues the transition to a more stable, democratic society, while trying to enhance its leadership position in the region and grow its influence in the European Union. Furthermore, in light of geopolitical events (growing Russian influence in the East and Turkish political turmoil), Romania is an anchor of stability that serves as a gateway to the EU market for Asian partners, such as China.

Similar to the other Central and Eastern Europe countries (CEECs), Romania is interested in development opportunities and seeks to boost its role in the EU decision-making process. Therefore, while remaining as a trustworthy partner in NATO and a reliable member of the EU, it concluded partnerships offering alternative ways to encourage growth.

The Belt and Road Initiative (BRI) is perceived as a useful opportunity, proposing a long-term, mutually advantageous relation of cooperation, a way to 'support the development and diversification of Romanian-Chinese bilateral relations and, by this, the enhancement of the EU-China strategic partnership'[53]. Thus, the country joined the 16+1 initiative at the Warsaw meeting in 2012. Within this context, most of the CEECs looked forward to overcoming the lasting consequences of the economic recession.

China advanced its diplomacy and economic interests to meet Romania's needs for recovery and development, proposing the One Belt, One Road Initiative, in response to which a competition started among CEECs to attract the most benefits

52 | In history, the image of Romania changed, depending on whose sphere of influence the country was under at a certain time. This caused difficulties in perceiving Romania as part of Europe *per se* and as part of the broader narrative discourse about Europeanisms. For example, in his formidable book, Goldsworthy discusses how writers perceived the Romanian capital, Bucharest, as the 'Paris of the East', a place where elements of the West mysteriously combine with remains of the Ottoman indirect rule, all these contributing to 'the first impressions of the city as an alien, Oriental capital, with an end-of-empire, decadent European feel.' (pp. 188-192). Source: Goldsworthy, V. (1998). *Inventing Ruritania: the Imperialism of the Imagination.* Yale University Press.

53 | Romanian Ministry of Foreign Affairs. (2015, July 13). *Press Room.* From Ministry of Foreign Affairs. Retrieved from https://www.mae.ro/en/node/32901.

from Chinese FDIs. The Initiative is controversial in the eyes of the European Union, which is reluctant concerning the cooperation under the BRI umbrella as it presumably does not respect the EU standards and regulations, eroding values and unity.

4.2. Investment Opportunities in Romania

We turn now to the strategy of Romania in promoting such investment opportunities and analyse the main sectors that could be advantaged by the co-operation within the Belt and Road Initiative.

Since 2012, Romania has held a discourse-based approach towards the BRI, emphasising the 'need to have an integrated synergetic, long-term approach'[54] via respecting previous agreements while engaging in deeper collaboration with China. Romania also upheld its interests in Chinese investments by hosting the second meeting of the 16+1 Initiative group, where the Bucharest Guidelines were settled[55].

In order to attract investments, Romania advertises several strong comparative advantages, such as its geostrategic position, the abundance of natural resources (non-renewables like oil and gas, as well as renewables like wind and hydro-energy), the potential of the energy sector in both renewables and non-renewables, the human capital in the ICT industry, as well as other assets like the port of Constanța (the largest container port in the Black Sea) and the transport network on the Danube river leading directly to the heart of Europe.

In the framework of BRI, the cooperation with Romania is mainly focused on several sectors: energy, ICT, infrastructure and transport, commerce and technology. The discussion hereafter will be limited to the first three sectors aforementioned.

The cooperation entails corridors of multiple types: pipelines, logistics stations, internet infrastructure, etc. Each of these corridors arguably comes with opportunities and drawbacks.

54 | Romanian Ministry of Foreign Affairs. (2015, November 19). *Consultations between State Secretary for Strategic Affairs Daniel Ioniță and EU Special Representative for Central Asia Peter Burian*. Retrieved from https://www.mae.ro/en/node/34744.

55 | The meeting reviewed the achievements of CEECs-China cooperation, while establishing the directions of cooperation in the future, respecting the EU-China strategic partnership and mutual interests. See Government of Romania. (2013, November 27). *The Bucharest Guidelines for Cooperation between China and Central and Eastern European Countries*. From Government of Romania website: http://gov.ro/en/news/the-bucharest-guidelines-for-cooperation-between-china-and-central-and-eastern-european-countries. In May 2017, at the Doha Forum, Romania reiterated the 'willingness to participate in cooperation projects aimed at increasing the transport of goods, services and people across the Black Sea' and to become 'a focal point for exchanges in both direction, between Europe and Asia'. Please see Romanian Ministry of Foreign Affairs. (2017, May 16). *Press room*. Romanian Ministry of Foreign Affairs. Retrieved from https://www.mae.ro/en/node/41802.

4.2.1. Energy Sector

In the energy sector, the BRI provides Romania with the chance to achieve greater energy efficiency by investing in connectivity with neighbouring countries, EU member states, and external BRI partners. Moreover, energy efficiency means taking one more step towards energy security and decreasing dependency on imports from Russia.

During a meeting in Brussels, Corneliu Badea, President of the Romanian Center for Energy, insisted that the geopolitical position of Romania can transform it into a major player in regional cooperation to achieve the objectives of the European Union's energy sector[56]. One example of a partnership that strengthens the commercial links with China in the energy sector is the €7 billion investment made by China General Nuclear Power in the construction of two nuclear reactors at Cernavoda, a Romanian power plant, that encourages the imports of Chinese nuclear technology in the country and the EU[57].

Likewise, in 2016, one of the largest private Chinese energy companies, CEFC (China Energy Company Limited), bought the majority of shares in KazMunayGas International (KMGI), the company which operates the largest oil refinery in Romania (Petromidia) and plays a major role in the Black Sea region with the purpose of creating a supply corridor from China to Kazakhstan and Romania. Such cooperation does not only diversify the supply sources and diminish Romania's energy dependence, but it also increases its energy security and promotes regional stability by rallying common interests around the connectivity useful for all parties involved.

4.2.2. Information and Communications Technology (ICT)

To achieve global connectivity, ICT is vital. In this respect, Romania can definitely boost its potential and role in the sector by taking advantage of the high-quality internet infrastructure already in place and the availability of a productive and less expensive labour force, compared to Western Europe.

56 | The European Union set the 2020 Energy Agenda and is in the process to create the Energy Union, with the aim to secure energy supplies, encourage competition, reduce consumption, and promote interconnectivity — the free flow of energy across national borders of member states. More at https://ec.europa.eu/energy/en/topics/energy-strategy-and-energy-unions. Also Deacu, E. (2017, June 6). *China va investi 11 miliarde de dolari in România prin proiectul One Belt One Road [China will invest 11 billion dollars in Romania through the One Belt One Road project]*. Adevarul. Retrieved from http://adevarul.ro/economie/afaceri/china-vainvesti-11-miliarde-dolari-romania-proiectul-onebelt-one-road-1_5936dbda5ab6550cb8d8517c/index.html.

57 | Mosoianu, A. (2016, November 6). China estimează investiția in Cernavodă 3-4 la 7,2 miliarde euro, aproape dublu față de cifra avansată inițial. 'Va stimula exporturile nucleare chinezești in UE și NATO' [China estimates the investment in Cernavoda 3-4 at €7.2 billion, almost double the figure initially advanced. 'Will boost Chinese nuclear exports to EU and NATO']. *Profit*. Retrieved from https://www.profit.ro/povesti-cu-profit/energie/china-estimeaza-investitia-in-cernavoda-3-4-la-7-2-miliarde-euro-aproape-dublu-fata-de-cifra-avansata-initial-va-stimula-exporturile-nucleare-chinezesti-in-ue-si-nato-15674481.

Numerous advancements can be induced via investments in the ICT industry, from better commercial links, good management of logistics, and transportation of goods and providing services, to more information security and innovation by sharing know-how and promoting connectivity of information across the BRI's global membership.

Romania already has a top-notch telecommunication infrastructure, allowing high-speed internet (one of the best in the world) and wide service coverage. Again, due to its geographical position, the country can represent an important hub of connectivity between CEE and Asian countries by becoming a provider of data security and firewall services.

A relevant example of one successful Romanian solution to IT security issues is the renowned BitDefender, a company that operates global cybernetic security, providing software for data protection and anti-virus for more than 500 million clients worldwide. Another proof of Romania's credibility in the quality of the Romanian ICT industry is the €10 billion investment made by the Chinese multinational Huawei Technologies in Bucharest, to open a Global Service Centre that provides technical support, network management, logistics, and customer support[58].

4.2.3. Physical Infrastructure

Finally, the development of physical infrastructure in Romania is crucial to the engagement within the BRI. The Chinese are admirably known for their infrastructural projects around the globe and Romania needs to be able to maximise this opportunity. First, Romania must rethink its national road network so as to transform it more efficiently. Second, it needs to build a network of highways, railways and adjacent roads to raise itself up to the standards set by the most developed Western states. Unfortunately, Romania has the worst conditions of infrastructure in the whole EU, as shown by the EU Transport scoreboards in 2016. This is an incentive to pay special attention to the infrastructure field as the long-term benefits can be large – attracting business, FDIs,

58 | This leads to another opportunity offered by BRI, i.e. the possibility to engage in cross-border knowledge exchanges, in parallel with supporting the private-public partnerships for higher education in the field and create specialised human capital — international groups of expertise — in the ICT industry. The projects under BRI's umbrella need specialists with a coherent vision about their implementation and goals in the long term. Also see Huawei Technologies. (2017, March 30). *From Huawei Enterprise Global Service Centre Launched*. Retrieved from http://e.huawei.com/en-HK/news/global/2017/201703301551.

and encouraging mobility at the national level. The Chinese party should invest in road infrastructure since it would be a disadvantage for businesses if imports arrive quickly in Constanta, the national port, and then spend unnecessary and costly lengths of time on land[59].

4.3. Challenges

As for any project involving different actors, interests and cultures, the BRI also comes with challenges that are specific to each sector and country.

4.3.1. Perception and public image of China in Romania

One of the very first challenges that Romania has to address is the negative perception its citizens have towards the BRI and China. Arguably, perception is one of the most critical aspects both in international relations and in business. How will Romanians react to deeper cooperation with China? How does the EU or Russia see this issue? What is the general opinion on Chinese business in the Romanian imagination?

Despite historically good diplomatic ties and improvements in cultural cooperation in recent years (e.g. establishment of Confucius Institutes in several Romanian cities), Romanians do not have a very favourable image of China. For example, the Romanian media often questions the real, presumably negative, motives behind the Chinese investments in Romania, creating an image that is worsened by the limited knowledge of China among the general population.

In addition, anti-communist feeling, which is quite powerful in Romania, negatively influences the political image of China. There are, nonetheless, assets that can be more easily used in building soft-power: culture, history, trade opportunities, and technological advancement[60].

Unfortunately, the reach of significant initiatives like 16+1 remains at the governmental or top-business levels, constraining the ability of the BRI to inspire promising results for the public

59 | Although Chinese imports represent only 5.3% of the total imports of Romania in 2018, the numbers have steadily increased over the past two decades. Moreover, there emerges the opportunity for Romania to become the gateway to the EU, once again accentuating the need for better infrastructure when it comes to connectivity. See World Integrated Trade Solution. (n.d.). *Romania Product Imports by country in US$ Thousand 2018*. Retrieved from https://wits.worldbank.org/CountryProfile/en/Country/ROM/Year/LTST/TradeFlow/Import/Partner/by-country/Product/Total.

60 | Song, W. (2017). *China's Relations with Central and Eastern Europe: From "Old Comrades" to New Partners.* Routledge Contemporary China Series.

and obtain wide support.

China and the BRI are not only critically perceived, but also insufficiently visible. Chinese public diplomacy related to the BRI must address the positive political and geostrategic implications for Romania to counterbalance the robust rhetoric that emphasises the 'evils' of Chinese projects – which raises environmental and social concerns as well as worry over China's hegemonic ambitions.

4.3.2. Balancing EU-China interests

Second, the co-operation with Romania also has to be in accord with its regional agreements and membership in the European Union. Since the EU attitude is quite reluctant towards BRI cooperation with CEECs[61], the governments of these countries have to ensure that their intentions are ably conveyed, understood, and advertised. On the one hand, the BRI has been presented as an attempt to boost Chinese influence in CEE, raising concerns about the degradation of democracy in the targeted countries. On the other hand, the Initiative might fuel economic growth and help the CEE states – Romania included – to better integrate with the European bloc, meaning that the real issue is how to balance the political and economic implications.

The existing European legal frameworks must be respected and that might engender difficulties in negotiating business projects within the BRI. Nonetheless, Romania cannot give up standards acquired from international agreements and has to pursue sustainable development by respecting social and environmental regulations. Chinese attempts to circumvent them might scandalise the public and have a further negative impact on the image of the BRI.

4.3.3. Technical challenges in physical and technology infrastructure

61 | Grieger, Gisela. (2017). *China, the 16+1 cooperation format and the EU.* Brussels: European Parliamentary Research Service.

The development of physical infrastructure in Romania provides a sure way to gain visibility and popularity. Better infrastructure also advantages the business sector and builds connectivity across the country. But since Romania is only a segment of the larger co-operation with the CEECs, the BRI has to harmonise the infrastructural projects – for example, the construction of railroads with similar features. This cross-border connectivity touches upon state sovereignty, an even more sensitive issue when it comes to ICT connectivity. Only if a credible level of cybersecurity is achieved (ensuring the safety of connections, sensitive information and secure transactions) and proper network topology is chosen, could ICT connectivity truly be established.

4.3.4. Corruption

Probably the greatest challenge to any initiative involving political variables is the recurrent theme of corruption. The way in which corruption in Romania is handled could prove to be a determinant of the success of the co-operation in the framework of the BRI. According to Transparency International's Bribe Payer Index 2006[62], Chinese enterprises are inclined to make use of bribery to pursue their interests. Chinese business corruption, on one hand, combined with the high levels of corruption in Romania on the other, could have disastrous effects. Taking into consideration the Initiative's public image and the possibility of escalating conflicts with the EU, the co-operation under the BRI has to keep in check these corruptive tendencies.

62 | The Bribe Payers Index is a ranking of 30 of the leading exporting countries according to the propensity of firms with headquarters within their borders to bribe when operating abroad. It is based on the responses of 11, 232 business executives from companies in 125 countries. See Transparency International. (2006). *Transparency International. Bribe Payers Index 2006: Analysis Report.* Retrieved from https://www.transparency.org/files/content/pressrelease/BPI_2006_Analysis_Report_270906_FINAL.pdf.

5. Suggestions for the Belt and Road Initiative and Improving Relations with the Chinese Youth

In the previous parts, we have briefly analysed the political system in Romania and the historical context, which still influences developments in the country, discussed the opportunities and the challenges affecting the youth and tried to forecast the direction of developments in various areas.

Now we turn to a list of suggestions that might improve the overall opportunities offered by BRI in Romania and limit the downsides. Briefly, we suggest that it is of prime importance to improve the overall image of China in Romania and indicate the factors that might contribute to that, such as investments in sectors attracting a lot of attention or projects that alleviate national challenges.

The first step to improving the possible outcomes of Romania's inclusion in the BRI is to invest more in popularising the initiative and the image of China. As shown in Section IV, China is still viewed in an unfavourable light by the majority of the population; this could harm collaboration with Chinese businesses. The image of China can only be elaborated through fine public diplomacy, long term planning, focused on popularising the win-win of co-operation.

The BRI is poorly reflected in the Romanian media and is largely unknown amongst the general public. How could the Initiative evoke the interest of Romanians and private business partners, as well as of public authorities, in the absence of proper dissemination of information about the advantages involved? For this, not only should the Chinese Embassy in Romania become more visible, but the cultural institutes should also engage more actively in social life, trying to overcome barriers and accomplish the cultural rapprochement desired by the BRI[63].

Grass-root and other forms of diplomacy usually familiarise the general public with small cultural elements that gather in time to create a subtle, but continuous presence; this can also be achieved by getting in touch with famous personalities from Romania and making them promoters of the BRI and Chinese culture. A good example here is that among the viewers of

63 | Wu, X. (2017, 5 12). Culture an essential piece for Belt and Road Initiative. *China Daily.* Retrieved from http://www.chinadaily. com.cn/business/2017-05/12/ content_29312197.htm.

one of Chinese rapper VAVA[1]'s[64] videos on YouTube, there were numerous French people who accessed the page due to one French vlogger who had previously mentioned the artist in his videos. The cultural rapprochement should also be a priority for the Confucius Institutes across Romania, along with the better promotion of scholarships for studies in China and public events with greater reach.

Besides direct action for image revision, China can take advantage of associating with successful projects and industries that bring about positive developments in areas of focus for Romanians. This, of course, includes the ICT sector, one of the most dynamic technology markets in the European Union[65]. Investment opportunities in this field are manifold.

First, private-public partnerships can be concluded for higher and technical education that prepares the future labour force. Similar projects are already underway for German companies in Brasov, where they financially support technological sections in several high schools. Investments in training the youth for the ICT sector will be positively perceived among the population, while the image of the investor will improve. Simultaneously, by investing in the ICT infrastructure, Romania can grow its role in regional telecommunication and information integration.

Second, innovation centres and research & development facilities can be established, making use of both the quality of graduates in the field, but also of Chinese talents. In 2007, only 20% of Romanian enterprises reported innovation activities (investments in R&D, acquisition of patents, the introduction of innovative marketing or organisational models etc.), while the European average was 38.9%[66]. Also, at the European level, 50% of the total productivity growth is due to investments in ICT, an opportunity for deepening Sino-European co-operation.

Third, creating an informational society by investing in e-business and e-government. The proven Chinese experience in the field of e-banking and practice of mobile payment apps can be brought to Romania, where the trend is still new. E-commerce has recorded steep growth with the value of Romanian

64 | Finalist in the Chinese show *The Rap of China (2017 edition)*.

65 | Mazurenc, M., Niculescu-Aron, I., & Mihaescu, C. (2006). *An Overview of the Romanian ICT Sector in the Context of European Emerging Markets*. Bucharest: Academy of Economic Studies.

66 | Baltac, V. (2009). *ICT - R&D Internationalisation: Romanian Experiences*. ATIC -Information Technology and Communications Association of Romania, 7.

e-commerce market reaching €1.8 billion in 2016, a 30% increase compared to 2015. Due to a rise in the number of smartphone users and online payments, national companies and online shopping platforms like Elefant.ro and Emag.ro have become regional players in a span of only a few years. At the governmental level, initiatives to achieve the European plan for Smart Cities[67] demand more efficient means of public transport, energy consumption, and waste disposal, which would benefit from Chinese expertise in managing urban areas.

ICT is not the only field where BRI should make improvements to tap into Romania's maximum potential. As discussed in Part IV, the energy sector and infrastructure can also improve the image of the BRI among Romanian youth. From reducing the time and costs of transport by constructing modern highways and reducing strikingly high unemployment through hiring Romanian labour, to improving the quality of public transportation by investing in the railway system and making it easier for businesses to access places of significance, all these are indicators that the BRI is not a unilateral project of a zero-sum game, but a win-win collaboration framework that gives advantages to all parties involved.

Romania was one of the first three countries to recognise the independent People's Republic of China, and relations between the two states have always been friendly. In order for the BRI to truly materialise in Romania, it is crucial to capitalise on this relationship and emphasise the close links that have bound China and Romania. A crucial first step in ensuring this is to involve Romania in as many of China's overseas projects as possible to ensure China's initiatives positively affect Romania and solidify China's assistance in the country.

Establishing more Confucius Institutes at varying universities in Romania could achieve this, and research-based Confucius Institutes (that have yet to appear in Romania) could be introduced to allow for a deeper academic understanding of Sino-Romanian relations. Similarly, Romania could be made a member of the Asian Infrastructure Investment Bank to allow for greater Chinese use of funds on Romanian infrastructure

67 | European Commission. (2017, 10 24). *Smart Cities*. Retrieved from https://ec.europa.eu/digital-single-market/en/about-smart-cities.

projects and on developing joint business ventures in Romania. Chinese investment projects in Romania worth €118 million allow young Romanians to pursue their interests and have long-lasting careers in a country marred by high youth unemployment.[68] In this regard, the Chinese acquisition of Rompetrol in 2015, one of Romania's largest oil companies, should be publicised to highlight the economic link China and Romania are forging.

Moreover, diplomatic and cultural exchanges between the two countries could be heightened by expanding embassies and consular missions in the other country. More honorary Chinese consular positions (i.e. no genuine diplomatic role, only the title) could be provided to local businessmen in Romania so as to boost China's domestic representation. Honorary consuls are often required to represent Chinese businesses in Romania and host cultural events in order to attain such a role. By giving out more honorary consul titles, more Chinese firms can access the Romanian market, as they have an honorary consul responsible for smoothening this process. This is particularly relevant to the smartphone and technology industries, as well as crucial in creating a broader cultural appreciation of China among young Romanians.

The BRI is a multi-faceted project – one must consider the economic issues related to it, but similarly, diplomatic and cultural efforts must be made to emphasise the long and prosperous relationship the two nations have had. This is best done by increased visits of Chinese dignitaries to Romania, by extended ministerial talks between Chinese and Romanian counterparts, and by Chinese state visits to Romania. Joint military drills and political visits are the best way of expressing to the populace the strong Sino-Romanian bond, as proven by the joint Sino-Russian naval drills in the Baltic over July of 2017[69].

Bilateral trade between China and Romania could also be increased through numerous avenues. The EU has no free trade agreement (FTA) with China, and because the EU prohibits member countries from forming their own trade agreements,

68 | The Diplomat Bucharest. (2016, June 15). China has five investment projects in Romania worth over 118 million Euro. *The Diplomat Bucharest*. Retrieved from http://www.thediplomat. ro/articol.php?id=7164.

69 | Higgins, A. (2017, July, 25) China and Russia Hold First Joint Naval Drill in the Baltic Sea. *The New York Times*. Retrieved from https://www.nytimes. com/2017/07/25/world/ europe/china-russia-baltic-navy-exercises.html.

Romania is incapable of striking an FTA with China. A long-term goal would be either for this restriction to be removed or for the EU to sign a free trade agreement with China, thus encouraging Chinese business owners to sell their goods in Romania at a more affordable price.

China also benefits from the fact that its government controls key assets and industries, in particular the oil industry. Because of the China National Petroleum Corporation's power, it could, should China wish, provide oil to Romania at a cheaper rate. The company is state-owned and making sizeable profits, meaning the Chinese government could use it as a financial incentive to strengthen Sino-Romanian ties. This can be done through a permanent agreement stipulating the cheaper sale of gas. In doing so, Romania could form a much stronger bond with China, and the presence of Chinese oil firms in Romania would likely increase. Not only would this lead to increased job opportunities for young Romanians, but also to more affordable lives (due to cheaper heating and gas prices, and so on), meaning Romanians are in turn able to spend more on other Chinese goods and are likely to have an increasingly favourable view of China. This is plausible as trade between the two nations is currently only at US$4.916 billion and Romania is only China's 60th export destination[70]. Similarly, trade could be pursued in under-developed areas, such as arms exports and consumer goods.

Ultimately, for the BRI to truly be embraced by Romania's youth, a two-pronged approach must be taken. The strong diplomatic relations between the two countries must be emphasised through political, cultural and social links, while economic opportunities must be provided to the youth (through increased job opportunities, investment in the ICT sector and cheaper products) in order to create economic inter-connectivity between the two countries and spark a Romanian interest in Chinese goods and Chinese support. In doing so, the Belt and Road project will succeed in providing Romania's youth with opportunity, prosperity and success.

70 | Ministry of Commerce of the People's Republic of China. (2017, March 24). *Romania-China Trade Report 2016*. Retrieved from http://english. mofcom.gov.cn/article/statistic/ lanmubb/chinaeuropeancoun try/201704/20170402560883. shtml.

The Saint Sava Temple and its surrounding area in Belgrade, Serbia.

CHAPTER 15

by   ĐINA GLAVČIĆ-KOSTIĆ
University of Belgrade

SARA KLJAJIC
Middlebury College, Bachelor of Economics (Class of 2020)

NIKOLA STOJANOVIC
Yenching Academy of Peking University

(Ordered by last name)

ESTONIA

LATVIA

LITHUANIA

POLAND

CZECH
REPUBLIC

SLOVAKIA

HUNGARY

SLOVENIA

CROATIA

ROMANIA

BOSNIA &
HERZEGOVINA

MONTENEGRO

BULGARIA

NORTH
MACEDONIA

ALBANIA

SERBIA

SERBIA

1. Overview of the Political Structure of Serbia

As an integral element of the phenomenon of constitutionalism, the political system is regarded as a powerful and indisputably useful apparatus of the state. It has become an inherent symbol of the state's stability, political justice, and security.

The Serbian system is *de jure* - according to the letter of the law – a rationalised parliamentary system[1] in which the president has disproportionately narrowed authorisations – more than his equal in a parliamentary system but less than the president in a presidential regime. However, when the president is simultaneously the leader of the dominant political party in the parliament, the Serbian system includes the features of a mixed-presidential system.

1.1. Constitution: A 'Political Building'

The history of constitutionalism in Serbia commenced in 1835 after Serbia had gained *de facto* independence due to the First and Second Serbian Uprisings against Ottoman rule that occurred two decades earlier. Significant constitutional examples culminated in 1888 in the Constitution, made and governed by the omni-legitimate power of Radicals[2], which was considered one of the most modern constitutional documents of that era. It had the highest quality of the letter of the law, as well as an introduction of guarantees for political and other relevant rights and liberties of the Serbian people[3].

For most of the 20th century, Serbia was included in numerous federal structures[4]. The Constitution from 2006 came about after the final political divorce between the two federal units remaining from Yugoslavia, following the referendum in Montenegro, held on 21st May 2006[5].

The Constitution from 2006 is a codified and democratically oriented constitution. Parallel to establishing legally the strongest guidelines to allowed behaviour of both functionaries and legal and natural persons, it underwrites democratic principles. It grants a rule of law, free elections, and judicial independence, delegates sovereignty to the people, and underwrites numerous rights and liberties of its citizens

1 | Marković, R. (2016). *Ustavno pravo* (21st ed., Ser. 1200) *[Constitutional right]*. Belgrade, Serbia: Pravni fakultet Univerziteta, Centar za izdavaštvo i informisanje, pp. 69-72.

2 | The People's Radical Party (Serbian: Народна радикална странка).

3 | Marković, *supra* note 1, pp. 101-103.

4 | Kingdom of Serbs, Croats and Slovenes (1918.), renamed into Kingdom of Yugoslavia (1929.), after Second World War: Federal People's Republic of Yugoslavia (1946.), Socialist Federal Republic of Yugoslavia (1963.) etc. (see: Marković, *supra* note 1, pp. 93-149).

5 | Traynor, I. (2006, May 22). Montenegro vote finally seals death of Yugoslavia. *The Guardian*. Retrieved from https://www.theguardian.com/world/2006/may/22/balkans.

in accordance with relevant international and European documents. It is explicitly committed to 'European principles and values'[6].

1.2. Division of Power

In Serbia, power is separated between the main decision-making organs[7]: a legislative body (the legislative branch), the executive branch, and a system of courts (the judiciary branch). All three are elements where crucial decisions shall be watchfully examined and concluded in the form of specific legal acts[8].

1.2.1. Legislative branch

Since the judicial branch holds proclaimed independence, the remaining two branches carry the responsibility of the insightful determination of state politics. The legislative body, resting on a principle of representative democracy, gifts and enacts the country with the most important *regulas* (rules).

In the Republic of Serbia, the legislative body is the National Assembly and has been given the legislative as well as constitutional power by The Constitution.[9] The National Assembly is a democratic institution in which the loudest legitimate political confrontation takes place. It is the official mediator between the people as a multi-will entity and the state as an artificial creature with a single will (legal person). Metaphorically speaking, the process is like making a unison voice out of a choir — to perceive a single, dominant melody in the buzzing of a thousand equally intense voices.

The total number of seats (250), being a constitutional design, shall suit the country's budget capabilities and demographic count. According to the universally acknowledged and verified 'square root formula', Serbia exceeds the desirable proportion between the total population and number of MPs[10].

6 | Article 1 and Article 4, *Constitution of The Republic of Serbia.*

7 | Marković, *supra* note 1, p. 77.

8 | Lukić, R. (1994). Pravna Organizacija Drzave, Vrsta državnih organa. In Uvod u pravo [Legal Organisation of the State, Type of state bodies In Introduction to law]. Belgrade, Serbia: Grafika PIROT, pp. 123-126.

9 | 'The National Assembly shall be the supreme representative body and holder of constitutional and legislative power in the Republic of Serbia.' (see: Article 98, *Constitution of Serbia*).

10 | Auriol, E. & Gary-Bobo, R. J. (2007). *On the Optimal Number of Representatives.* Retrieved from http://www.econ.uiuc.edu/~skrasa/seminar_april.pdf.

1.2.2. Electoral system

The main parameters for observing and assessing the quality of an electoral system are intuitively sequential: who can affect the content of the legislative body; who has the ability to compete as a candidate; to what extent does it embrace population in percentage; and its precision in transmitting the people's will. These democratic methods allow the people's will to be delineated through the result of the elections.

The ability to participate in the most vital political processes is general and equal: there are no special requirements. The acquisition of suffrage (ability to vote for) and electability (ability to be voted for) in Serbia coincides with the acquisition of legal capacity. This is a minimum requirement in the modern legal system in comparison with other specific requirements that the Serbian legal system practised through history, e.g. one's precisely determined educational level or the amount of property tax that one was obliged to provide for the state. In Serbia, it is not obligatory to exercise the suffrage.

During the last century, Serbia reverted to the multi-party system under the assumption that a more diverse scope of interests is to be expressed and social solutions are to be proposed in accordance with Western democratic tradition.[11] Serbia thus stands behind the proportional system, which protects minor parties and allows them to enter the parliament after breaking the electoral census – an official threshold of 5% of the total received votes[12].

1.2.3. Executive branch

The executive body in Serbia is an arrangement of two political elements: passive, the president, and active, the Government. The Government determines the state's inner and outer politics: bringing about vital decisions in coordination with the law and directing the state's motion to the future while the president personifies and 'expresses state's unity'.[13]

The president holds standardised authorisations such as

11 | Narodna Skupstina Republike Srbije [National Assembly of The Republic of Serbia]. (n.d.). *Izbori i izborni system [Elections and Electoral System]*. Retrieved from http://www.parlament.rs/narodna-skupstina-/uloga-i-nacin-rada/izbori-i-izborni-sistem.906.html

12 | Marković, *supra* note 1, pp. 234-246.

13 | Mijović, M. (2002). *Mislilo: knjiga misli 1000 mudraca [Thought: a book of thoughts of 1000 sages]*. Belgrade, Serbia: Narodna knjiga – Alfa, p. 60.

representing the Republic of Serbia in the country and abroad. Authorisations are mostly expressed through decisions of minor importance (awarding, granting amnesty, etc..), but there are also those such as the promulgation of law ('legal veto') and the proposal of a prime minister candidate to the National Assembly which are of greater importance[14].

The Government, on the other hand, consists of the prime minister, vice-presidents, and ministers. Due to the 'incompatibility of functions', ministers do not derive from the Parliament/National Assembly. In other words, ministers cannot be Parliamentary members and cannot perform both functions simultaneously. For that reason, it is the candidate for the prime minister who selects the names of ministers and gives the final approval to the MPs who are in power. It is, in effect, the total engagement of MPs in the process.

After the Government is successfully designated, political responsibility through interpolation combined with the vote of confidence and vote of no confidence makes it liable to the will of the Assembly. Depending on the level of contentment of MPs in opposition over the Government's conduct and work, a group counting 60 MP's (vote of no confidence) can activate a series of questions regarding either particular matter of the Government's recent actions, or work of the Government in general. After a moderated dispute takes place, optional vote over dismisses of the Government's composition can be induced, demanding majority to be put in effect[15]. This is a mechanism which strengthens the principle of 'checks and balances' by introducing the most direct sanction of responsibility of the Government before the parliament.

1.2.4. Judicial branch

Judicial power naturally belongs to courts (of general and of special jurisdiction), which falls under a systematic and hierarchical net of jurisdiction. 'And all judges shall judge led by the law, and not by the fear of my empire/governance!', says a provision from Dušan's legal code[16]. It was brought in May of 1349, in the first Empire of Serbia governed by Dušan Silni,

14 | Article 112-121, *Constitution of The Republic of Serbia*. Retrieved from https://www.ilo.org/dyn/natlex/docs/ELECTRONIC/74694/119555/F838981147/SRB74694%20Eng.pdf.

15 | Article 18-19, Government Law. *Zakon o Vladi [Government Act]*. Retrieved from https://www.paragraf.rs/propisi/zakon_o_vladi.html.

16 | Mirković, Z. (2017). *Srpska pravna istorija [Serbian legal history]*. Belgrade, Serbia: Pravni fakultet Univerziteta, Centar za izdavaštvo i informisanje, p. 26.

which solemnly granted judicial independence[17].

Today, Serbia opposes legal insecurity in the same way. Relocation, unless with the consent of the judge, is strictly prohibited. Influence on a judge's decisions and a judge's engagement in political actions are also prohibited[18]. There is also the principle of unique judicial power in the whole territory of the Republic of Serbia, aiming to provide legal security through many layers. Moreover, judges operate not only according to the Constitution, Law, and other general acts, but also to 'generally accepted rules of international law and ratified international contracts'[19].

The constitution introduces the existence of the institutions which hold very specific tasks as follows:

> **The Constitutional Court** which, on the grounds of continental law, has been given the 'surname' of the negative legislator. It is an official protector of the highest principles of legal order: legality and constitutionality, as well as legal remedy for 'wounded' rights and liberties through an institute of a constitutional complaint.

> **Independent regulatory bodies** such as the Civic defender, a 'legal transplant'[20] adopted from Scandinavian legal tradition and supported by the word of Constitution. They are established to, among the rest, monitor the work of the bodies of public administration.

The Republic of Serbia has a relatively stable, universally acknowledged constitutional basis. Nevertheless, the Republic is in the process of creating a new constitutional architecture. The official announcement of these constitutional changes will be further elaborated in section two of this paper. All reforms shall enable the state to fit into the modus of effective, direct and successful contact with its legitimate 'titulars' - the people, who are the very initial source of the state's power. It is within the scope of this change to determine national necessities, and consequently, to locate the least arduous and turbulent path to highly desired prosperity and progress in Serbia.

17 | *Ibid.*

18 | Article 152, *Constitution of The Republic of Serbia*. Retrieved from https://www.ilo.org/dyn/natlex/docs/ELECTRONIC/74694/119555/F838981147/SRB74694%20Eng.pdf.

19 | *Ibid.*

20 | Term first used by Alan Watson. See Votson, A. (2000). *Pravni transplanti [Legal transplants]*. Belgrade, Serbia: Pravni fakultet Univerziteta, Centar za izdavaštvo i informisanje.

2. The Political Direction of the Country

This part of the paper interviews Prof. Milan Jovanovic, a prominent expert in political systems, to analyse the political direction of Serbia. The interview was conducted by the authors in October 2017. This part will observe the current political situation, compare it to the optimal one to state the differences and provide the conclusion. According to Professor Jovanovic, the future political situation in Serbia will see the change of the current Constitution, the rise in political culture, no changes in the political spectrum, and a further move towards European Union integration.

2.1. Change of the Constitution

In 2018, Serbia was going to see changes to its Constitution. According to Prof. Milan Jovanovic, the constitution would emphasise judicial independence, but should also aim towards the change of the electoral system.

Q *Will the independence of the judiciary be emphasised in the constitution?*

A (Professor Jovanovic): 'The assembly and government should be excluded from the choice and funding of courts, the choice of judges and prosecutors. This would balance the relationship between the three branches of the government and preserve the usual balance of reciprocal influence and control.

The absence of consensus causes frequent changes in the constitution, as shown by statistics. It proves we have not yet constituted ourselves as a political community. The most important changes in the constitution will meet the obligations of the EU accession process. The most realistic thing is that changes will happen only in the area of the independence of the judiciary and, possibly, in provisions that would encourage relaxation of relations with Kosovo and Metohija [the southwest part of Kosovo].'

Q *Does the electoral system need to change?*

A (Professor Jovanovic): 'We need to abandon the voting mode for the electoral lists where we vote for the party, more precisely for the party leader. In order to avoid depersonalised elections, some forms of preferential voting should be introduced, and citizens should be able to elect their concrete MPs and councillors by name and surname, and the candidates will thus fight for our votes armed with qualities they own.

'And to reduce the leverage of power and encourage acting within the limits of the constitutional power, direct election of the president of the state should be abolished and he should be elected by the parliament so that the government would become the only active factor of the executive power.'

2.2. The Rise of Political Culture

Q *How do we improve the trust of the people in political institutions?*

A (Professor Jovanovic): 'Citizens, to a certain extent, expect to hear their wishes from the authorities, to recognise their projections in public policies – this is a process of mutual seduction, but in this relation, the measure is a reality.

'We need the political engagement of citizens who need to be more informed, educated, and also, not to, as bearers of sovereignty, leave the area of public decision-making to politicians without control. The expansion of media, internet, and use of social networks creates a place for populist nets.'

'Generational change in the political class would contribute to new patterns of behaviour and action but that takes time and cultural change.'

2.3. Political Spectrum: Changes and Tendencies

Q *What can you say about the balance between the government and the opposition party?*

A (Professor Jovanovic): 'The parties forming the government have strong support in the electorate. In addition, they also show a higher level of coalition capacity. They have additional stability by their cooperation with all relevant parties of national minorities, and a series of local and regional parties and movements.

'However, there is a disoriented and fragmented opposition that does not have a consistent programme that would oppose government policy and gain voters for political reversal. There is simply a lack of an energetic and promising alternative.

'There is a crowd on the axis of the political spectrum towards the centre, making many parties colourless and unrecognisable on the political scene. The power of extreme poles: extreme left and right, in the electorate, is minor.

'Serbia was the only one of all former republics of the Socialist Federal Republic of Yugoslavia to retain an extremely multi-ethnic character. This eloquently says that the right has no perspective.'

2.4. EU Integration

Serbia will keep strengthening its candidacy to the European Union, but the process will only conditionally affect the country's democratisation in the future. In the first quarter of the 21st century, Serbia is surrounded by the semi-European environment of the Balkans (Croatia, Romania, Bulgaria). The country was granted the status of a candidate country in 2012 by the European Council[21]. Accession negotiations started in 2014. In order to become a full member of the EU, Serbia needs to go through ambitious political and economic reform. For example, it would need to address issues such as the rule of law, specifically judicial reform and anti-corruption policies, independence of key institutions, freedom of expression, anti-discrimination policies, and the protection of minorities. Since the negotiations started, Serbia has met the minimum conditions, which implies that EU assistance can be now addressed through sector approach and multi-annual planning

21 | European Commission. (2016, November 9). Commission Staff Working Document Serbia 2016 Report. *COM(2016) ed., Vol. 715, p. 91, Rep.* Retrieved from https://ec.europa.eu/neighbourhood-enlargement/sites/near/files/pdf/key_documents/2016/20161109_report_serbia.pdf.

perspective over the 2014-2020 period[22].

Q *Will the future realisation of EU membership improve the level of democracy in Serbia?*

A (Professor Jovanovic): 'EU membership in itself will not raise the level of democracy, nor solve all of our problems in the development of democracy. Even countries of the EU with all normative guarantees are not deprived of deformation in the functioning of politics. Democracy depends on politics at the national level. It is therefore of particular importance to understand that the responsibility of preserving, building, and strengthening democracy rests on the national plane – citizens, parties, and political institutions. There is no (legal) transplantation or copying, democracy is a process that involves institution building.'

In conclusion, Serbia is passing through numerous changes so analysing its future political route is not an easy task. Taking into account the country's political history, this part heavily relied on the interview with Prof. Jovanovic. While Professor Jovanovic is known to be a distinguished expert in this area, we are aware that our arguments have limits and that predictions may vary. Overall, Serbia will experience a strengthening of judicial independence and progress towards the European Union integration, but most likely it will not change its current government-opposition relation nor its level of democracy. The country needs to go a long way to gain a higher level of trust of its citizens in political institutions.

22 | European Commission. (2014). *Instrument for Pre-accession assistance (IPA II), Indicative Strategy Paper For Serbia (2014-2020)* (p. 44, Issue brief). Retrieved from https://ec.europa.eu/neighbourhood-enlargement/sites/near/files/pdf/key_documents/2014/20140919-csp-serbia.pdf.

3. Challenges and Opportunities Faced by Young People in Serbia

23 | Potočnik, D, &, Williamson, H. (2015). *Conclusions of the Council of Europe International Review Team.* Council of Europe, Република Србија Министарство Омладине и Спорта [Republic of Serbia Ministry of Youth and Sports]. Retrieved from www.mos. gov.rs/public/ck/uploads/files/YP_Serbia.pdf. p. 11.

24 | *Ibid.,* p. 11.

25 | Stojanović, B. (2017). *Alternativni Izveštaj o Položaju i Potrebama Mladih u Republici Srbiji [Alternative Report on the Position and Needs of Youth in the Republic of Serbia].* 1st ed., vol. 150, Krovna Organizacija Mladih Srbije-KOMS, pp. 2-184. Retrieved from http:// koms.rs/wp-content/uploads/2017/08/ Alternativni-izves%CC%8Ctaj-o-poloz%CC%8Caju-i-potrebama-mladih-u-Republici-Srbiji-Boban-Stojanovic%CC%81-1. pdf. p. 14.

26 | Potočnik, D, &, Williamson, H. (2015). *Conclusions of the Council of Europe International Review Team.* Council of Europe, Република Србија Министарство Омладине и Спорта [Republic of Serbia Ministry of Youth and Sports]. Retrieved from www.mos. gov.rs/public/ck/uploads/files/YP_Serbia.pdf. p. 11.

27 | *Ibid.,* p. 42.

28 | Stojanović, B. (2017). Alternativni Izveštaj o Položaju i Potrebama Mladih u Republici Srbiji [Alternative Report on the Position and Needs of Youth in the Republic of Serbia]. *1st ed., vol. 150, Krovna Organizacija Mladih Srbije-KOMS,* pp. 2–184. Retrieved from http:// koms.rs/wp-content/uploads/2017/08/ Alternativni-izves%CC%8Ctaj-o-poloz%CC%8Caju-i-potrebama-mladih-u-Republici-Srbiji-Boban-Stojanovic%CC%81-1. pdf. p. 4.

Serbian law defines 'youth' as all people aged 15 to 29 years old[23]. In 2017, Serbia had 7.2 million people[24] categorised as 'youth', which is approximately 17.35% of the total population[25]. To evaluate the current situation of young people in Serbia, one must first understand the Serbian political and economic situation. Young people in Serbia still feel the consequences of the country's communist and socialist past as well as the breakup of Yugoslavia. These issues, followed by the macroeconomic instability, formed the paternalistic, traditional regime among youth in the country[26]. What this means is that young people face many challenges, the biggest three being the high unemployment rate, lack of youth autonomy, and ineffective educational system. It is worth mentioning, however, that in recent years, there have been significant developments made in youth policy.

3.1. Youth Unemployment

In Serbia, youth inactivity rates are the highest among the population. 'Young people are the most disadvantageous and the least employable in the labour market'[27]. According to the government, the youth in Serbia is viewed as a 'vulnerable group', because 22.75% of all unemployed people in the country belong to the young population[28].

29 | The World Bank. (2018). *Unemployment, youth total (% of total labor force ages 15-24) (modeled ILO estimate) - Serbia.* Retrieved from https://data.worldbank.org/indicator/ SL.UEM.1524.ZS?end=2020&locations=RS &start=1991&view=chart.

30 | Stojanović, B. (2017). *Alternativni Izveštaj o Položaju i Potrebama Mladih u Republici Srbiji [Alternative Report on the Position and Needs of Youth in the Republic of Serbia].* 1st ed., vol. 150, Krovna Organizacija Mladih Srbije-KOMS, pp. 2–184. Retrieved from http://koms. rs/wp-content/uploads/2017/08/Alternativni-izves%CC%8Ctaj-o-poloz%CC%8Caju-i-potrebama-mladih-u-Republici-Srbiji-Boban-Stojanovic%CC%81-l.pdf. p. 88.

31 | Đurović, A., Stevanović, A., & Jevtović, B. (2017). Mladi i izazovi savremenog tržišta rada. *Progovori o pregovorima [Youth and the challenges of the modern labour market. Talk about negotiations],* 16. Retrieved from http:// eupregovori.bos.rs/progovori-o-pregovorima/ uploaded/Bilten_Broj_27_28_2017.pdf. p. 4.

32 | Potočnik, D, &, Williamson, H. (2015). Conclusions of the Council of Europe International Review Team. Council of Europe, Република Србија Министарство Омладине и Спорта [Republic of Serbia Ministry of Youth and Sports]. Retrieved from www.mos.gov.rs/ public/ck/uploads/files/YP_Serbia.pdf. p. 42.

33 | *Ibid.,* p. 11.

34 | World Economic Forum. (2016). *Human Capital Report 2016 - Serbia.* Retrieved from http://reports.weforum.org/human-capital-report-2016/economies/#economy=SRB.

35 | Potočnik, D, &, Williamson, H. (2015). Conclusions of the Council of Europe International Review Team. Council of Europe, Република Србија Министарство Омладине и Спорта [Republic of Serbia Ministry of Youth and Sports]. Retrieved from www.mos.gov.rs/ public/ck/uploads/files/YP_Serbia.pdf. p. 13.

36 | *Ibid.,* p. 12.

37 | Ninamedia Research. (2016, December 2). *Survey on Position and Needs of Youth in the Republic of Serbia.* Retreived from www.mos. gov.rs/public/ck/uploads/files/Istrazivanje%20 polozaja%20i%20potreba%20mladih%20 teren%20decembar%202016_ENG%20(2).pdf. p. 33.

This view is further supported by official RZS (Republic Institute of Statistics of Serbia) data from March 2017, showing that 31.2% of the youth is unemployed[29], 2.5 times higher than the overall unemployment rate in the state[30]. After graduating, a young person needs two years on average to find employment[31]. Youth opportunity in the labour market is worsened by the grey-market and non-regulated work, commonly found in Serbia[32]. While it is true that there has been a slight decrease in the unemployment rate in recent years, the situation is still far from good.

As a consequence of facing a precarious labour market[33], young people either decide to leave the country or heavily depend on family sources. As the World Economic Forum report shows, Serbia has the second-highest annual number of experts that leave the country in the world[34].

Unfortunately, a large number of young people consider leaving the country as well. For example, in 2015, 64.2% of people said they considered going abroad. When asked about the reasons for making such a decision, they mentioned the low living standard, lack of job prospects, safer life abroad, hope for better education, and corruption within the country as their main reasons[35]. Moreover, young people heavily rely on family sources, some even until their thirties[36]. This is shown by the fact that 38.4 % of young people aged 25-30 and more than one half in the range 20-24 live with their parents[37].

Many people hold the opinion that opportunities for youth employment would significantly increase if Serbia entered the European Union. In that case, young people would gain easy access to enter other EU countries' labour markets and have the same rights as their peers in those countries. Equally important, the biggest reason usually given for the high percentage of youth unemployment in the country is inadequate skills that do not match the labour market. To improve the status of youth in the labour market, the country needs to invest in practice and internship programmes, and develop career guidance and counselling systems[38].

3.2. Education System and Quality of Teaching

The education system in Serbia is unresponsive to the needs of the labour market, resulting in a mismatch between demand and supply. Serbia ranks only 43rd out of 65 countries in the Programme for International Student Assessment (PISA) from the Organisation for Economic Cooperation and Development (OECD)[39]. Education in Serbia does not follow modern market demands and is rather traditional in nature.

The curricula do not follow market demand, which results in a lack of knowledge and skills sought by employers[40]. Most students rely on outdated heavy printed books as their main source of information, instead of using online sources. The data shows that Serbia is one of the least developed countries in Europe when it comes to internet usage, with only 39% of the population having access to it[41]. In other words, a significant amount of young people do not have immediate access to information. This lack of access contributes to the exclusion of youth in the political decision-making process, which will be further discussed in the next section.

Having a limited budget, the government does not invest enough in research and development, textbooks are outdated, and professors' wages are low, which in turn lowers the quality of teaching and education overall. Educational reforms are characterised by a slow pace[42], and even the reforms that are in place have been blocked by the local educational authorities and universities, who do not want to regulate their enrolment quotas in accordance with the labour market need[43].

3.3. Youth Autonomy and Decision-Making Freedom

Youth autonomy is defined as the possibility of young people to

38 | Đurović, A., Stevanović, A., & Jevtović, B. (2017). *Mladi i izazovi savremenog tržišta rada. Progovori o pregovorima [Youth and the challenges of the modern labour market. Talk about negotiations]*, 16. Retrieved from http://eupregovori.bos.rs/progovori-o-pregovorima/uploaded/Bilten_Broj_27_28_2017.pdf. p. 5.

39 | European Commission. (2014). *Instrument for Pre-accession assistance (IPA II), Indicative Strategy Paper For Serbia (2014-2020)* (p. 44, Issue brief). Retrieved from https://ec.europa.eu/neighbourhood-enlargement/sites/near/files/pdf/key_documents/2014/20140919-csp-serbia.pdf. p. 5.

40 | Đurović, A., Stevanović, A., & Jevtović, B. (2017). *Mladi i izazovi savremenog tržišta rada. Progovori o pregovorima [Youth and the challenges of the modern labour market. Talk about negotiations]*, 16. Retrieved from http://eupregovori.bos.rs/progovori-o-pregovorima/uploaded/Bilten_Broj_27_28_2017.pdf. p. 5.

41 | Potočnik, D, &, Williamson, H. (2015). *Conclusions of the Council of Europe International Review Team*. Council of Europe, Република Србија Министарство Омладине и Спорта [Republic of Serbia Ministry of Youth and Sports]. Retrieved from www.mos.gov.rs/public/ck/uploads/files/YP_Serbia.pdf. p. 13.

42 / *Ibid.*, p. 12

43 | *Ibid.*, p. 42

live, think, and work independently. In Serbia, youth autonomy is low because of limited financial resources and the strong distrust of young people towards political institutions. Over a third of people do not have independent sources of funds but depend on parents instead[44]. While one could argue that the youth can execute autonomy through non-governmental bodies, i.e. youth organisations, with the Ministry of Youth and Sport reporting an impressive 836 youth organisations and associations registered in the country[45], limited financial funds prevent the organisations from realising their full potential.

Research has shown that young people have no confidence in any of the institutions mentioned in the research, especially political parties and political institutions[46]. When it comes to the level of political activity, only 52.2% of young men and 47.8% of women consider themselves to be active[47].

There is a significant exclusion of young people from decision-making processes as well. Not a single minister in the Serbian government, minister's assistant, or Secretary of State falls under the category of youth. While it is true that young people are less qualified and experienced to hold a ministerial position, this number is significantly higher in other European countries. In the current convening of the National Assembly, there are only three MPs younger than 30 (1.2%)[48]. Additionally, the number of youth in the National Assembly is only decreasing year by year[49].

3.4. Recent Youth Policy Developments

To combat traditional and paternalistic attitudes in Serbia, which are common among young people who live in post-socialist states, several crucial youth policy reforms have taken place since 2007[50]. In 2007, a crucial year regarding youth policy development, the Serbian government formulated the National Youth Policy and established The Serbian Ministry of Youth and Sport. In 2011, the government introduced the Law for Youth[51]. These changes opened a new era of youth policy in Serbia by requiring the opening of local youth offices across the country and distributing a significant amount of human and

44 | *Ibid.,* p. 12.

45 | *Ibid.,* p. 25.

46 | Stojanović, B. (2017). *Alternativni Izveštaj o Položaju i Potrebama Mladih u Republici Srbiji [Alternative Report on the Position and Needs of Youth in the Republic of Serbia].* 1st ed., vol. 150, Krovna Organizacija Mladih Srbije-KOMS, pp. 2–184. Retrieved from http://koms.rs/wp-content/uploads/2017/08/Alternativni-izves%CC%8Ctaj-o-poloz%CC%8Caju-i-potrebama-mladih-u-Republici-Srbiji-Boban-Stojanovic%CC%81-1.pdf. p. 85.

47 | Potočnik, D, & Williamson, H. (2015). *Conclusions of the Council of Europe International Review Team.* Council of Europe, Република Србија Министарство Омладине и Спорта [Republic of Serbia Ministry of Youth and Sports]. Retrieved from www.mos.gov.rs/public/ck/uploads/files/YP_Serbia.pdf. p. 13.

48 | *Ibid.,* p. 41.

49 | *Ibid.,* p. 41.

50 | *Ibid.,* p. 19.

51 | *Ibid.,* p. 19.

financial resources for the implementation of policy initiatives[52].

The latest improvement was made on 27th February 2015, when the government adopted the 'National Youth Strategy for the period from 2015 to 2025,' and an action plan for the first three years (until 2017)[53]. Even though the direct effects of the strategy have yet to be observed, the current SNS government party has begun to put more emphasis on youth education and skill development[54]. For instance, Prime Minister Ana Brnabic has said: 'This government will put emphasis on enabling young people to work, create and implement their ideas and initiatives through youth entrepreneurship. We want young people to develop through education a driving spirit and gain more courage to embark on new endeavours.'[55]

There is a great emphasis on the development of entrepreneurship and career guidance centres, but it is also noticeable that young people are finally gaining recognition as a 'power of the future' and potential of the country[56]. Although there are many policies that have been or are being undertaken with a goal of improving the status of young people in the country, many had setbacks and pitfalls, which prevented their proper execution[57].

In conclusion, there are some positive youth policy developments being made, but there is still work to be done in Serbia to enable its young people to enjoy the same privileges as their peers in the European Union or more developed countries. As mentioned, people are increasingly saying that they want to leave the country. From the youth's perspective, Serbia does not have an optimistic future. On the other hand, the newly implemented policies have still not taken full effect, which implies that the views of Serbian youth might change over time. Serbia is currently thought to be at a politically important turning point after selecting Brnabic as the first female Prime Minister. Brnabic herself has been educated abroad and plans to actively use her experience abroad to implement new youth policies and improve the overall situation in the country.

52 | *Ibid.*, p. 14.

53 | Stojanović, B. (2017). *Alternativni Izveštaj o Položaju i Potrebama Mladih u Republici Srbiji [Alternative Report on the Position and Needs of Youth in the Republic of Serbia].* 1st ed., vol. 150, Krovna Organizacija Mladih Srbije-KOMS, pp. 2–184. Retrieved from http://koms.rs/wp-content/uploads/2017/08/Alternativni-izves%CC%8Ctaj-o-poloz%CC%8Caju-i-potrebama-mladih-u-Republici-Srbiji-Boban-Stojanovic%CC%81-1.pdf. p. 31

54 | *Ibid.*, p. 55.

55 | *Ibid.*, p. 48.

56 | *Ibid.*, p. 49.

57 | Potočnik, D, &, Williamson, H. (2015). *Conclusions of the Council of Europe International Review Team.* Council of Europe, Република Србија Министарство Омладине и Спорта [Republic of Serbia Ministry of Youth and Sports]. Retrieved from www.mos.gov.rs/public/ck/uploads/files/YP_Serbia.pdf. p. 14

4. Serbia in 16+1 Cooperation

The point of departure for the Sino-Serbian bilateral relationship and the basis for Serbia's involvement in the 16+1 initiative is a traditional friendship between the two governments and peoples. During his visit to Serbia, in a signed article for the Serbian newspaper *Politika*, President Xi Jinping maintained that: 'Over the last 60 years, the two peoples have always been united in each other's special feelings for each other and true friendship across time and space.'[58] The overall tone for the positive development of Beijing's relationship with Serbia was rooted in joint historical struggles and common memories. Chinese and Serbians, among others, opposed Fascist aggression on the Eastern and Western fronts during the World War II and shared the 'traumatic experience of the United States' bombardment of the Chinese embassy in Belgrade during the NATO campaign against Serbia in 1999'[59]. Even before Chinese President Xi Jinping first announced the vision of the New Silk Road in 2013, China and Serbia had already been frequently characterised as reliable partners with a high level of mutual trust and support.

Given that the Chinese government adopted strategic partnerships diplomacy as one of the major foreign policy tools over the past two decades, aiming to build a new type of state-to-state relations based on non-alliance, non-confrontation, and not directed against any third party[60], it came as no surprise that China and Serbia officially elevated their mutual relationship to the level of strategic partnership by a Joint Statement signed by Presidents Boris Tadic and Hu Jintao during the former's visit to Beijing in August 2009[61].

Serbia was the first among the Central and Eastern Europe (CEE) countries to be added to the portfolio of China's strategic partners. Explaining the underlying rationale for Beijing to establish a strategic partnership with Serbia, former Chinese ambassador to Serbia Zhang Wanxue highlighted similar historical experiences, complementary economic structures, equivalent views on numerous international and territorial issues, and joint opposition to foreign interferences in internal affairs of countries[62]. Moreover, reflecting China's importance for Serbia, shortly after the establishment of the strategic

58 | Stanzel, A., Kratz, A., Szczudlik, J., & Pavlićević, D. (2016). China's Investment in Influence: the Future of 16+1 Cooperation. *European Council on Foreign Relations*, pp. 12-13. Retrieved from https://www.ecfr.eu/page/-/China_Analysis_Sixteen_Plus_One.pdf. pp. 12-13.

59 | Stanzel, A., Kratz, A., Szczudlik, J., & Pavlićević, D. (2016). China's Investment in Influence: the Future of 16+1 Cooperation. *European Council on Foreign Relations*, pp. 12-13. Retrieved from https://www.ecfr.eu/page/-/China_Analysis_Sixteen_Plus_One.pdf. p. 13.

60 | Strüver, G. (2017). China's Partnership Diplomacy: International Alignment Based on Interests or Ideology. *The Chinese Journal of International Politics*, 10(1), p. 35.

61 | Pavlićević, D. (2011). The Sino-Serbian Strategic Partnership in a Sino-EU Relationship Context. *Briefing Series*, p. 6.

62 | Arežina, S. (2013) *Odnos NR Kine sa Jugoslavijom i Srbijom od 1977. do 2009. Godine [The relationship of the People's Republic of China with Yugoslavia and Serbia from 1977 to 2009]* (Unpublished doctoral dissertation). University of Belgrade, Faculty of Political Sciences, Belgrade, Serbia, p. 365.

relationship, Belgrade adopted a 'four pillars' foreign policy doctrine, proclaiming Beijing as one of the principal places of reference and international partnership in its diplomatic agenda (together with Brussels, Moscow and Washington).

4.1. Sino-Serbian Political Cooperation Under the 16+1 Framework

Following the establishment of the 16+1 initiative, China's multilateral framework for cooperation with 16 countries of Central and Eastern Europe (CEE), Serbia positioned itself as one of China's principal partners in this mechanism. Serbia hosted the third meeting of heads of government of China and Central and Eastern Europe countries (CEEC) in Belgrade in 2014; it led the efforts to establish a China-CEEC association on transport and infrastructure cooperation.

Describing the prominent role of Belgrade in the 16+1 initiative, the Chinese ambassador to Serbia Li Manchang jokingly referred to the platform as '15+1+1, due to the high number of agreements and projects agreed upon and implemented by China and Serbia over recent years'[63]. Moreover, as a willing partner, Serbia was eager to join China's key foreign policy projects - 'Going out' and the Belt and Road Initiative (BRI) - being one of the 29 countries to be represented by the head of state during the Belt and Road Forum in Beijing in May 2017.

Sino-Serbian ties further deepened in June 2016, during President Xi Jinping's visit to Serbia, when the two countries decided to upgrade their relationship to the status of a comprehensive strategic partnership. In general, comprehensive strategic partnership agreements are regarded as more important than strategic partnerships. As such, they are often guided by a desire to place future relations on more solid footing, including rather detailed agendas for bilateral collaboration; regular consultations both during and 'before big events, including meetings of the UN Security Council, the UN General Assembly, WTO, and climate change conferences'[64]; and the 'establishment of specific communication channels

63 | Stanzel, A., Kratz, A., Szczudlik, J., & Pavlićević, D. (2016). *China's Investment in Influence: the Future of 16+1 Cooperation.* European Council on Foreign Relations, pp. 12-13. Retrieved from https://www. ecfr.eu/page/-/China_Analysis_ Sixteen_Plus_One.pdf. p. 12.

64 | Strüver, G. (2017). China's Partnership Diplomacy: International Alignment Based on Interests or Ideology. *The Chinese Journal of International Politics*, 10(1), p. 45.

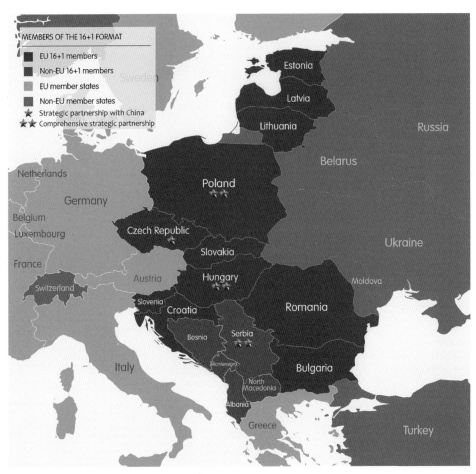

Figure 1: Members of the 16+1 Format[65]

65 | Modified from Stanzel,
A., Kratz, A., Szczudlik, J., &
Pavlićević, D. (2016). *China's
Investment in Influence: the
Future of 16+ 1 Cooperation.*
European Council on Foreign
Relations, pp. 12-13. Retrieved
from https://www.ecfr.eu/
page/-/China_Analysis_
Sixteen_Plus_One.pdf.

66 | Strüver, G. (2017). China's Partnership Diplomacy: International Alignment Based on Interests or Ideology. *The Chinese Journal of International Politics, 10(1),* p. 45; Feng, Z., & Huang, J. (2014). China's strategic partnership diplomacy: engaging with a changing world. *European Strategic Partnerships Observatory Working paper,* p. 15; Emmott, R. (2016 July 15). EU's statement on South China Sea reflects divisions. *Reuters.* Retrieved from https://www.reuters.com/article/southchinasea-ruling-eu-idUSL8N1A130Y.

67 | Stanzel, A., Kratz, A., Szczudlik, J., & Pavlićević, D. (2016). *China's Investment in Influence: the Future of 16+1 Cooperation.* European Council on Foreign Relations, p. 13. Retrieved from https://www.ecfr.eu/page/-/China_Analysis_Sixteen_Plus_One.pdf. p. 13.

68 | Pavlićević, D. (2011). The Sino-Serbian Strategic Partnership in a Sino-EU Relationship Context. *Briefing Series,* p. 6.

69 | As a country striving to join the European Union, Serbia is expected to harmonize its foreign policy with the stances of the EU. However, in 2016, Serbia aligned with only 66% of the EU foreign policy declarations and resolutions, rejecting those related to China and Russia. See Perkučin, J. & Novaković, I. (2017). Analiza usaglašavanja Srbije sa spoljnopolitičkim deklaracijama i merama Evropske unije tokom 2016 [Analysis of Serbia's harmonization with foreign policy declarations and measures of the European Union during 2016]. *CFSP and Serbian Accession to the European Union.* Retrieved from https://www.isac-fund.org/wp-content/uploads/2017/05/CFSP-Analiza-usaglasavanja-2016.pdf. Also Emmott, R. (2016 July 15). EU's statement on South China Sea reflects divisions. *Reuters.* Retrieved from https://www.reuters.com/article/southchinasea-ruling-eu-idUSL8N1A130Y.

70 | Centre for Insights in Survey Research. (2015). *Survey of Serbian Public Opinion.* Retrieved from http://www.iri.org/sites/default/files/wysiwyg/serbia_november_2015_poll_public_release.pdf.

to facilitate regular exchanges between the heads of state and high-level representatives of different government units'[66]. Serbia was the first among the CEE countries to elevate its relationship with China to a level of comprehensive strategic partnership; it still remains the only CEE country in the Balkans region with strategic relations with Beijing.

The essence of political cooperation between Beijing and Belgrade is mutual respect for each other's core interests. Serbia reaffirms its commitment to the One China policy, safeguarding China's position on issues related to Taiwan, Xinjiang, and Tibet[67]; whereas China supports Serbia in the diplomatic battle to protect its sovereignty over Kosovo. Furthermore, Belgrade adopted a state policy to boycott any initiatives criticising China in international forums and continuously refuses to join EU initiatives that condemn the state of human rights in a number of countries, fully reflecting China's stance[68]. For instance, in 2016, Serbia refrained from joining the EU Declaration on the South China Sea that emphasised a need to maintain 'a legal order of the seas and oceans'[69]. Finally, unlike occasional negative campaigning by the media and think tanks on China in some other European countries, the public perception of China in Serbia is very positive. In a recent survey of Serbian public opinion, 89% of respondents claimed that Serbia's interests are best served by maintaining strong relations with China[70].

4.2. Sino-Serbian Economic Cooperation Under the 16+1 Framework

The CEE region plays an important role in the construction of the Silk Road Economic Belt, linking the European and Asian markets and facilitating cooperation between the EU and China. Generally speaking, CEE countries share numerous similar economic indicators. With cheap and abundant labour, lower acquisition prices, higher demand for preferential lending, cost-effective human capital, high concessions for Chinese investors, and easy access to EU technology and market, CEE countries offer relatively good investment conditions[71]. Given its abundant foreign reserves and industrial restructuring, Chinese investment in CEE countries is a 'window of opportunity' to produce much-added value and to realise the localisation, and even 'Europeanisation', of the production, circulation, sales, and branding of Chinese products. Finally, due to the favourable geographical position, China can use CEE as a springboard to access Russian and Turkish markets[72].

Owing to differences in the economies, populations, and market potentials, the benefits of economic cooperation with China are not equally distributed among the 16 CEE countries. Trade is highly concentrated in five countries – Poland, the Czech Republic, Hungary, Slovakia, and Romania – constituting 80% of the total CEE-China trade. Similarly, these five CEE countries have attracted 95% of Chinese FDI in recent years[73]. On the other hand, Serbia attracted only 1.75% of total Chinese investment in CEE in 2014 (US$29.71 million).

Nonetheless, if we limit our analysis to only non-EU CEE countries, the small and underdeveloped markets of southeast Europe, Serbia again begins to occupy a central position. Out of the US$3 billion trade exchanges between southeast European CEE countries and China in 2015, half of it was conducted with Serbia. Likewise, Chinese FDI in the Balkans remains limited to Serbia[74]. Overall, Serbia is offering excellent opportunities for Chinese investors, especially in infrastructure, energy, agriculture, communication technologies, and tourism. Incentives include a large supply of low-cost but skilled labour, government subsidies, and other sorts of preferential policies[75]. Furthermore, due to its preferential trade agreements with neighbouring countries, the EU, Russia, Belarus, Kazakhstan, and Turkey, as well as its tax-free access to markets of approximately 800 million people, Serbia is a strategic investment destination for Chinese enterprises.[76]

Cooperation in the domain of infrastructure has been the most

71 | Zeneli, V. (2017). What Has China Accomplished in Central and Eastern Europe?. The Diplomat. Retrieved from https://thediplomat. com/2017/11/what-has-china-accomplished-in-central-and-eastern-europe/.

72 | Liu, Z. (2012). New Circumstances for China's Investment in Central and Eastern Europe. China International Studies, 37, p. 148.

73 / Zeneli, V. (2017). What Has China Accomplished in Central and Eastern Europe?. The Diplomat. Retrieved from https://thediplomat. com/2017/11/what-has-china-accomplished-in-central-and-eastern-europe/.

74 | Ibid.

75 | Stanzel, A., Kratz, A., Szczudlik, J., & Pavlićević, D. (2016). China's Investment in Influence: the Future of 16+ 1 Cooperation. European Council on Foreign Relations, pp. 13-14. Retrieved from https://www. ecfr.eu/page/-/China_Analysis_ Sixteen_Plus_One.pdf. pp. 13-14.

76 | Ibid., p. 14.

Figure 2: China's Proposed 'Land-Sea Express Route'[77]

77 | Modified fromKynge, J., Beesley, A., Byrne, A. (2017, February 20). EU Sets Collision Course with China Over 'Silk Road' Rail Project. *Financial Times*. Retrieved from https://www.ft.com/content/003bad14-f52f-11e6-95ee-f14e55513608.

78 | Stanzel, A., Kratz, A., Szczudlik, J., & Pavlićević, D. (2016). *China's Investment in Influence: the Future of 16+ 1 Cooperation.* European Council on Foreign Relations, pp. 13-14. Retrieved from https://www.ecfr.eu/page/-/China_Analysis_Sixteen_Plus_One.pdf. p. 13.

prominent aspect of Sino-Serbian economic ties. The combined value of four joint infrastructure projects in Serbia is already higher than in any other CEEC[78]. There are three principal factors making infrastructure development the focal point of Sino-Serbian economic cooperation. First, Serbia occupies a geographically advantageous position at the crossroads of the EU, south-east Europe, and Asia. Hence, it has the potential to become one of the key regional logistics hubs for air, rail, road, and water transportation; a valuable partner in the BRI; and an irreplaceable component of the China-Europe land-sea express route (passageway linking the port of Piraeus in Greece and Hungary).

Moreover, as a country outside the European Union, Serbia is not fully bound by EU legislation and special provisions which have, in the past, complicated Chinese infrastructure projects in other CEE countries. For instance, when conducting business in the EU pursuant to public procurement, a company has to hire a high proportion of its workforce from the EU labour market[79]. Conversely, in Serbia, Chinese companies undertaking infrastructure projects are permitted to hire a notably higher share of Chinese labour (51% or more). In a similar fashion, the domestic legislation allows the Serbian Government to conclude contracts with investors directly, avoiding sometimes lengthy and complex public open procurement procedures, which are mandatory in the EU member states[80]. This is illustrated by the US$2.5 billion Belgrade-Budapest high-speed railway project, agreed to be financed 85% by China's Export-Import Bank and built by state-owned China Railway and Construction Corporation (CRCC). While the Serbian side would gladly start the construction work, the Hungarian side had to wait for the European Commission's approval due to an investigation launched against Hungary in 2017 for breaching the EU laws of public tenders[81].

Finally, Serbia's domestic strategy of reindustrialisation seems fairly well-aligned with the BRI. To improve its basic infrastructure, Serbia is in need of an estimated €9-10 billion over the next 10 years[82]. As the Western development institutions are not able to address these needs for various reasons, a valuable opportunity in the sectors of infrastructure development emerges for Chinese enterprises, which lack neither capital nor expertise. In fact, Sino-Serbian infrastructure cooperation can serve as a noteworthy example of win-win cooperation under the framework of the BRI, with the Serbian national strategy of reindustrialisation working in synergy with China's own developmental goals. By improving its transportation infrastructure, Serbia would close the infrastructure gap with Western Europe, allow for a 'more cost-efficient transfer of Chinese goods from several cargo nodes northward to the European market'[83], bolster Chinese exports to Europe, and support the 'going out' policy for Chinese SOEs (state-owned enterprises)[84].

Chinese enterprises have been, up to this point, contracted to undertake several major projects in Serbia. China's state-owned China Road and Bridge Corporation (CRBC), largely financed with a loan from China's Exim Bank, constructed a

79 | Jones, D. (2014) On the Road away from Mandalay: Heading West along the "Silk Road" as China Moves Its Investments into Europe, around Russia. *Journal of Business and Economics 5 (6)*, p. 792.

80 | Arežina, S. (2013). *Odnos NR Kine sa Jugoslavijom i Srbijom od 1977. do 2009 [The relationship of the People's Republic of China with Yugoslavia and Serbia from 1977 to 2009]*. (Unpublished doctoral dissertation). University of Belgrade, Faculty of Political Sciences, Belgrade, Serbia, p. 369.

81 | Szunomár, A. (2017 December 6). Cooperation between China and Central and Eastern Europe: Promising Start, Doubtful Outlook. *China-US focus*. Retrieved from https://www.chinausfocus.com/finance-economy/cooperation-between-china-and-central-and-eastern-europe-promising-start-doubtful-outlook.

82 | Stanzel, A., Kratz, A., Szczudlik, J., & Pavlićević, D. (2016). *China's Investment in Influence: the Future of 16+ 1 Cooperation*. European Council on Foreign Relations, pp. 13-14. Retrieved from https://www.ecfr.eu/page/-/China_Analysis_Sixteen_Plus_One.pdf. p. 14.

83 | Pavlićević, D. (2014). China's Railway Diplomacy in the Balkans. *China Brief, 14 (20)*.

84 | *Ibid.*

US$260 million road bridge over the Danube River in Belgrade, making it the 'first infrastructure project financed and delivered by China in Europe'[85]. Moreover, The China Railway and Construction Corporation (CRCC) is going to construct the high-speed railway line between Serbia and Hungary, increasing the average train speed from the current 45 mph to 125 mph and reducing the time by rail between Belgrade and Budapest from eight to three hours. Serbian Railways have also reached a US$100 million agreement with Huawei to modernise Serbia's railway telecommunication infrastructure. In April 2016, Hesteel, a major Chinese state-owned iron and steel manufacturing conglomerate, acquired Smederevo steel mill for €46 million, saving one of Serbia's most important industrial assets from bankruptcy.

Moreover, Serbia received a 'US$608 million loan, with an interest rate as low as 2.5% from China's Exim Bank, for the construction of a new unit at the coal-fired Kostolac power plant, the first new power station to be built in Serbia in approximately three decades'.[86]

4.3. Cooperation with China in Light of Serbia's European Union Integration

Chinese initiatives in Central and Eastern Europe have not always been warmly greeted by all the actors operating in this region, especially the European Union. Brussels expressed several concerns over China's involvement in CEEC. For instance, the EU is unsure about Chinese companies' determination to uphold the EU's legislation and technical standards, particularly in the fields of infrastructure construction and financial cooperation[87]. Also, Brussels is concerned that China's 'non-Western receipt for the economic success' could undermine CEE countries' confidence in European principles and values, such as democratic reforms, liberal market regulations, etc.[88] Finally, fears have been expressed that a comprehensive and multifaceted Sino-Serbian cooperation would provide this country with a viable foreign policy alternative, potentially derailing it off the European Union accession path.

85 | *Ibid.*

86 | Pavlićević, D. (2015). China's New Silk Road Takes Shape in Central and Eastern Europe. *China Brief, 15 (1).*

87 | Pavlićević, D. (2017). 16+1 Promoting Belt and Road Initiative: Challenges and Potential Policy Responses. . In Huang Ping & Liu Zuokui (Ed.). *How the 16+1 Cooperation promotes the Belt and Road Initiative.* Beijing: China Social Sciences Press, p. 57.

88 | Eszterhai, V. (2017). "16+1 Cooperation" Promoting Belt and Road Initiative: Better after the Belt and Road Forum for International Cooperation. In Huang Ping & Liu Zuokui (Ed.). *How the 16+1 Cooperation promotes the Belt and Road Initiative.* Beijing: China Social Sciences Press, pp. 94-95.

China has already adopted several policies and initiatives to reduce the EU's anxieties. Namely, the 16+1 cooperation is always defined as a form of 'sub-regional' cooperation that supplements, rather than replaces, the China-EU Comprehensive strategic partnerships. Moreover, the transparency of the 16+1 initiative was increased, and since the Belgrade Summits in 2014, the EU has regularly sent its representatives to attend China-CEEC Annual Leaders Meetings.

With regards to Serbia, China has explicitly expressed its support for this country's EU accession path. For instance, in the Joint Statement Between the People's Republic of China and the Republic of Serbia on Establishing Strategic Partnership in 2009, China indicated its understanding for 'Serbia's efforts to integrate into the European family'; similarly, in the China-Serbia Joint Statement on Establishing a Comprehensive Strategic Partnership in 2016, 'both sides agreed that China-CEEC cooperation and China-EU comprehensive strategic partnership complement each other'.

Moreover, the bulk of Chinese infrastructure projects in Serbia actually increases Serbia's interconnectedness with other EU countries. In other words, contrary to Brussels' fears, China is practically reinforcing Serbia's EU integrations by embedding it deeply into the European network of infrastructure corridors. Hence, to reduce Brussels' often unsubstantiated anxieties, China needs to emphasise louder and more regularly that its initiatives and investment projects support Serbia on its EU accession path; that they complement similar EU projects in the Balkans; that they make Serbia more ingrained into the European road and rail networks; and, consequentially, that they boost mutually beneficial ties between China and the European Union, which ultimately reinforces European unity.

5. Serbian Youth and the Belt and Road Initiative

One of the proclaimed goals of the BRI is to 'expand and advance practical cooperation between countries along the Belt and Road on youth employment, entrepreneurship training, and vocational skill development'[89]. The BRI is still in its early stages. Therefore, it is too soon to quantify and qualify the overall impact of Chinese investments on Serbian youth, especially in terms of spillover effects such as SME development, youth job creation, and economic growth[90]. Nonetheless, there is a shared belief among the general public that these projects will boost employment and economic opportunities for young people, both in the short and long term, thus helping to mitigate one of the most pressing youth issues in Serbia.

One should not underestimate the indirect benefits of infrastructure development on young people either. The Belgrade-Budapest high-speed railway is going to provide young people with a cheaper and more comfortable route to visit their friends in the EU; the reduction of travelling time from eight to three hours is going to increase the number of foreign tourists visiting Serbia, further improving prospects of employability for the youth. Moreover, good transportation infrastructure increases the availability of comparatively cheaper Chinese products in the Serbian market and lowers the living expenses of young people.[91]

5.1. The Role of Youth in Enhancing People-to-People Bonds

People-to-people exchanges, as one of the five pillars and cooperation priorities of the BRI, are particularly relevant for the CEE countries. Despite China enjoying a predominantly positive image, there is a major lack of understanding of Chinese culture and values among the Serbian public. Cultural exchanges have become even more important, knowing that differences in national customs and business cultures complicate joint business endeavours of Chinese and Serbian counterparts[92]. Hence, Serbian youth must play a major role in bridging the cultural gap, reinforcing mutual learning, and building understanding, trust, and respect among the two nations.

89 | National Development and Reform Commission, Ministry of Foreign Affairs, and Ministry of Commerce of the People's Republic of China, with State Council authorization. (March 2015). *Vision and Actions on Jointly Building Silk Road Economic Belt and 21st-Century Maritime Silk Road.* Retrieved from https://engyidaiyilu.gov.cn/qwyw/qwfb/1084.htm.

90 | Zeneli, V. (2017). What Has China Accomplished in Central and Eastern Europe?. *The Diplomat,* Retrieved from https://thediplomat.com/2017/11/what-has-china-accomplished-in-central-and-eastern-europe/.

91 | Tanjug. (2017, December 12). "Alibaba" stiže u Srbiju ["Alibaba" arrives in Serbia]. *Blic,* Retrieved from http://www.blic.rs/.

92 | Zakić, K. (2015), Mogu li srpski i kineski menadžeri uspešno da saraduju: kritički osvrt [Can Serbian and Chinese managers cooperate successfully: a critical review]. *Medunarodni Problemi, Vol. LXVII,* pp. 2-3.

Under the bilateral partnership framework, Serbia and China reached several agreements in the fields of elementary, secondary, and higher education. Serbia hosts two Confucius Institutes (Belgrade and Novi Sad) with the Chinese government sponsoring Chinese language teachers in Serbia. A key point in the educational cooperation between the two countries occurred in 2011 when the Serbian Ministry of Education initiated a pilot project of Chinese language classes in 70 schools across the country with approximately 2,500 students[93].

In line with China's commitment to providing 10,000 government scholarships to countries along the Belt and Road every year from 2009 to 2015, the Chinese Ministry of Education awarded 67 scholarships to students from Serbia to enrol into a Chinese language program, Master's, or PhD degree.[94] Yet, the number of Serbian students in China is still low, especially compared to the number in European countries. One way to further increase academic flows between the two countries is to proliferate direct, unmediated ties between Serbian and Chinese universities – by establishing partnerships and joint study programmes, and facilitating student and staff exchanges. Serbia's University of Belgrade, University of Novi Sad, and University of Niš, being the country's best public high education institutions, particularly stand out as possible partners for their Chinese counterparts.

Good bilateral relations led Serbia to become the first European country to establish a visa free-entry regime with China. Signed during the fifth China-CEE Summit in Riga in November 2016, this bilateral visa-free regime for visits lasting up to one month[95] further reinforced bilateral relations between the two countries, increased the volume of tourism and opened more channels for youth interactions and mutual learning. Accordingly, the number of Chinese tourists coming to Serbia multiplied in 2017. As Serbian domestic tourist agencies still lack a sufficient number of tour guides with adequate Chinese language skills, organising language and culture workshops could be a meaningful advancement for both sides.

93 | Zakic, K., Stakic, A. J., & Jurcic, A. (2016). Managing cooperation programmes between Serbia and China in the field of higher education. *Economic and Social Development: Book of Proceedings*, p. 324.

94 | United Nations Development Programme. (2017). *The Economic Development Along the Belt and Road.* Retrieved from https://www.cn.undp.org/content/china/en/home/library/south-south-cooperation/the-economic-development-along-the-belt-and-road-2017.html#:~:text=It%20suggests%20that%20with%20a,BRI's%20quest%20for%20economic%20prosperity.

95 | Stanzel, A., Kratz, A., Szczudlik, J., & Pavlićević, D. (2016).*China's Investment in Influence: the Future of 16+ 1 Cooperation.* European Council on Foreign Relations, pp. 12-13. Retrieved from https://www.ecfr.eu/page/-/China_Analysis_Sixteen_Plus_One.pdf. p. 12.

Finally, the youth has a potentially indispensable role in enhancing the visibility of the BRI in Serbia. In practice, the attractiveness of the BRI to the general public is often shadowed by the lack of widespread and accessible information about the initiative. Noting that cultural and 'academic exchanges between China and the CEE countries started late, and that quantity, quality, scale, and influence are still weak'[96], numerous authors proposed a list of ways to improve the situation. These included: strengthening cooperation in the areas of law, culture, education, and science; boosting inter-party exchanges; reinforcing think-tank exchanges, joint conferences and research projects; and establishing special financing funds for further cultural and educational exchanges[97]. In fact, since 2012, the Chinese government has set aside approximately two million yuan annually for the China-CEE Relations Research Fund, aiming to support academic exchanges and cooperation[98].

Nonetheless, to increase its media presence, and to improve the public reach of its initiatives in Serbia, China needs to begin focusing on cooperating with the non-governmental. So far, cooperation under the 16+1 framework has been mostly targeting governmental bodies and semi-official government-affiliated research institutions which, in the case of Serbia, do not have as significant an impactful a role over public opinion. Bilateral cooperation needs to further expand to encompass civil society associations, youth groups, and non-governmental organisations, which play a dynamic and influential role in Serbian society and can serve as useful channels for amplifying China's voice.

By offering grants and project financing to civil society and youth groups, the BRI framework can gain a new and more impactful channel of communication with the Serbian youth; increasing their enthusiasm and involvement in the initiative. In addition, another way for Chinese initiatives to better reach young people and reinforce China's media presence in Serbia is to invest more efforts in building China's identity in the digital space – i.e. to increase the amount and quality of China-related content on the social media platforms widely used by Serbians,

96 | *Ibid.*, p. 5.

97 | *Ibid.*, p. 5.

98 | *Ibid.*, p. 5.

Sunset over Belgrade and ships in the harbour, Serbia.

such as Facebook and Instagram. Hence, to fully utilise the creative potential of Serbian youth in enhancing the visibility of the BRI, China has, in addition to holding China-CEEC Young Political Leaders' Forums and Bridge of the Future China-CEEC youth camps under the 16+1 framework, to recalibrate its approach so it fits the specific composition and unique features of Serbian society.

A train passing through a lake near Mlynky village in the Slovak Paradise national park, Slovakia.

CHAPTER 16

by   LENKA ČURILLOVÁ
Comenius University

JAZMÍNA OBEDOVÁ
Ringling College of Art and Design

(Ordered by last name)

ESTONIA

LATVIA

LITHUANIA

POLAND

CZECH
REPUBLIC

HUNGARY

SLOVENIA

ROMANIA

CROATIA

BOSNIA &
HERZEGOVINA

SERBIA

MONTENEGRO

BULGARIA

NORTH
MACEDONIA

ALBANIA

SLOVAKIA

SLOVAKIA

1. The Brief History of Slovakia

The following article is an overview of the political structure of the Slovak Republic, internationally known as Slovakia. A major focus of the article will be on the discussion about the country´s evolutionary history, followed by an investigation of the current political structure and government. The extent of Slovakia's real sovereignty, major state and non-state actors such as the European Union, and the overall power dynamics in the country will be discussed.

Slovaks commonly believe that the very first Slovak state was created some 1,200 years ago during the 9th century. This state was called Veľká Morava (Great Moravia) and was the first sign of a large population of Slavic people forming as a single community.[1] Veľká Morava had existed for almost a century before being conquered and destroyed by nomadic Hungarians in the 10th century. The Hungarians ruled the Slovaks as part of the Hungarian Empire, which existed from then until 1918 in one form or another.[2] In 1918, the Austro-Hungarian Empire finally collapsed into Austria, Hungary, Czechoslovakia and several Balkan states, after being defeated during World War I.

The Slovaks started a movement for a sovereign Slovak state and for the Slovak language, which continued until the Austro-Hungarian Empire collapsed. On October 28, 1918, a sovereign state of Czechoslovakia was created as the Austro-Hungarian Empire was crumbling. According to the theory behind the creation of Czechoslovakia, there was only to be a single Czechoslovak identity to unite the Czechs and Slovaks. However, Slovakia started a complex rebirth on cultural, political, and educational fronts. In 1919, the first university, Comenius University, was built in Bratislava. the construction of this university was a milestone and it proved to be a great asset to education reforms and development in the Slovak region.[3]

However, the region kept struggling on the economic front as the newly-formed Slovak industries were no match for their Czech counterparts, which were not only years ahead in terms of the level of development thanks to the heavy investment from the Empire over the years but also growing bigger and more competitive. This, combined with the Great Depression

1 | Plaštiak, K. (2017, September 19). Veľká Morava – zem našich predkov, na ktorú môžeme byť právom hrdí. [Great Moravia - the land of our ancestors, of which we can be rightfully proud]. Retrieved from: https://zaujimavysvet.webnoviny.sk/velka-morava-zem-nasich-predkov-na-ktoru-mozeme-byt-pravom-hrdi/.

2 | European Commission. (2000). Portrait of the Regions Volume 7 Slovakia. Luxembourg: Office for Official Publications of the European Communities, p.1.

3 | Comenius University in Bratislava. (n.d.). The Comenius University History. Retrieved from: https://uniba.sk/en/about/history/.

during the 1930s, was a catastrophe for the Czechoslovak economy since it triggered a huge emigration crisis in the country.

In the meantime, an existential crisis appeared for Czechoslovakia with the rise of Nazi Germany. Adolf Hitler suggested two options for the future for Czechoslovakia: either division between Poland and Hungary or the formation of a new Slovak state which was to be completely under German control. Jozef Tiso, who later became the first president of Slovakia, chose the lesser of two evil options and so, for the first time since the 9th century, a somewhat sovereign Slovak state was created on March 14, 1939.[4] That state later became the Slovak Republic. The state was a dictatorship under a single political party rule. Foreign relations were limited and only served German interests. Thus, Slovakia introduced the same German restrictions for the Jewish population, starting with creating concentration camps for their persecution.[5] Slovakia at this point was merely a puppet state of Germany.

On May 1, 1945, Slovakia was liberated from the Nazis by the Red Army of the Soviet Union. After the liberation, the old Czecho-Slovakia was reconstituted which came under the influence and later occupation of the Soviet Union.

In 1969, the Czechoslovak Socialist Republic was formed, and it existed until 1990. As a socialist state, the Czechoslovak Federation, now comprised of the Czech Socialist Republic and the Slovak Socialist Republic strived to become a model communist state.

However, the regime was extremely unpopular among the population since it gravely restricted people's freedom. For instance, the state actively suppressed religion and any democratic movement against the regime. The economic situation was worse than other communist countries, let alone non-communist countries. The promotion of atheism, limited freedom of movement, and the strong influence of the Soviet Union in the state further angered the people.

4 | The Slovak Spectator. (n.d.). *Nazi-allied Slovak State Emerged 80 Years Ago*. Retrieved from: https://spectator.sme.sk/c/22074748/anniversary-Slovak-state-Nazi-totalitarian-history-Slovakia.html.

5 | US Department of State. (n.d.). *The JUST Act Report: Slovakia*. Retrieved from: https://www.state.gov/reports/just-act-report-to-congress/slovakia/.

Finally, on November 16, 1989, the Velvet Revolution spread across the Czechoslovak Federation. The revolution is called 'velvet' because of its non-bloody and rather peaceful nature that caused the end of the communist regime in the state. Soon, the word 'Socialist' was dropped and the country was now called the Czech and Slovak Federal Republic. Government officials and political powers then begun discussions about the division of the state into two sovereign states. On July 17, 1992, Slovakia declared independence to become a sovereign state of the Slovak Republic. By December 31, 1992, the federation was officially dissolved to form the Slovak Republic and the Czech Republic.[6]

The state of democracy in Slovakia remained questionable as most of the communist leaders were still leading the now democratic Slovak Republic. However, since the country was under a constitution that was followed by all democratic institutions and democracy is enforced throughout the country, the Slovak Republic is technically fully democratic.

The constitution of the Slovak Republic splits the governing system into three branches.[7] The constitutional and legislative power, executive power, and judiciary. The nature of government is officially a parliamentary democracy. The word 'parliament' suggests that the constitutional and legislative powers are held by the National Council of the Slovak Republic.[8] The National Council is the only constitutional and legislative body of the Slovak Republic. It represents the sovereignty of the state and the people.

Members of the National Council are elected directly through a general election by secret ballot by all Slovak citizens. There are 150 seats in the parliament, each member elected for a term of four years.[9] Executive power is held by both the president and government of the Slovak Republic. The government encompasses 14 ministries including the ministries of Finance, Health, Education, Defence, Economy and so on.[10] The prime minister is the head of the government and is among the group of most important and powerful people in the country, along with the president. While the position of the

6 | European Commission. (2000). *Portrait of the Regions Volume 7 Slovakia.* Luxembourg: Office for Official Publications of the European Communities, p.1.

7 | Constitute Project. (n.d.). *Slovakia's Constitution of 1992 with Amendments through 2014.* Retrieved from https://www.constituteproject.org/constitution/Slovakia_2014.pdf.

8 | *Ibid.*

9 | *Ibid.*

10 | Greenpages. (n.d.). *Government Ministries.* Retrieved from http://greenpages.spectator.sme.sk/en/c/government-ministries.html?page=1.

president is mostly ceremonial, they are able to sign new laws and legislation as well as to veto any.

The judiciary is represented by the Constitutional Court and other courts. The Constitutional Court (Supreme Court in other countries) is above all the other courts and is an independent judicial authority for the protection of the constitution and provision of justice to the people.[11] However, all courts of the Slovak Republic, even the Constitutional Court, fall under the European Court of Justice.

The government of the Slovak Republic is a coalition of the three major political parties in the country. During the elections in 2016, no single party was able to win a significant majority to form an independent government. The major political parties in the government include SMER (Direction – Social Democracy), SNS (Slovak National Political Party) and MOST-Híd (which gets its name from the Slovak and Hungarian words for 'bridge').[12] The ministers for all ministries in the government are elected through a secret ballot election by all members of the National Council.[13] All political parties nominate a member for each ministry, thus, all members, regardless of their parties, have a chance to lead a ministry.

All parties, both in the government and opposition, can propose new laws or amendments to existing laws. The president, with veto power, is a part of all the decision making and may use his powers to return legislation back to the National Council for further discussions and amendments or sign it into implementation. However, if the president refuses to sign the legislation for a second time, the legislation will proceed to implementation without his signature and approval. This shows that the government is the strongest part of the political structure as it is independent of the influence of any single member. However, the government is not independent of the Constitutional Court of the Slovak Republic and is bound to follow verdicts given by the court. This keeps the government in check and ensures it upholds the constitution.

The Slovak Republic joined the European Union (EU) in 2004.

11 | Constitute Project. (n.d.). *Slovakia's Constitution of 1992 with Amendments through 2014.* Retrieved from https://www.constituteproject.org/constitution/Slovakia_2014.pdf.

12 | National Democratic Institute. (2016). *2016 Parliamentary Elections Signal A Shift to the Right in Slovakia.* Retrieved from: https://www.ndi.org/Parliamentary-Elections-Signal-Political-Change.

13 | Constitute Project. (n.d.). *Slovakia's Constitution of 1992 with Amendments through 2014.* Retrieved from https://www.constituteproject.org/constitution/Slovakia_2014.pdf.

As part of the EU, Slovakia is no longer a completely sovereign state and must follow the laws of the EU. Thus, a Slovak citizen can challenge a verdict by the Slovak Constitutional Court in the European Court of Justice. The European Court of Justice holds the authority to overturn or validate any verdict given by the Constitutional Court of Slovakia.

2. Political, Economic and Social Reforms in Slovakia

This section will discuss the potential political, legal or economic reforms and a few of the important movements in country's politics, followed by an evaluation of the current state and support of these policies, reforms and movements as per the author's opinion.

One of the biggest issues that have swept across Slovakia recently is the Istanbul Convention (IC), a convention for the protection of women's rights. With the '#metoo' campaign taking the world by storm, the discussion of women's rights in Slovakia has come to the forefront. Slovakia is a signatory of the Istanbul Convention although its ratification into the constitution has been postponed several times. There are people in the government who fear the ratification of such law could create disputes in society.

The biggest opposition to the Convention comes from conservative groups. According to them, the biggest issue with the Istanbul Convention is that it allegedly introduces 'gender ideology' into national and international laws[14]. They also believe that by ratifying the Convention, Slovakia would have to introduce non-stereotypical gender roles into schools, which could stop parents from raising their children according to their

14 | China-CEE Institute. (2018, May 29). *The Instanbul Convention Controversy.* Retrieved from https://china-cee. eu/2018/05/29/slovakia-social-briefing-the-istanbul-convention-controversy/#.

traditional religious beliefs.[15] According to them, ratification would also bring sexual education to schools, and those who would disagree with such classes would be punished. The referendum failed because it did not attract the necessary voter turnout of 50%.[16]

Other aspects of the education system of Slovakia have also received a lot of attention and reforms from the government.

The country is struggling to satisfy the demand for a skilled workforce by local industries, and there are no other options but to invite more workers from abroad, especially from countries such as Ukraine, Serbia, Bosnia & Herzegovina, etc. Universities also reported increased demand for degrees in humanities subjects. By minimising classes of grammar schools and investing in vocational schools, the government aims to control the flow of foreign workers into the country and thus eventually decrease the rate of immigration to avoid any possibility of increased unemployment in the country.

In addition to the unemployment issue, Slovakia is also facing problems regarding religion. Being a constitutionally secular state, Slovakia should be independent of religious beliefs and influences. However, because of a pact with the Vatican, the Catholic Church receives funds from the government and holds some influence on the governance of the country. The discussion of independence from religious influences of any kind started after the failed referendum about reforms to family laws in 2015. The real purpose of the referendum was to limit the legal rights of the LGBT community in Slovakia, as well as to influence the teaching of sex education in schools. Many churches and religious groups supported this referendum, including the Catholic Church.[17] This religious view is gaining more influence in Slovakia and has polarised the country by emphasising the role of religion in politics.

At the same time, many have formed petitions and referendums calling for complete separation of religion and politics in the country. This has caused a surge in hate speech throughout the country, mainly from the supporters of the

15 | *Ibid.*

16 | Rousek, L. (2015, February 8). Slovakia Referendum on Gay-Adoption Ban Fails. *Wall Street Journal.* Retrieved from: https://www.wsj.com/articles/slovakia-referendum-on-gay-adoption-ban-fails-1423355269.

17 | China-CEE Institute. (2018, May 29). *The Instanbul Convention Controversy.* Retrieved from https://china-cee.eu/2018/05/29/slovakia-social-briefing-the-istanbul-convention-controversy/#.

failed referendum, as well as the churches.

The refugee crises in Europe sent panic throughout the country. This panic polarised the nation and created new groups of political parties, ones that openly sympathise with the Nazis and the first Slovak Republic, which was created and controlled by Nazi Germany.[18] Such groups are led by political leaders with racist agendas and views and are followed and supported by panicked and uninformed Slovaks. Such groups believe in white Slovaks as the only true citizens of Slovakia, and Jozef Tiso, the leader and president of the first Slovak state, as the only true president of Slovakia. The government, especially then Prime Minister Robert Fico, who was also the leader of the biggest political party, failed to take the necessary steps to avoid the rise of such populist groups. As the refugee crisis continued, the rhetoric of the prime minister started suggesting the refugees might be a matter of national security for Slovakia, especially refugees from Islamic countries such as Afghanistan and Pakistan.[19]

However, the issue is bigger than just refugees. Even progressive Slovaks have reportedly supported such extremist parties in elections to make a change in the political climate of the country. The traditional parties have been hit with countless scandals of corruption. Thus, other parties that offered a war against corruption and policies against refugees got public support regardless of their ideologies. One such party is the Kotleba – People's Party Our Slovakia, which was able to secure a presence in the government from the parliamentary election of 2016.[20] This was a significant moment in Slovak government structure, as it was the first time such an extremist party entered parliament since the end of WWII when a Nazi political party won their way to the National Council.

This has caused a divide within the country where now several politicians, including the president, and countless NGOs have begun campaigns against such fascist groups. As time passed, the government finally addressed the corruption scandals and the refugee crisis was also put under control in Central Europe. Once the fear and chaos were gone, it enabled the public to

18 | Cameron, R. (2016, March 6). Marian Kotleba and the Rise of Slovakia's Extreme Right. *BBC News*. Retrieved from https://www.bbc.co.uk/news/world-europe-35739551.

19 | Estevens, J. (2018). Migration Crisis in the EU: Developing a Framework for Analysis of National Security and Defence Strategies. CMS 6, 28.

20 | Filipec, O. (2018). People's Party – Our Slovakia: An Anti-System Party?. *Conference Paper: Current Trends and Public Administration.*

focus more on the actual performance of Marian Kotleba's political party. Kotleba, the leader of Kotleba – People's Party Our Slovakia, had been a leader of the region Banska Bystrica for four years after the general elections in 2013.[21] However, in the same region, it was a very different picture in the elections of November 2017. Kotleba lost his seat in Banska Bystrica, once considered as his stronghold. The results for other politicians of his party were also similar and the party has since failed to achieve any major political success.

As mentioned previously, corruption scandals are one of the biggest issues in Slovakia. The issue is so sensitive and extreme that it was single-handedly responsible for the rise of Kotleba and his party. Corruption scandals of the government have led to serious disruptions in major cities. It is no longer uncommon to see marches and protests against the government in all major cities. There were many major protests against corruption organised by students, and it is believed that the fight against corruption in the streets would become more intense if the government kept ignoring the issues. Unsurprisingly, the campaign against corruption gained intensity when the government failed to investigate the death of young journalists Jan Kuciak and his fiancée, who was shot in their house by an unknown perpetrator in February 2018.[22] Kuciak had been investigating shady financing and tax evasion of companies connected to Slovak oligarchs and their connection to governing parties.[23] His last work discovered a huge misuse of funds from the European Union by the Italian mafia in Slovakia and even the signs that connect mafia and people close to the leading parties.[24] As of now, there does not seem to be an end to the political chaos in Slovakia as the government is crippled by corruption.

21 | Cameron, R. (2016, March 6). Marian Kotleba and the Rise of Slovakia's Extreme Right. *BBC News*. Retrieved from https://www.bbc.co.uk/news/world-europe-35739551.

22 | BBC. (2018, March 16). Slovakia Protests: 65,000 Join Bratislava Anti-Government Protests. *BBC News*. Retrieved from: https://www.bbc.co.uk/news/world-europe-43437993.

23 | *Ibid.*

24 | *Ibid.*

3. Opportunities and Challenges Faced by Young People in Slovakia

25 | Council of Europe. (2012). *Child and Youth Participation in the Slovak Republic - A Council of Europe Policy Review.* Retrieved from https://rm.coe. int/168046c7fa.

26 | Iuventa. (n d.). *Slovak Youth Institute - In A Nutshell.* Retrieved from: https://www. iuventa.sk/en/IUVENTA/ Slovak-youth-iustitute-short-introduction.alej.

27 | European Union. (n.d.). *Further Expansion.* Retrieved from https://europa.eu/ european-union/about-eu/ history/2000-2009_en.

28 | European Comission. (n.d.). *What is Erasmus+?.* Retrieved from https://ec.europa.eu/ programmes/erasmus-plus/ about_en#:~:text=support%20 HTML5%20video.- ,Erasmus%2B%20is%20the%20 EU's%20programme%20to%20 support%20education%2C%20 training%2C%20youth,just%20 have%20opportunities%20for%20 students..

29 | European Commission. (n.d.). *Erasmus + Youth Exchanges.* Retrieved from https:// ec.europa.eu/programmes/ erasmus-plus/opportunities/ individuals/young-people/youth-exchanges_en#:~:text=Youth%20 exchanges%20allow%20 groups%20of,shared%20 projects%20for%20short%20 periods.&text=On%20a%20 youth%20exchange%2C%20 you,plays%2C%20outdoor%20 activities%20and%20more.

3.1. Existing Opportunities for Young People in Slovakia

Generally, there are many opportunities available for young people for their personal and professional development, but most of the opportunities are centred in the capital Bratislava. There are fewer chances for development in schools in other parts of the country, especially small towns and rural areas. But there are some. For instance, in almost all small towns, there are centres of leisure-time activities where children and high school students can spend their time in the afternoons participating in workshops, discussions, dancing or music clubs. In addition, young people are becoming more and more active and want to create a change in their community. Therefore, 'youth parliaments' have been set up in several cities.[25] Young people who join these organisations seek to create opportunities for themselves and others.

Regarding opportunities at the national level, The Slovak Youth Institute is an institution managed by the Ministry of Education, Science, Research and Sport and offers plenty of educational and informational activities for youths and youth workers around the country.[26]

Many opportunities have become accessible to Slovaks after joining the European Union (EU) in 2004.[27] In 1998, the first batch of Slovak students got a chance to study abroad thanks to the Erasmus (EuRopean Community Action Scheme for the Mobility of University Students) programme that aims to boost cultural exchanges and mobility among students in the EU.[28] Moreover, Erasmus offers not only well-known study exchange programmes during university studies, but also internships, volunteer programmes, short-term youth exchanges, training courses and many other opportunities.[29] Nowadays, all those opportunities are included in the Erasmus+ programme. Almost all universities and many high schools participate in the programme and introduce such opportunities to their students. Many NGOs have been established (e.g. Keric, Mladiinfo, ADEL, Sytev, Youthfully Yours, SPACE, etc.) that help young people facilitate the application processes of joining volunteer programmes or one-to-two-week educational projects.[30] INEX, a volunteer programme, helps young people participate in workcamps abroad. Young people who would like to start their own business can spend three-to-six months job-shadowing more experienced entrepreneurs abroad thanks to the programme 'Erasmus for Young Entrepreneurs'.[31]

Slovak Academic Information Agency (SAIA) is another institution that young people refer to when they want to go abroad to gain experience, knowledge, and language skills.[32] It is an organisation that provides advisory services about the opportunities to study abroad, particularly for university students. They have a broad database that contains information about various academic exchange programmes, scholarships and research programmes, summer schools and many more.[33]

There are also many educational and extracurricular opportunities for young people available in Slovakia. In the eight university cities and regional centres in Slovakia, there is a very well-developed network of student organisations and clubs such as AIESEC, ESN, IAESTE, BEST, etc. Almost every university offers their students opportunities to practise their writing or speaking skills by establishing student magazines, radio or TV programmes. Many NGOs also offer diverse educational programmes on business administration, financial literacy, project management, mentoring programmes, career counselling, outstanding student awards, young innovative entrepreneurs programmes and unique 'mini-Erasmus' programmes for high-school students who can spend a week at a university (such a programme is organised by NGOs such as Future Generation Europe, Nexteria, JA Slovakia, JCI, Leaf, etc.).[34] An interesting example is 'The Duke of Edinburgh's International Award' that runs activities through licensed Independent Award Centres around the country.[35] Young people between 14-24 set their own goals as a challenge for their personal development together with mentors who help in setting goals and providing support.[36]

In addition, there are occasional events such as workshops, information days, networking and discussion opportunities. Many organisations, institutions and companies also provide

30 | Keric. (n.d.). *What's Keric*. Retrieved from https://www.keric.sk/en/about-us/what-s-keric.

31 | Erasmus for Young Entrepreneurs. (n.d.). *The Programme and Its Benefits*. Retrieved from https://www.erasmus-entrepreneurs.eu/page.php?cid=20.

32 | Slovak Academic Information Agency. (n.d.). *About Us*. Retrieved from https://www.saia.sk/en/pages/about-us/.

33 | Slovak Academic Information Agency. (n.d.). *SAIA Programmes*. Retrieved from https://www.saia.sk/en/main/saia/.

34 | Future Generation Europe. (n.d.). *Mini Erasmus*. Retrieved from https://futuregenerationeurope.eu/en/mini-erasmus-2/.

35 | The Duke of Edinburgh's International Award Slovakia. (n.d.). *About the Award*. Retrieved from https://www.dofe.sk/en/about-award/.

36 | *Ibid*.

usually unpaid internships for students to gain practical experience. Young people who are neither in education nor in employment (NEET), who constituted 15.3% of Slovak youths in 2017, can apply for small grants from the labour offices to start their own business.[37]

To sum up, the overview focuses on the opportunities which now exist, not those that are still missing. There is definitely still a huge gap and we should develop more opportunities especially in smaller towns and rural areas. We would conclude that there are various opportunities for young people, but not many of them are aware of their existence or they are not motivated to take advantage of them since they still think that a university diploma is sufficient for their future development.

3.2. Young People's Views Towards Slovakia

Many public opinion surveys were held to ask about people's satisfaction with the country. The Youth Council of Slovakia conducted one with 700 young people aged between 15-24 in 2017. The most interesting finding was that 29% of young people thought that they would move abroad within three years.[38]

Young Slovaks deeply mistrust the state institutions with 80% of them not trusting the government and parliament, and 67% of them not trusting the courts.[39] Another alarming finding was that 43% of the respondents deemed that they had no influence on what was happening in the country and society.[40] Despite the grim attitudes towards institutions, 69% of the respondents appreciated the opportunities brought by youth and student activities.[41]

Another interesting fact is between 2015 to 2016, around 14% of young people attending Slovak schools attended university in the Czech Republic as there is no language barrier between the two nations.[42] The main reasons why young people choose a university abroad is better university quality abroad. Compared to neighbouring countries such as the Czech Republic, Hungary, and Poland, Slovakia has 'the lowest number

37 | Eurostat. (2018, June 15). *Young People Neither in Education Nor Employment.* Retrieved from https://ec.europa.eu/eurostat/web/products-eurostat-news/-/DDN-20180615-1.

38 | Mladez. (2017, September, 27). *Prieskum Rady mládeže Slovenska: Táto krajina nie je pre mladých [Youth Council of Slovakia Survey: This Country is Not For Young People].* Retrieved from http://mladez.sk/2017/09/27/prieskum-rady-mladeze-slovenska-tato-krajina-nie-je-pre-mladych/.

39 | *Ibid.*

40 | *Ibid.*

41 | *Ibid.*

42 | The Slovak Spectator. (2018, July 5). Young Slovaks Want to Study Abroad. *The Slovak Spectator.* Retrieved from https://spectator.sme.sk/c/20864305/young-slovaks-want-to-study-abroad.html.

43 | *Ibid.*

44 | Manpower Group. (2018). *2018 Talent Shortage Survey Slovakia*. Retrieved from https://cdn2.hubspot.net/hubfs/2942250/Local%20Infographics/2018_TSS_Infographics-Slovakia.pdf.

45 | *Ibid.*

46 | *Ibid.*

47 | OECD Data. (n.d.). *Youth Unemployment Rate*. Retrieved from https://data.oecd.org/unemp/youth-unemployment-rate.htm#indicator-chart.

48 | *Ibid.*

49 | Eurostat. (2017). *Share of Young Adults Aged 18-34 Living with Their Parents by Self-defined Current Economic Status - EU-SILC Survey*. Retrieved from http://appsso.eurostat.ec.europa.eu/nui/show.do?dataset=ilc_lvps09&lang=en.

50 | European Union. (n.d.). *Further Expansion*. Retrieved from https://europa.eu/european-union/about-eu/history/2000-2009_en.

51 | OECD. (n.d.). *Slovak Republic and the OECD*. Retrieved from https://www.oecd.org/slovakia/slovak-republic-and-oecd.htm.

52 | North Atlantic Treaty Organization. (2004, April 2). NATO Welcomes Seven New Members. *NATO Update*. Retrieved from https://www.nato.int/docu/update/2004/04-april/e0402a.htm.

53 | European Central Bank. (n.d.). *Slovakia (Since 1 January 2009)*. Retrieved from https://www.ecb.europa.eu/euro/changeover/slovakia/html/index.en.html.

of universities in the international rankings'.[43] Universities abroad also tend to have more choices of study programmes. Overall, studies suggest there is a talent shortage in Slovakia.[44] In 2018, more than 50% companies found it difficult to recruit employees.[45] Other than the lack of applicants, other major reasons cited by employers and human resources specialists include the lack of hard and soft skills such as leadership abilities and communication skills.[46]

Youth unemployment was particularly alarming. In 2017, the youth unemployment rate for men and women were 18.1% and 20.1% respectively.[47] Although this represented a drop from 2016 (the youth unemployment rates for men and women were 19.7% and 26.3%), there was still a large number of young people who were not able to find a job, particularly those who reside in less developed parts of Slovakia.[48] Besides youth unemployment, in 2017, 59.9% of young people between 18 to 34 years-old were still living with their parents, which placed Slovakia in second place for the most number of young people who are still somewhat dependent on their parents.[49]

3.3. How Slovakia Supports the Young People

The paradox is that many of the abovementioned opportunities at the beginning of this chapter are either funded by the EU or provided by NGOs, albeit some of them are also directly financed by Slovak ministries or the government too. Politicians proclaim that the economic situation of the country is improving. Since the fall of communism in 1989, Slovakia experienced a very turbulent period of transition from federalism towards independence, from communism to democracy, from a planned economy to market economy. A period of structural reforms, revitalisation and stabilisation resulted in economic and political integration with the world. Slovakia joined the Organisation for Economic Co-operation and Development (OECD) in 2000, the EU and NATO in 2004, and the Euro Zone in 2009.[50][51][52][53]

Through the promulgation of more protective business laws, the business environment in the country has improved as seen

54 | GHK. (n.d.). *Study on Volunteering in the European Union Country Report Slovakia*. Retrieved from https://ec.europa.eu/citizenship/pdf/national_report_sk_en.pdf.

55 | Ministry of Interior. (2017). *Registers and Evidences of the Ministry of Interior of the Slovak Republic*. Retrieved from http://www.ives.sk/registre/.

56 | Council on Foundations. (n.d.). *Nonprofit Law in Slovakia*. Retrieved from https://www.cof.org/content/nonprofit-law-slovakia#:~:text=A.-,Tax%20Exemptions,(7a)%20and%2050).

57 | OECD. (2018, May). *Slovak Republic - Economic Forecast Summary*. Retrieved from http://www.oecd.org/economy/slovak-republic-economic-forecast-summary.htm.

in an increased inflow of foreign investors. The third sector (also known as the voluntary sector, comprising NGOs and non-profits) has grown rapidly as well, growing from 6,000 registered NGOs in 1993 to over 48,000 in 2017.[54,55] NGOs were favoured in terms of tax exemption and benefit in the form of tax assignation from income.[56] Economic performance is characterised by favourable forecast, with continuing economic growth expected.[57] However, there is still a lot of work to do, particularly in the areas of improving education and healthcare systems and support of small and medium enterprises.

4. Opportunities and Challenges Offered by the Belt and Road Initiative

The Belt and Road Initiative (BRI) would undeniably bring more opportunities to Slovakia. The main benefits would be the economic growth and opportunities that are brought by the infrastructure projects and multilateral engagement of the 60-plus participating countries.

It could be rated a considerable success because several projects have been implemented: an electric rail line from Djibouti to Addis Ababa launched in October 2016.[58] A few months later, in January 2017, the first direct train from China arrived in London.[59] In November 2017, the first train from the Chinese town of Dalian arrived in Bratislava, etc.[60] Not only are trains more reliable, they are also cheaper than air freight and faster than maritime shipment.[61,62]

However, the BRI is a long-term project and those who will receive the full benefit of this tremendous initiative are young people. With the BRI, youths could expand their business

58 | BBC News. (2016, October 5). Ethiopia-Djibouti Electric Railway Line Opens. *BBC*. Retrieved from https://www.bbc.co.uk/news/world-africa-37562177.

59 | Kentish, B. (2017, January 18). First Direct Train Service from China to the UK Arrives in London. *The Independent*. https://www.independent.co.uk/news/uk/home-news/first-direct-train-china-uk-arrives-east-london-yiwu-city-barking-channel-tunnel-a7533726.html.

opportunities into new markets and trade either in goods or services abroad. Nowadays, most Slovak companies export to countries within the EU, particularly because of the establishment of free trade agreements, reduction of transaction costs and bureaucracy. It is assumed that BRI will reduce transportation costs, lower existing barriers to the market, encourage cooperation and even access to investors and transfer of technology. It will be a step towards improving and deepening business relationships between Slovak and foreign entrepreneurs and can also bring new investment opportunities to Slovakia.

Moreover, the BRI is presented as more than just a trade development project. It will integrate three continents (Europe, Asia and Africa) and the 60 countries involved not only in terms of economic cooperation, but also by promoting cooperation in forms of increased academic and cultural exchanges and tourism. Plus it will foster intercultural dialogue as well as understanding and tolerance among countries in the BRI.[63]

4.1. Potential Challenges Brought by the BRI from Slovakia's Perspective

Such an ambitious megaproject has never been attempted before, consequently, some questions have to be answered to eliminate risks, which could endanger its successful implementation and curtail further benefits.

The countries along the BRI have diverse legal systems and laws regarding requirements on customs, safety standards, taxation, labour law, etc. Such differences could cause disputes on various levels and between different stakeholders, i.e. state-state, business-state, etc. It is also important to set up clear rules on who will bear responsibility and legal obligations for health, safety or disasters.[64]

BRI should not be an initiative of only one country, but a 'partnership agreement based on multiple bilateral and multilateral trade agreements'.[65] China currently has separate

60 | Railway Gazette International. (2017, November 17). Dalian to Bratislava Freight Service Arrives. *Railway Gazette International.* Retrieved from https://www.railwaygazette. com/freight/dalian-to-bratislava-freight-service-arrives/45503. article.

61 | BBC News. (2016, October 5). Ethiopia-Djibouti Electric Railway Line Opens. *BBC.* Retrieved from https:// www.bbc.co.uk/news/world-africa-37562177.

62 | Kentish, B. (2017, January 18). First Direct Train Service from China to the UK Arrives in London. *The Independent.* https://www.independent. co.uk/news/uk/home-news/ first-direct-train-china-uk-arrives-east-london-yiwu-city-barking-channel-tunnel-a7533726.html.

63 | Wong, C. (2017, May, 12). One Belt, One Road, One Language?. *The Diplomat.* Retrieved from https://thediplomat. com/2017/05/one-belt-one-road-one-language/.

64 | Ngai, H. (2017). *An Effective Dispute Resolution Mechanism Necessary for the Chinese "The Belt and Road" Initiative. Opportunities for the Youth Along the Belt and Road.* Zhejiang University Press, pp. 78-83.

65 | Ascensio, H. (2016). Building International Infrastructure Networks: Some Lessons from the Experience of the European Union in the Energy Sector. In *International Law Perspective of the Belt and Road Initiative.* Zhejiang University Press.

agreements with many countries along the road and many of those countries are members of some other international organisations such as the EU, World Trade Organisation (WTO), Energy Charter Treaty, etc. It is also important to keep in mind the laws regarding environmental sustainability, particularly The Paris Agreement that aims to cut down carbon dioxide emissions. That can result in complications with the enforcement of agreements and unclear rules. It should be decided whether to apply existing international treaties or to create a new set of rules[66]. A potential challenge that faces small countries such as Slovakia is that the voice or opinions of countries with a small number of inhabitants or small economic power are not taken into consideration equally.

Many countries along the road have weak economies and/or unstable political systems. Therefore, conditions on lending should be strict to avoid abuse and the chance of jeopardising the whole project[67]. Another challenge for the BRI could be political and economic instability in certain countries along the Silk Road, instability caused by factors such as internal conflicts, terrorism, separatist groups, etc. All those are issues that must not be neglected, and some measures must be adopted to avoid those political risks.

Another challenge concerns the geopolitical tensions in the region. For example, the most logical and fastest trade route linking Asia to Europe would be via Russia, Ukraine and also Slovakia. The Slovak politicians even promoted it as a great economic opportunity for Slovakia because a train container terminal could be built to serve the route and improve the geostrategic significance, logistics, and business opportunities of Slovakia. However, due to the conflict between Russia and Ukraine, the drawn path of the new Silk Road does not currently connect with Slovakia. Liu Jiangyong, an International Relations professor at Tsinghua University, says: 'In Beijing's experience, a safe and sustainable geopolitical situation can be achieved only through peaceful co-operation between maritime and land-locked countries.'[68] Hopefully, BRI will also address the possibilities arising from the potential improvement of the geopolitical relationships in the region.

66 | Wang, G. (2016). The Belt and Road Initiative in the Context of Globalization, in *A New International Legal Order*. Brill Publishers, pp. 279-294.

67 | Ryan, C. (2016). Economics, Geo-Politics, Culture and Environment. *International Law Perspective of the Belt and Road Initiative.* Zhejiang University Press.

68 | Fu, Y. (2015, August 25). Time for China and the US to Build Inclusive World Order. *The Telegraph*. Retrieved from https://www.telegraph.co.uk/sponsored/china-watch/politics/11816946/china-us-relations-world-order.html.

The key to the success of this project is cooperation and mutual understanding. In this case of cooperation among 60 countries with different cultures, religions, political structures, and business practices, we were confident that more frequent exchanges can accelerate mutual understanding and better awareness. Such heightened awareness can be beneficial for enterprises who want to expand their businesses worldwide.

4.2. The Role of Young People in the Belt and Road Initiative

China focuses intensively on promoting the BRI among young people. A series of Bedtime Stories videos about the BRI was filmed and introduces how the Initiative will be financed, who can join, etc. in a very simple and comprehensible way. Some songs were produced, targetting children and young people with lyrics such as: 'We are breaking barriers, we are making history, the world we are dreaming of starts with you and me... ideas start to flow, and friendships starts to form. Then things impossible all become the norm'.[69] Another song that young people sing goes: 'mutual benefit, joint responsibility and shared destiny'. All those promotion efforts have already yielded visible results. Half of the respondents/young people from China expressed in a survey that the BRI will have a big influence in their lives[70].

In Europe, this project gets little attention from media and as we found out from a small survey of our friends, few people have even heard about it. The role of us young people is of course important. 'Give young people a greater voice. They are the future and they are much wiser than we give them credit for,' said anti-apartheid activist Archbishop Desmond Tutu of South Africa.[71] It sounds already like a cliché since many quotes and thoughts like this one have been said. Nevertheless, not much has been done in this regard. The propensity to view young people as serious decision-makers with power is still rather occasional and considered to be odd by the public.

69 | Hollingsworth, J. (2017, May, 12). 'Belt and Road' for Kids: Are These Videos China's Latest Bid to Win Over Young Western Generation?. *South China Morning Post*. Retrieved from http://www.scmp.com/news/china/society/article/2094080/belt-and-road-kids-are-these-videos-chinas-latest-bid-win-over.

70 | *Ibid.*

71 | Schnall, M. (2010, April 5). The Elders Speak: Desmond Tutu, Jimmy Carter And Mary Robinson. *Huffpost*. Retrieved from https://www.huffpost.com/entry/desmond-tutu-jimmy-carter_b_476610?guccounter=1.

5. Suggestions for the Belt and Road Initiative

We have previously mentioned the challenges and there are more that should be addressed in order to implement the BRI successfully and allow the countries to fully enjoy the benefits brought by it. As mentioned in one of the songs written to promote the Initiative, the BRI is 'China's idea but it belongs to world'.[72] Although China is the largest stakeholder in the project and is putting in the most effort to accelerate the process, a mechanism to facilitate genuine cooperation and decision-making of all countries affected should be established. A long-term strategy is required from all the countries and without institutionalisation, it will not be efficient.

Cooperation with other international and regional organisations or networks could be strengthened to prevent resources duplication and provide synergies in certain fields. Erasmus (EuRopean Community Action Scheme for the Mobility of University Students) for young entrepreneurs is an exchange programme financed and facilitated by the EU for newly established entrepreneurs who can spend several months with the more experienced entrepreneur abroad. Nowadays, the programme operates only within the European Economic Area (EEA), yet it has the potential to expand and allow exchanges within all countries along the BRI too. It could greatly contribute to cross-border business development and interconnectivity between continents. For example, in the first phase, the registered hosting entrepreneurs could be contacted and have business ideas, created by young entrepreneurs, introduced to them.

The BRI is presented in a way that it will also increase cultural exchanges. There are plenty of cultural exchanges within Europe, but very few are cross-continental. Asia-Europe Foundation focuses on enhancing Asia-Europe relations by organising seminars, workshops, conferences in the fields of education, culture, economy and more. Cooperation between the Foundation and the BRI could be negotiated in facilitating the previously mentioned exchanges. Not only cultural exchanges, but also higher interactions between universities and exchanges of researchers and students would be welcome and beneficial for the implementation of the BRI.

72 | Hollingsworth, J. (2017, May, 12). 'Belt and Road' for Kids: Are These Videos China's Latest Bid to Win Over Young Western Generation?. *South China Morning Post*. Retrieved from http://www.scmp.com/news/china/society/article/2094080/belt-and-road-kids-are-these-videos-chinas-latest-bid-win-over.

The BRI should publicise more about its projects to promote its ideas and benefits. In particular, Europeans should be informed more about this Initiative. First of all, we recommend establishing a portal where all information about the BRI could be found and will be regularly updated since the current presentation is inadequate and confusing. The website could serve also as a networking platform for finding business partners. Another possible step could be to establish a think tank that would play a role in research, advocacy, exchange of best practices, and conducting events or campaigns.

In our view, the establishment of venture capital funds for innovative cross-continental start-ups could be favourable, especially for social enterprises – those that address problems in society, help achieve Sustainable Development Goals (SDG) and support a green economy in the long run.

5.1. Encouraging Youth Participation in the Belt and Road Initiative

We were members of several working groups and involved in the creation of diverse policy recommendations. Therefore, from our experience, we would recommend the following suggestions to encourage more young people to get involved in the Initiative. Incorporation of young people cannot be conducted randomly or occasionally through an invitation to conferences or forums. It is necessary to institutionalise our involvement in the future BRI structures through the establishment of:

- A Belt and Road Initiative Youth Council comprising one youth representative from each country on the Silk Road. Members of the BRI Youth Council could work closely on specific topics in working groups.

- A Board selected among representatives that would act as a permanent liaison within other BRI structures and BRI Youth Council. The Board could consist of several positions responsible for different fields.

Cargo transported by railway in Slovakia.

- Annual conferences of the BRI Youth Council with other relevant stakeholders invited in order to consult progress made and develop future initiatives or activities.

Each youth representative would act as the national contact point and help spread information to other young people and organise local events relevant to the opportunities offered by the BRI. We would also recommend creating and conducting surveys in each country among youths to find out relevant information, problems and needs on how they view the opportunities. Such information would be a valuable point of reference to improve the Initiative for young people. Furthermore, another kind of involvement that young people in the BRI could have is through the formation of specific internships or job shadowing programmes in newly industrialised economies. Under those circumstances, young people would get new experiences in a new country and probably also have a say in the decision-making process.

Piran city in Slovenia (photo provided by the author).

CHAPTER 17

by   NINA PEJIČ
Researcher at Center of International
Relations, University of Ljubljana.

URBAN ŠPITAL
University of Ljubljana

(Ordered by last name)

SLOVENIA

ESTONIA

LATVIA

LITHUANIA

POLAND

CZECH
REPUBLIC

SLOVAKIA

HUNGARY

ROMANIA

CROATIA

BOSNIA &
HERZEGOVINA

SERBIA

MONTENEGRO

BULGARIA

NORTH
MACEDONIA

ALBANIA

SLOVENIA

1. Political Structure of Slovenia: An Overview

1.1. A Brief History of Slovenia

Slovenia lies at the crossroads of Central Europe, at the north of the Adriatic Sea, locked between Hungary in the east, Croatia in the south, Italy in the west, and Austria in the north. The history of the region dates to prehistoric times as evidenced by such findings as the oldest known musical instrument, wooden wheel, and wooden axle discovered in the Ljubljana Marshes.[1] Celtic and Illyrian tribes resided in the area of what is currently known as Slovenia until 100 B.C., when the Romans established their presence[2]. The first Slavic ancestors of the Slovenes settled the region in the sixth century and became part of the tribal union of the Slavic King Samo in 623 A.D.[3]. Two more notable entities existed in the approximate area of today's Slovenia: Carantania and Carinthia, but the area was *de facto* owned by feudal lords from European nobility including the Dukes of Spanheim, Counts of Celje, Counts of Gorizia and the House of Habsburg[4].

Although the oldest document in the Slovene language, the Freising Manuscripts, date to the 10th century A.D., the first mention of a Slovene ethnic group outside of the area dates back to the 16th century[5]. With the Protestant Reformation during this era, Primož Trubar wrote the first books in Slovene, which resulted in numerous books printed in Slovene, including a Jurij Dalmatin's translation of the Bible[6]. Several Slovene towns that had flourished after the reformation, were destroyed in the wars of the 15th, 16th and 17th centuries between the Habsburg Monarchy, Ottoman Empire and the Venetian Republic, after which they never managed to recover. This also led to several peasant revolts, which left a big imprint on Slovene folklore and are celebrated up to this date. In the 18th century, the area experienced a rare period of peace that brought economic recovery, scientific advances, and resulted in the first political movement for United Slovenia – *Zedinjena Slovenija*.[7]

In the 19th century, Slovene national awakening brought several political, financial and cultural institutions to the forefront and formed a kind of loose national infrastructure. The standardised

1 | UKOM. (2003). *World's Oldest Wheel Found in Slovenia*. Retrieved from: http://www.ukom.gov.si/en/media_room/background_information/culture/worlds_oldest_wheel_found_in_slovenia/.

2 | Granda, S., Dular, A. & Grdina, I. (2008). *Mala zgodovina Slovenije [A Small History of Slovenia]*. Ljubljana: Celjska Mohorjeva Družba, (27).

3 | *Ibid.*, pp. 43-51.

4 | *Ibid.*, pp. 43-51.

5 | *Ibid.*, p. 53.

6 | *Ibid.*, p. 58.

7 | Granda, S., Dular, A. & Grdina, I. (2008). *Mala zgodovina Slovenije [A Small History of Slovenia]*. Ljubljana: Celjska Mohorjeva Družba, (27), pp. 61-94.

Slovene literary language had emerged, and the idea of a common political entity, known as Yugoslavia, captivated the minds of liberals, progressives and anticlericals.[8] Slovenia was one of the main battlegrounds of World War I with hundreds of thousands of Slovenes conscripted, and civilians resettling in refugee camps on both sides of the front. During that war, the Slovene People's Party launched a self-determination movement, which resulted in more than 200,000 signatures of support by early 1918. On 6 October the National Council of Slovenes, Croats and Serbs came to power and merged with Serbia on 1 December 2018 to form the Kingdom of Serbs, Croats and Slovenes that later renamed itself to the Kingdom of Yugoslavia in 1929.[9]

During World War II, Slovenia was annexed in its entirety by Germany, Italy and Hungary after the invasion on 6 April 1941. The Liberation Front of the Slovene Nation was organised on 27 April 1941, while about 21,000 members of the Slovene Home Guard collaborated with the occupying powers.[10] Yugoslavia was liberated in 1945 and became a federal socialist state, which Slovenia joined as a socialist republic. The Communists perceived the opponents of the Liberation Front as anti-revolutionary and that resulted in a parallel civil war after 1942, which ended with largely ideologically motivated massacres shortly after the end of World War II.[11]

Due to the Tito-Stalin split, economic and personal freedoms in Yugoslavia were broadened more than in the rest of the Eastern Bloc[12]. The Slovenian borders with Yugoslavia were mostly established in 1947, with the exception of the Free Territory of Trieste, which was contested until 1975, when it fell under Italian control with the ratification of the Treaty of Osimo.[13] With the unique placement of Slovenia at the border of Austria and Italy, it enjoyed a high degree of autonomy that resulted in heavy industrialisation with diverse markets and the highest economic standard among all of the socialist republics in Yugoslavia.

In the 1980s, Slovenia started to experience the rise of political and intellectual movements like the *Neue Slowenische Kunst*

8 | *Ibid.*, p. 96.

9 | *Ibid.*, p. 128. For more insight into Slovene history, we suggest reading Luthar, O. (2008). *The Land Between: A History of Slovenia*. Internationaler Verlag der Wissenschaften.

10 | *Ibid.*, p. 143.

11 | *Ibid.*, p. 156.

12 | National Public Library. (n.d.). *World Heritage Enclycopedia Edition – Slovenia.* Retrieved from http://www.nationalpubliclibrary.org/articles/eng/Slovenia.

13 | *Ibid.*, pp. 198-222.

14 | *Ibid.,* pp. 198-222.

15 | To illustrate, Slovenes at that time formed 10% of Yugoslav population, but produced one fifth of the GDP. See *Ibid.,* pp. 232-241.

16 | Uradni list Republike Slovenije. (1990). *Odlok o razglasitvi ustavnih amandmajev k ustavi Socialistične Republike Slovenije [The Decree About the Proclamation of Constitutional Amendments to the Constitution of the Socialist Republic of Slovenia].* Retrieved March 16, 2017, from https://www.uradni-list.si/Data/File/vip_akti/1989-02-1075.pdf.

17 | Created by an agreement of the Slovenian Democratic Union, the Social Democrat Alliance of Slovenia, Slovene Christian Democrats, the Farmer's Alliance and the Greens of Slovenia, and led by Jože Pučnik.

18 | Uradni list Republike Slovenije. (1991). *Deklaracija o neodvisnosti [Declaration of independence].* Retrieved May 25, 2017, from https://www.uradni-list.si/glasilo-uradni-list-rs/vsebina/1991-01-0007/deklaracija-ob-neodvisnosti.

19 | Ramet, S. P. (1993). *Slovenia's Road to Democracy.* Europe-Asia Studies, 45(5), pp. 869-886.

20 | Uradni list Republike Slovenije. (1991). *Deklaracija o neodvisnosti [Declaration of independence].* Retrieved May 25, 2017, from https://www.uradni-list.si/glasilo-uradni-list-rs/vsebina/1991-01-0007/deklaracija-ob-neodvisnosti.

21 | Government of the Republic of Slovenia – Public Relations and Media Office. (2001). *Path to Slovene State.* Retrieved February 05, 2018, from http://www.slovenija2001.gov.si/10years/path/.

(New Slovenian Art, a collective), *Nova Revija* (*New Review* magazine), *Mladina* (*Youth* magazine), etc.[14] This coincided with the Yugoslav economic crisis of the 1980s, where many Slovenes felt economically exploited, due to harsh economic measures and a large but mostly ineffective federal administration.[15] Several clashes occurred between the civil society and the communist regime, which led to the 'Slovene Spring'. Unions began organising into parties and a crackdown on journalism resulted in mass demonstrations. In 1989, the Slovenian Assembly introduced several amendments to the 1974 constitution. In particular, by passing amendment XCI, the Assembly changed the state's official name from the Socialist Republic of Slovenia to the Republic of Slovenia on 8 March 1990.[16]

The first multiparty parliamentary elections and presidential elections were held on 8 April 1990. The Democratic Opposition of Slovenia[17] won the election with Milan Kučan winning the presidential elections on 22 April 1990[18]. Political and economic reforms to establish a liberal democratic political system with a market economy began under the formal leadership of Lojze Peterle.[19] On 23 December 1990, a plebiscite on Slovenian independence was held with more than 88% of Slovenes voting in favour of the separation from Yugoslavia.[20] Slovenia finally became independent on 25 June 1991, starting a short 10-day war, in which Slovenes successfully deterred the Yugoslav military presence, resulting in the last Yugoslav soldier leaving Slovenia on 26 October 1991. The new Slovenian Constitution was passed on 23 December 1991 and the political processes of a now independent Slovenia began. On 22 May 1992 Slovenia became a member of the United Nations, joined NATO on 29 March 2004, and finally became a member of the EU on 1 May 2004.[21]

1.2. The Ideology of the Slovenian Constitution and Government System

Based on the Slovene Constitution, Slovenia is a unitary democratic republic, a parliamentarian state, with rule of law, with its legitimacy deriving from the self-determination of the

Slovene nation[22]. Its parliament is composed of two houses, the National Assembly of the Republic of Slovenia and the National Council. The Government is elected in the National Assembly. There are two levels of courts in Slovenia, courts of the first instance and appeal courts, with litigation starting in district courts and county courts, while higher court and Supreme Court serve as courts of appeal[23]. All of the courts in Slovenia are completely independent. On the local level, Slovenia is divided into 212 municipalities.[24]

As already hinted in the general overview of the history of Slovenia, the system of rule, governance, and the role of institutions more or less evolved from the previous systems, mainly from the Socialist Republic of Slovenia. The political system develops on two levels, by implementing the legislation forwarded from the European Union (EU) and adopted in the National Assembly.[25] The biggest milestone was the creation of the Constitution of the Republic of Slovenia.

The Slovenian Constitution is a modern normative constitution inspired by the Italian and German constitutions.[26] It has 10 chapters with a preamble and 174 articles. There are elements of corporatism evidenced by the existence of (1) the National Council that represents local, social, cultural and economic interests; (2) elements of socialism defining Slovenia as a social state with rights to employment, union freedom, etc.; and, (3) pluralism with determined elections as a prerequisite for a pluralistic society.[27]

22 | Constitution of the Republic of Slovenia, 2017, Article 1-4.

23 | Government of the Republic of Slovenia. (2017). *Politicni Sistem [Political System of the Republic of Slovenia]*. Retrieved from: http://vlada.arhiv-spletisc. gov.si/o_sloveniji/politicni_ sistem.

24 | *Ibid.*

25 | Brezovšek and Haček. (2012). *Politični sistem Republike Slovenije [Political System of the Republic of Slovenia]*. Ljubljana: Založba FDV, p. 127.

26 | *Ibid.*, p. 77.

27 | *Ibid.*, pp. 78-80.

2. Political Direction of Slovenia

2.1. Constitution, Political, Economic and Legal Reforms

At the time of the writing of this paper, on 14th March 2018, the Slovenian Prime Minister Dr Miro Cerar offered his resignation.[28] It came after an accumulation of problems.[29] The main issues were: an on-going strike of teacher unions; an episode where equipment was overpaid for through public procurement due to an administrative issue[30]; and a decision of the Slovenian Supreme Court that a referendum on the implementation of the 'Second Rail' project wasn't conducted properly.[31] With the resignation of the prime minister all work on constitutional reforms, political or economic projects that were not agreed upon, stopped. The last time that the Slovenian constitution was amended was in 2016 when Article 70a was added, enshrining the right to drinkable water in the Slovenian constitution.[32] There were no publically expressed ambitions to make further constitutional changes to the Slovenian constitution.

In 2015, the Republic of Slovenia reaffirmed its new foreign policy strategy titled 'Slovenia: Safe, Successful and Respected in the World'.[33] In it, Slovenia reaffirmed its foreign policy objectives. At the time of writing (14th March 2018), there was not any political party that would oppose this strategy and have enough support to change it or stop its implementation. However, there were movements that advocated policies like discontinuation of Slovenian membership in the North Atlantic Treaty Organisation (NATO) or the EU, which will be elaborated on further later in this paper. The Slovenian Government sets its reform priorities in six fields: health, youth employment, traffic infrastructure, enhancement of the legal system, ecology, and digitalisation.[34] Ultimately, the implementation of these reforms heavily depends on the overall political climate and electoral politics of Slovenia.

As previously mentioned, the ideological divide from World War II was, for a long time, the main reason for division within Slovenia, to such an extent that one of the priorities of Borut Pahor, in his first term as the President of the Republic in 2012 was reconciliation. This sentiment is still present in generations

28 | Potič, Z. (2018). Miro Cerar odstopil, odnesel ga je drugi tir [Miro Cerar resigned, he was taken away by the second rail]. Delo. Retrieved from https://www.delo.si/novice/politika/miro-cerar-je-odstopil-odnesel-ga-je-drugi-tir.html.

29 | With regular elections coming up in less than three months after his resignation, it was also a strategic decision.

30 | The equipment was a model of the politicised 'second rail' project, which added to its importance.

31 | The interpretation of the Slovenian Supreme Court was the Government is not entitled to use public funds for a referendum campaign even in instances when the Government isn't the entity calling for a referendum. See Vukelić, M. (2018). Zakona o drugem tiru vlada še ne more izvajati [The second rail project cannot yet be implemented by the government]. Delo. Retrieved from https://www.delo.si/novice/politika/ustavno-neskladna-referendumska-zakonodaja-nic-o-drugem-tiru.html.

32 | Supreme Court of the Republic of Slovenia. (2017). Spremembe in dopolnitve ustave [Amendments to the Constitution]. Retrieved from http://www.us-rs.si/o-sodiscu/pravna-podlaga/ustava/spremembe-in-dopolnitve-ustave/.

that lived in Yugoslavia or experienced loss first-hand during or after World War II, and there are still parties who try to fuel these divisions; however, they are slowly losing traction. It is interesting that there are political parties on both the political left and right, liberals and conservatives, that are trying to make their pitch outside of the discourse of the divisions that still last from World War II.

For example, the narrative of Prime Minister, Dr Miro Cerar, was seen by the Slovenian public as a new face trying to bridge the divides from the past. The goal of reconciliation will be hard to repeat with falling support from both the left and the right, culminating in almost 30 per cent decrease in projected voter share. Such would also be the case of the new political party Prstan, which is trying to fill the gap of a typical liberal party in Slovenia, as well as the case of Andrej Šiško and his party, *Zedinjena Slovenija* (United Slovenia), who was also at least publically trying to put Slovenian nationalism at the forefront. Slovenia is a curious case about political parties since there is a visible trend, where a party with a front-running prodigy won every election between 2011 and 2017[35].

There were new and renewed political parties forming across the political spectrum. One such was Marjan Šarec, who was the runner up at the elections for the president of the Republic of Slovenia with 46.91% of the votes and whose party, The List of Marjan Sarec, bears his name.[36] Other new notable political party movements as of 2018 included the conservative political activist Aleš Primc, who formed an election coalition with the previous mayor of Maribor, Franc Kangler[37]. parliament member Dr Bojan Dobovšek is with the anti-corruption focused Good Country. And another parliament member, Andrej Čuš, took over the Slovenian Green Party from his father.

Another interesting phenomenon present in Slovenia regarding the forming of nationalistic movements was the condemnation of them. Slovenia was on the path of the so-called Balkan route of the refugee crisis that was dealt with by the EU, which led to the rise of anti-immigrant movements similar to those

33 | Ministry of Foreign Affairs of the Republic of Slovenia. (2015). *Slovenija: varna, uspešna in v svetu spoštovana [Slovenia: safe, successful and respected in the world]*. Retrieved from http://www.mzz.gov.si/fileadmin/pageuploads/Zakonodaja_in_dokumenti/dokumenti/strategija_ZP.pdf.

34 | Gole, N. (2017, January 26). Prioritete vlade v 2017: zdravstvo, infrastruktura, mladi [Government priorities in 2017: health, infrastructure, youth]. *Delo*. Retrieved from https://old.delo.si/novice/politika/vlada-zacrtala-prednostne-naloge-za-letosnje-leto.html.

35 | State Election Commission of the Republic of Slovenia. (2017). *Volitve predsednika Republike Slovenije 2017 [Elections of the President of the Republic of Slovenia 2017]*. Retrieved from http://www.dvk-rs.si/index.php/si/arhiv-predsednika-rs/volitve-predsednika-rs-leto-2017.

36 | *Ibid.*

37 | Maribor is the second largest city in Slovenia.

in much of the EU. While the Slovenian Government itself implemented some policies that are considered anti-immigrant, like the introduction of a barbed wire fence on the border with Croatia, the most vocal anti-immigrant party in the National Assembly was the Slovenian Democrat Party. This led to a situation where every political party in the National Assembly stated in November 2017 that they would not be willing to go into coalition with the Slovenian Democrat Party under the leadership of Janez Janša, rejecting views that are recognised as far-right.[38]

With very encouraging economic forecasts for Slovenia, a projected GDP rise of 4.3% for 2018, and almost monthly news of investment into the country, the political constellation was not expected to drastically change in terms of the general left-right orientation in the National Assembly.[39]

38 | Mekina, B. (2017). *Koalicija proti nestrpnosti [Coalition against intolerance]. Mladina.* Retrieved from http://www.mladina.si/182933/koalicija-proti-nestrpnosti/.

39 | OECD. (2017). *Developments in Individual OECD and Selected Non-Member Economies – Slovenia.* Retrieved from http://www.oecd.org/eco/outlook/economic-forecast-summary-slovenia-oecd-economic-outlook.pdf.

3. Opportunities and Challenges for the Youth in Slovenia

Young people present a very particular part of Slovenian society, facing both country-specific obstacles and opportunities connected with the three main transitions: completion of schooling and entry into the labour market; the transition from economic dependence to economic independence and self-responsibility; and the transition of moving from home to their own home and creation of their own family. This chapter will introduce the social situation of youth in Slovenia – it will identify the developmental (personal, career, business, education etc.) opportunities for young people in the country, the challenges, which hinder the youth's potential in Slovenia, and the measures that Slovenian governments have taken to tackle them. This chapter is based on the descriptive-analysis and interpretation of the secondary sources and two interviews with the youth associations' representatives.[40]

Who is the youth in Slovenia? Although the United Nations defines youth as individuals aged 15 to 24 years old, Slovenia (Statistical Office of the Republic of Slovenia – SURS), as well as some other European countries, define youth as young people aged 15 to 29 years old.[41] At the beginning of 2016, there were 327,000 young people in Slovenia (15-29 years), which represents almost 16% of the population. The share of young people in Slovenia is decreasing – 20 years ago, they represented more than 20% of the population.[42]

3.1. Opportunities for the Youth

A representative of youth pointed out that in terms of new developmental opportunities for young people, Slovenia has some very good results. 'Our Global Human Capital Index ranks us number 9 globally (right after Norway, Finland, Switzerland, the US, Denmark, Germany, New Zealand and Sweden),' she said. 'Our Sustainable Development Goals (SDG) Index rank is also 9 (which means that we are in the top 10 countries on the planet when it comes to the implementation of the Agenda 2030 for sustainable development), Slovenia and Norway are the most child-friendly countries on the planet (End of Childhood Index), etc.'[43] In general, she stated, 'some of our key strengths are [a] highly educated population, high level of gender equality, participation of youth in decision-making, well-recognised youth sector and state support of the youth organisations, also awareness of global citizenship and active volunteering'.[44]

Moreover, the official governmental agencies, news outlets (state-owned and commercial)

40 | Interview no. 1 was conducted with Ms Kaja Primorac, former President of the United Nations Association of Slovenia Youth Section (until 2017) and an intern at the Office of the Advocate of the Principle of Equality. Interview no. 2 was conducted with Ms Sabina Carli, United Nations Youth Delegate of Slovenia with the mandate in 2017-2018.

41 | Since this is a period of life characterised by transience and interdependence, it is very difficult to unambiguously determine the 'age brackets' for the youth category, especially in Slovenia, which has a very high percentage of youth still dependent on their parents – more than 75% of young people live in the household of their parents. Such a high percentage is only characteristic for two other countries in Europe, Italy and Greece. See OECD. (2016). *Society at a Glance 2016: OECD Social Indicators. OECD Publishing.* Retrieved from http://dx.doi.org/10.1787/9789264261488-en; SURS. (2009). Mladi v Sloveniji. *Ljubljana: Statistični urad Republike Slovenije [Ljubljana: Statistical Office of the Republic of Slovenia].* Retrieved from http://www.mladi-in-obcina.si/wp-content/uploads/2017/06/mladi2009-SLO.pdf.

42 | SURS. (2017, August 7). *Deleža mladih (15-29 let) med celotno populacijo v EU in v Sloveniji še vedno upadata [The shares of young people (15-29 years) among the entire population in the EU and in Slovenia are still declining].* Retrieved from http://www.stat.si/StatWeb/News/Index/6799.

43 | Carli, S. (2017, October 27). *Interview no. 2 on the status of youth in Slovenia.*

44 | *Ibid.*

45 | Siol. (2016). *Kako lepo je danes v Sloveniji biti mlad [How nice it is to be young in Slovenia today]*. Retrieved from https://siol.net/novice/slovenija/kako-lepo-je-danes-v-sloveniji-biti-mlad-423658; Carli, S. (2017, October 27). *Interview no. 2 on the status of youth in Slovenia.*; Primorac, K. (2017, October 26). Interview no. 1 on the status of youth in Slovenia; Morel. (2017, May 4). *Slovenija: Vladna verzija poročila o posvetu o mladih [Slovenia: Government version of the report on the Youth Conference]*. Retrieved from http://www.morel.si/Notranja_politika/Slovenija_Vladna_verzija_porocila_o_posvetu_o_mladih/; Vlada Republike Slovenije. (2017). *Skupni imenovalec vlade in mladih: spodbujanje zaposlovanja, inovativnosti, kreativnosti in podjetnosti ter zagotavljanje dostopnih najemnih stanovanj [Common denominator of government and youth: promoting employment, innovation, creativity and entrepreneurship, and providing affordable rental housing]*.

46 / Zavod Republike Slovenije za zaposlovanje. (2017). *Registrirane brezposelne osebe po starostnih razredih, 2005-2017 [Registered unemployed persons by age groups, 2005-2017]*. Retrieved from https://www.ess.gov.si/trg_dela/trg_dela_v_stevilkah/registrirana_brezposelnost.

47 | Carli, S. (2017, October 27). *Interview no. 2 on the status of youth in Slovenia.*

48 | Morel. (2017, May 4). *Slovenija: Vladna verzija poročila o posvetu o mladih [Slovenia: Government version of the report on the Youth Conference]*. Retrieved from http://www.morel.si/Notranja_politika/Slovenija_Vladna_verzija_porocila_o_posvetu_o_mladih/; Vlada Republike Slovenije. (2017). *Skupni imenovalec vlade in mladih: spodbujanje zaposlovanja, inovativnosti, kreativnosti in podjetnosti ter zagotavljanje dostopnih najemnih stanovanj [Common denominator of government and youth: promoting employment, innovation, creativity and entrepreneurship, and providing affordable rental housing]*.

49 | SURS. (2016). *Število študentov v Sloveniji še naprej pada, a še vedno študira skoraj polovica mladih [The number of students in Slovenia continues to fall, but almost half of young people still study]*. Retrieved from http://www.stat.si/StatWeb/News/Index/5929.

and interviewees all emphasise that the number of opportunities for youth is increasing since 2013, especially in terms of positive effects in the labour market, education and the system of social protection.[45] Unemployment among the youth has dropped.[46] Why is the level of unemployment among youth decreasing? According to the interviewee: 'Slovenia has undergone a very major crisis in the period between 2008 and today, as a consequence of the global financial crisis and the resulting austerity measures across the EU. As a result, the unemployment rate is slowly but gradually decreasing, there are numerous job opportunities arising.'[47]

The Government of the Republic of Slovenia confirmed this view, while also emphasising that the increase in developmental opportunities for youth (with new financing and subsidies etc.) was one of the five priorities of their mandate and that 'the state has a key role in the process of inclusion of youth to the labour markets, with active policies and measures that will facilitate the start of their careers'.[48]

However, Slovenia has an unusually high percentage of young people included in tertiary higher education programmes, with almost 50% of youths studying to achieve a bachelor's degree. According to the number of students per inhabitants, Slovenia, therefore, ranks among the leading countries in the EU.[49] This means that partly, the 'real' percentage of unemployment among young people could be 'hidden' due to the high inclusion of young people into higher levels of education, as a way to extend their studies and enter the labour market later.

In terms of opportunities for education and personal development, there are numerous new opportunities for both formal and informal education. The former – formal education – is characterised by mandatory primary school, very accessible and free secondary

50 | Carli, S. (2017, October 27). *Interview no. 2 on the status of youth in Slovenia*.; Študentska organizacija Slovenije. (2017). *Študij [Study]*. Retrieved from http://www.studentska-org.si/studentski-kazipot/studij/.

51 | Any restaurant in a city which offers tertiary education, can apply for the system of subsidies. This means that a student in Slovenia (local or exchange student) can eat a three-course meal in these (privately-owned) restaurants for €2.50 on average – the rest of the meal is subsidised by the Slovenian Government. The system is digitalised and works through a mobile phone. See Univerza v Mariboru. (2017). *Subsidised students meals [Students coupons]*. Retrieved from https://www.um.si/en/international/erasmus/Pages/Subsidized-students-meals-(Students-coupons).aspx.

52 | Študentska organizacija Slovenije. (2017). *Študij*. Retrieved from http://www.studentska-org.si/studentski-kazipot/studij/.

53 | *Ibid.*

54 | The voluntary activities of young people in Slovenia are usually associated with the city of their residence and interest for the local environment (in all possible sectors). Moreover, young people in Slovenia have a developed sense of social justice or solidarity, since it demonstrated by the willingness to voluntarily help people in need as a consequence of natural disasters (especially earthquakes in the northwest of the country) and provide other forms of public welfare assistance. For example, one of the most successful youth projects in Slovenia called Simbioza works to educate elderly how to use computers and mobile phones and it has educated more than 11.000 people in three years. Moreover, this solidarity is also expressed through large volunteering in youth organisations such as scouts and young firefighters and youth sections of civil protection and army. In addition, young people are most often members of organisations from the fields of sports, recreation and culture – considerably less young people are members of political parties. However, youth high school and students' organisations play a great role here, supplementing for the lack of political involvement through more appealing conditions to the youth, helping them to influence society and contributing to the country's development. Through their participation in especially the Student Organisation of Slovenia, they gain an idea of what public decision-making is, which also decreases political apathy among the youth. See Office of the Republic of Slovenia for Youth. (2015). *Mladi v številkah – število mladih [Youth in numbers - the number of young people]*. Retrieved from http://www.slovenija25.si/mladi-obrazi/mladi-v-stevilkah/.

55 | Carli, S. (2017, October 27). *Interview no. 2 on the status of youth in Slovenia*.

56 | Office of the Republic of Slovenia for Youth. (2015). *Mladi v številkah – število mladih [Youth in numbers - the number of young people]*. Retrieved from http://www.slovenija25.si/mladi-obrazi/mladi-v-stevilkah/.

education, as well as the aforementioned high enrolment into tertiary levels of education, which is also a consequence of it being free for all citizens (undergrad and post-grad studies).[50] Furthermore, there are numerous other benefits for higher education students. For example, the food subsidies system, is unique in Slovenia, since it is the only country in Europe with coupons systems[51]. Basic health insurance is also covered by the state.[52]

Doctoral studies are, in most cases, subsidised for 50% of the tuition by the Government as well.[53] Growing space for opportunities is visible also in the field of informal education, which offers personal, educational and career improvements. There are 'a growing number of NGOs and private sector organisations offering courses, [and] volunteering opportunities.[54] We observe increased popularity of career and leadership development programmes. Entrepreneurship and business sector have attracted the attention of young people in the past couple of years especially through the start-up culture.'[55]

3.2. Satisfaction of Young People with Their Status, Their Country and Societal Systems

Among all the age groups young people are the most satisfied with their lives, since — on a scale from 0 to 10 — they rated their lives with an average score of 7.8 in 2015, which was 0.7 points more than the average in Slovenia and 0.2 points more than the average among the youth in the EU[56]. The average satisfaction with life was higher for young women than young men (by 0.3

points). Those enrolled in secondary and tertiary education were more satisfied with their lives (8.0 points)[57]. In 2015, the Statistical office conducted a survey among youth, which showed that 80% perceived themselves as generally satisfied with their lives.[58]

This means that young people in Slovenia remain optimistic about their future. However, concern about their future has increased in the following fields: employment security, housing, and in terms of 'the lack of trust towards the society as a whole and towards formal societal systems.'[59] In the same survey, young people marked great satisfaction (8.5 points) with their personal relationships (towards family, friends and colleagues), while their assessment of trust in the police gathered only 5.4 points. Even worse was their assessment of trust in the legal system (3.2 points), while trust in the political system was assessed at 2.0 points.[60]

Both of the first two categories (employment security and housing) are worrying, since, for example, 83% of young women in 2015 (aged 15-29) do not have a child, meaning that the fear of unemployment or unstable employment, lack of housing security and so on, might influence the fertility of Slovenia as a whole, which has grim consequences for the future of the Slovenian social system (free education, pensions, free healthcare etc.). The level of risk to fall into poverty among young people in 2016 was 14.6% and higher than that among the general population. Furthermore, the level of unemployment among the youth is also higher than total Slovenian unemployment (around 8% in 2016), meaning that youths are an especially vulnerable group in terms of the social assistance that they need.

For instance, Ms Carli mentioned that the development of youth is hindered by 'structural unemployment, unstable employment, the delayed transition to adulthood, ageing populations, lack of skills and competencies that employers are looking for, a mismatch between the supply and demand. The institutions are also witnessing growing numbers in rates of NEETs (Not in Employment, Education or Training),

57 | SURS. (2015). *Mladi v Sloveniji na splošno zadovoljni s svojim življenjem.* Retrieved from http://www.stat.si/StatWeb/News/Index/5345.

58 | *Ibid.*

59 | *Ibid.*

60 | *Ibid.*

young people who basically "gave up" and are discouraged to further participate in the labour market or continue with their education'.[61]

Ms Primorac[62] identified similar problems with the mentioned 'systemic' issues that 'should be dealt with by the government', such as housing, 'which is extremely expensive and therefore inaccessible', and precarious work, which results in 'huge struggles of young people while trying to gain their independence'. The second issue that she mentioned coincides with the data regarding the lack of trust of young people in societal structures in Slovenia, which she marks as 'challenges that stem out of the incompatibility of the values the youth possess and the rigidity of the existent social structures'. The biggest problem of this category, she marks as 'the political apathy of young people that is reflected in the voting turnouts. Active participation of youth is below the average in Europe, which is absolutely extremely problematic'.

Last, but not least, in the field of mobility and improving travel habits among young people, the government does provide subsidies to unemployed students for the usage of public transport;[63] however, the public transport system in Slovenia remains underdeveloped, as well as the connections to other countries, which are almost unavailable (with train or aeroplane), especially to those outside Europe.[64]

3.3. Remedies for These Barriers

According to the government, increasing developmental opportunities for youth was one of its priorities. In terms of improving conditions of employment, there are several different programmes available to young unemployed people such as the Guarantee for Youth (€300 million was allocated in the 2016-2020 period), on-the-job training, and subsidised employment to traineeships. In March 2017, the government adopted a joint staffing plan for public administration bodies for the years 2017 and 2018, and 237 jobs are foreseen to be created for trainee employment. The accelerated promotion of entrepreneurship among young people is another priority,

61 | Carli, S. (2017, October 27). *Interview no. 2 on the status of youth in Slovenia.*

62 | Primorac, K. (2017, October 26). *Interview no. 1 on the status of youth in Slovenia.*

63 | Morel. (2017, May 4). *Slovenija: Vladna verzija poročila o posvetu o mladih [Slovenia: Government version of the report on the Youth Conference].* Retrieved from http://www.morel.si/Notranja_politika/Slovenija_Vladna_verzija_porocila_o_posvetu_o_mladih/; Vlada Republike Slovenije. (2017). *Skupni imenovalec vlade in mladih: spodbujanje zaposlovanja, inovativnosti, kreativnosti in podjetnosti ter zagotavljanje dostopnih najemnih stanovanj [Common denominator of government and youth: promoting employment, innovation, creativity and entrepreneurship, and providing affordable rental housing].*

64 | Radiotelevizija Slovenija. (2017). *Slovenija šepa pri letalski infrastrukturi in kongresnem turizmu, čeprav ruši rekorde [Slovenia is lame in aviation infrastructure and congress tourism, although it is breaking records].* Retrieved from http://www.rtvslo.si/tureavanture/podobe-slovenije/slovenija-sepa-pri-letalski-infrastrukturi-in-kongresnem-turizmu-ceprav-rusi-rekorde/433635.

65 | For example, in the year 2016, the Ministry of the Economy has already launched half a million € of funds for launching youth cooperatives, and this year, the public tender will select the national youth network of a supportive environment for social entrepreneurship. Sources: Morel. (2017, May 4). *Slovenija: Vladna verzija poročila o posvetu o mladih [Slovenia: Government version of the report on the Youth Conference]*. Retrieved from http://www.morel.si/Notranja_politika/Slovenija_Vladna_verzija_porocila_o_posvetu_o_mladih/; Vlada Republike Slovenije. (2017). *Skupni imenovalec vlade in mladih: spodbujanje zaposlovanja, inovativnosti, kreativnosti in podjetnosti ter zagotavljanje dostopnih najemnih stanovanj [Common denominator of government and youth: promoting employment, innovation, creativity and entrepreneurship, and providing affordable rental housing]*.

66 | Morel. (2017, May 4). *Slovenija: Vladna verzija poročila o posvetu o mladih [Slovenia: Government version of the report on the Youth Conference]*. Retrieved from http://www.morel.si/Notranja_politika/Slovenija_Vladna_verzija_porocila_o_posvetu_o_mladih/; Vlada Republike Slovenije. (2017). *Skupni imenovalec vlade in mladih: spodbujanje zaposlovanja, inovativnosti, kreativnosti in podjetnosti ter zagotavljanje dostopnih najemnih stanovanj [Common denominator of government and youth: promoting employment, innovation, creativity and entrepreneurship, and providing affordable rental housing]*.

67 | Mlad.si. (2017, May 5). Delovni posvet Vlade RS z mladimi [Working consultation of the Government of the Republic of Slovenia with young people]. *Mlad.si*. Retrieved from http://arhiv.mlad.si/2017/05/delovni-posvet-vlade/; Vlada Republike Slovenije. (2017). *Delovni posvet Vlade RS: Spodbujanje zaposlovanja, inovativnosti, kreativnosti in podjetnosti ter zagotavljanje dostopnih najemnih stanovanj za mlade [Working consultation of the Government of the Republic of Slovenia: Promoting employment, innovation, creativity and entrepreneurship and providing affordable rental housing for young people]*. Retrieved from http://www.vlada.si/fileadmin/dokumenti/si/projekti/2016/prioritete/Mladi/PPTposvetVlade.pdf.

68 | *Ibid.*

for which the government will spend a total of €45 million in grants and €10 million in EU co-financing.[65] In addition, the Ministry of Agriculture, Forestry and Food has also announced a public tender to support young farmers in the framework of which €14 million will be earmarked. The objective of the call for tenders is to improve the age structure of holders of agricultural holdings and to improve the competitiveness of agricultural holdings.[66]

In terms of education, a law on scholarships has also been revised, which has remedied some of its shortcomings. Several student benefits have been extended, such as prolonged social insurance after graduating, prolonged hours of using student coupons with subsidies. The government also focused on the improvement of school camps. Last year they published funds for three learning production laboratories in the total amount of €2.4 million. In cooperation with experts in the field of modern digital technologies, they develop models of learning production laboratories, in which young people could produce different products and learn the practical skills of creative and innovative work.[67]

In terms of housing, only two pilot projects have been prepared, namely the pilot project for providing facilities for youth housing units and a pilot project for providing affordable housing for young people (first-time homebuyers). There is a third pilot project in preparation, aimed at designing a housing cooperative model to provide affordable rental housing for young people. According to the government, this measure should produce up to 400 affordable rentals for young people.[68]

4. Challenges and Solutions for the Belt and Road Initiative in Slovenia

69 | The head of the Chinese delegation was the First Deputy Prime Minister Zhang Gaoli. This was so far the highest ranked representative of the Chinese authorities to visit Slovenia, coming in the wake of the 25th anniversary of the establishment of diplomatic relations between the two countries. See Sta, M.N. (2017, April 14). Premier Miro Cerar danes na delovnem obisku v Sloveniji gosti prvega podpredsednika kitajske vlade Zhanga Gaolija [Today, Prime Minister Miro Cerar is hosting the First Deputy Prime Minister of China Zhang Gaoli on a working visit to Slovenia]. *Hudo.com.* Retrieved from http://novice. najdi.si/predogled/novica/4c92f 23511e7fc77a42273f511f4564d/ Hudo-Slovenija/Zanimivosti/ Premier-Miro-Cerar-danes-na-delovnem-obisku-v-Sloveniji-gosti-prvega-podpredsednika-kitajske-vlade-Zhanga-Gaolija.

70 | Sta, M.N. (2017, April 14). Premier Miro Cerar danes na delovnem obisku v Sloveniji gosti prvega podpredsednika kitajske vlade Zhanga Gaolija [Today, Prime Minister Miro Cerar is hosting the First Deputy Prime Minister of China Zhang Gaoli on a working visit to Slovenia]. *Hudo.com.* Retrieved from http://novice.najdi.si/predogled/ novica/4c92f23511e7fc77a422 73f511f4564d/Hudo-Slovenija/ Zanimivosti/Premier-Miro-Cerar-danes-na-delovnem-obisku-v-Sloveniji-gosti-prvega-podpredsednika-kitajske-vlade-Zhanga-Gaolija.

The Republic of Slovenia and the People's Republic of China have traditionally had very good relations, with political contact regularly held at various levels, both bilaterally and multilaterally. Which is why it was no surprise that in April 2017, when the Chinese delegation[69] visited Slovenia, the Slovenian Prime Minister Miro Cerar expressed satisfaction that Slovenia is part of the Chinese plan for the establishment of the 21st century maritime and land 'Silk Road'. He also estimated that the Chinese One Belt, One Road Initiative is an additional opportunity to promote cooperation and to strengthen the strategic partnership between the EU countries and China.[70] This section will discuss whether the BRI truly provides opportunities for youth in Slovenia and will critically analyse the challenges that the initiative might be posing in this regard. Furthermore, it deals with the role of young people in the implementation of the project initiative in Slovenia. The chapter is based on the descriptive-analysis and the interpretation of primary and secondary sources.

4.1. Potential Challenges Faced by the Belt and Road Initiative in Slovenia

A challenge awaiting the researcher trying to study the potential challenges of the Belt and Road Initiative (BRI) is that there is no contact person or body, which would coordinate all the developments regarding Chinese involvement in the country. It is clear that the first and the biggest challenge of Belt and Road in Slovenia is that government did not yet approve any sort of strategy, agenda or a vision, which one could see as the listed aims of Slovenia in the framework of the BRI. This, in turn, means that no ministry or governmental agency is responsible for the planning and implementation of the possible projects – or even to answer a set of simple questions about the challenges ahead, especially in terms of youth. In other words, there is nobody in charge of Belt and Road cooperation in the field of the participation of youth. This could lead to diminished or non-existent cooperation between Slovenia and China regarding young people.

The institutions and universities that could provide cooperation benefiting the youth are free to do so at their own costs and initiative. There is currently no structure that would facilitate the process of cooperation and no strategy for the overall goals that should be achieved with the cooperation. The Strategy for the Internationalisation of Slovenian Higher Education 2016-2020 does not mention potential cooperation within the Belt and Road framework.

On top of that, the lack of a responsible governmental body and strategy being presented to the public could cause a deficiency in knowledge about the BRI among citizens. The media present little information about the Belt and Road and what is reported is vague. There is a limited number of governmental documents referencing cooperation within the Belt and Road framework, and a meagre amount of media articles discussing its benefits.

Due to the shortage of information, there is a lack of interest among the youth to engage in the initiative. This could be the most prominent challenge faced by the BRI in the country but at the same time, it could be a great opportunity. While in this digital era, an interesting Instagram picture from China could have more effect than several articles promoting the implementation and benefits of the initiative. We strongly suggest a wide variety of approaches aimed at different target groups, bridging the divides between people from Slovenia and China. Addressing only one side without having a similar counterpart for the other cannot achieve the set results. An informative public diplomacy campaign coordinated with a governmental plan to create specific aims for Slovenia in regard to the Chinese Initiative would clarify most misunderstandings and prejudices, thus easing the path to its implementation.

The second challenge of the BRI is one of the possible disapproval of the Slovenian public that may be manifested by the civil society – in a referendum, as has happened before.[71] If new economic investments brought to Slovenia by the BRI also bring Chinese workers into the labour market, this will affect employment and wages in the receiving country, Slovenia.[72] Slovene public opinion will go one of two ways regarding the

71 | See the referendum on the construction of the second railway track Divača-Koper, which was subject to civil-society initiative. The law adopted by the National Assembly regulates the method of carrying out the investment into the railway, the construction and management of the track, determines financing and concessions for the management of the railway etc. The referendum against the law – and investment – was raised by the citizens, however, the law was accepted by a slight majority of referendum voters in the end. See: Radiotelevizija Slovenija. (2017). *Referendum o drugem tiru ni uspel, volivci ob nizki udeležbi zakon podprli [The referendum on the second track failed, and voters supported the law with low turnout].* Retrieved from http://www.rtvslo.si/slovenija/referendum-2017/referendum-o-drugem-tiru-ni-uspel-volivci-ob-nizki-udelezbi-zakon-podprli/433329.

72 | Radiotelevizija Slovenija. (2017). *Referendum o drugem tiru ni uspel, volivci ob nizki udeležbi zakon podprli [The referendum on the second track failed, and voters supported the law with low turnout].* Retrieved from http://www.rtvslo.si/slovenija/referendum-2017/referendum-o-drugem-tiru-ni-uspel-volivci-ob-nizki-udelezbi-zakon-podprli/433329.

import of foreign workers. The first way justifies the demand and the need for foreign workers since they are supposedly accepting those jobs that the domestic workforce deems inferior and refuses to perform.

The influence that foreign investment has on the parallel migration of workers and on employment and wages in the host country depends on whether the foreign workers are substitutes or complements for certain groups in the labour market. In the case of Chinese investments in the previously identified sectors – agriculture, tourism, green connectivity, infrastructure etc. – the workers brought to Slovenia would effectively substitute (the effect of substitution is greater than the effect of the income), since these are all highly valued sectors in Slovenia, and the demand for the jobs in these sectors by Slovenian workers is great. This means a reduction in wages for the remaining (local, Slovenian) group of domestic workers and lowered demand for this group of workers. Slovenian public opinion would therefore go in the direction of disapproval, believing that any new foreign worker entering the labour market will cause one domestic worker, who could become employed, to remain unemployed.[73]

This was also a part of the domestic Slovenian debate on the investment of China into Slovenian railways around 2011, where the public largely supported then Prime Minister, Borut Pahor, in his disagreement with the investment into railways carried out by Chinese companies with Chinese workers from construction to implementation.[74] One of the alternatives that might mitigate the negative effect on the Slovenian labour market of the Belt and Road investments is to take into account the sectors in which Slovenia is growingly recognised as successful and to employ locals who can bring a higher value to the investment itself.

Another potential obstacle is the language barrier. There is little knowledge of the Chinese language in Slovenia, although there are several language schools that can provide service in this regard, as well as the different Faculties of Arts.[75] Intertwined with the language barrier is the problem of cultural

73 | Dolinar, M. (2011). *Migracije in trg dela v Sloveniji: teoretična in empirična analiza [Migrations and the labour market in Slovenia: theoretical and empirical analysis]*. Graduate Thesis. University of Ljubljana: Faculty of Economics. Retrieved from http://www.cek.ef.uni-lj.si/u_diplome/dolinar4490.pdf. pp. 9-11.

74 | In this context, China has expressed its willingness to participate in the development of the Port of Koper and the railways network in Slovenia. Then Prime Minister of Slovenia, Borut Pahor, (who became president in 2012) complained that the Chinese government offered up to €10 billion for the purchase of the railways and the construction of expressways – provided that the construction is carried out in its entirety by Chinese workers. Pahor told reporters that he rejected the offer, as the arrival of 25,000 Chinese workers would cause a 'social bomb', which Slovenia could not endure. See Radiotelevizija Slovenija. (2017). *Na obisk v Slovenijo je prišel prvi podpredsednik kitajske vlade Džang Gaoli [First Deputy Prime Minister of China Zhang Gaoli paid a visit to Slovenia]*. Retrieved from https://www.rtvslo.si/gospodarstvo/na-obisk-v-slovenijo-je-prisel-prvi-podpredsednik-kitajske-vlade-dzang-gaoli/419858.

75 | Dnevnik. (2015). *Se že učite kitajsko? [Are you already learning Chinese?]*. Retrieved from https://www.dnevnik.si/1042722645.

understanding and the differences in business culture between China and Slovenia. This could be seen as an opportunity, for the promotion and funding of learning about Chinese culture.

The solution is to make the Chinese culture more known in Slovenia, which can be achieved by several tools. What would be especially beneficial would be to involve the youth in the promotion of cultural opportunities. For example, in order to target students, the creation of exchange programmes, stipends, language courses or even the availability of books written by Chinese academics would introduce Slovenes to Chinese culture (a good example to engage with is the East Asia Resource Library in Ljubljana). A good practical example is the Youth in Action programme of the EU, where youths originating from different countries are brought together to tackle common challenges. Such an action would be ideal for the BRI since it would mobilise the youth to simultaneously connect and solve issues that might arise during the implementation of the initiative.

4.2. Addressing the Challenges and Creating Opportunities for Youth in the Belt and Road Initiative

In terms of providing opportunities for youth in Slovenia, one should emphasise the most visible part of the general Slovenian-Chinese cooperation until now – economic cooperation, in the fields of both trade and investment.[76] China is the most important trading partner of Slovenia in Asia – and the scope of this connection has been steadily increasing in recent years. In 2016, trade in goods between the countries exceeded €1 billion, with Slovenian imports exceeding the exports by three times.[77] In December 2015, Chinese companies in Slovenia founded the Chinese Chamber of Commerce, which was a significant milestone in Slovenian-Chinese relations.[78] Therefore, the opportunities that the BRI can offer to young people can be of an economic nature, opportunities which are the most publically pronounced or visible in the eyes of the Slovenian citizens.

Among the areas where the BRI will contribute to the

76 | Delo. (2015). *Kitajski pogled na svet: en pas, ena pot – evroazijski Marshallow načrt? [China's worldview: one lane, one route - the Eurasian Marshall Plan?]*. Retrieved from http://www.delo.si/svet/globalno/kitajski-pogled-na-svet-en-pas-ena-pot-ndash-evroazijski-marshallov-nacrt.html.

77 | Radiotelevizija Slovenija. (2017). *Na obisk v Slovenijo je prišel prvi podpredsednik kitajske vlade Džang Gaoli [First Deputy Prime Minister of China Zhang Gaoli paid a visit to Slovenia]*. Retrieved from https://www.rtvslo.si/gospodarstvo/na-obisk-v-slovenijo-je-prisel-prvi-podpredsednik-kitajske-vlade-dzang-gaoli/419858.

78 | Radiotelevizija Slovenija. (2017). *Na obisk v Slovenijo je prišel prvi podpredsednik kitajske vlade Džang Gaoli [First Deputy Prime Minister of China Zhang Gaoli paid a visit to Slovenia]*. Retrieved from https://www.rtvslo.si/gospodarstvo/na-obisk-v-slovenijo-je-prisel-prvi-podpredsednik-kitajske-vlade-dzang-gaoli/419858.

79 | Vlada Republike Slovenije. (2017). *Premier dr. Cerar s kitajskim podpredsednikom vlade Gaolijem o nadaljnji krepitvi sodelovanja med državama [Prime Minister Dr. Cerar with Chinese Deputy Prime Minister Gaoli on further strengthening cooperation between the two countries].* Retrieved from http://www.vlada.si/predsednik_vlade/ sporocila_za_javnost/a/premier_dr_cerar_s_kitajskim_ podpredsednikom_vlade_gaolijem_o_nadaljnji_krepitvi_ sodelovanja_med_drzavama_828/.

80 | For example, Prime Minister Cerar expressed hope that Luka Koper, Slovenian main port, will become a key entry point for Chinese products to Central Europe. See Radiotelevizija Slovenija. (2017). *Slovenija šepa pri letalski infrastrukturi in kongresnem turizmu, čeprav ruši rekorde [Slovenia is lame in aviation infrastructure and congress tourism, although it is breaking records].* Retrieved from http://www.rtvslo.si/tureavanture/podobe-slovenije/ slovenija-sepa-pri-letalski-infrastrukturi-in-kongresnem- turizmu-ceprav-rusi-rekorde/433635.

81 | Ministry of Foreign Affairs of the Republic of Slovenia. (2015). *Slovenija: varna, uspešna in v svetu spoštovana [Slovenia: safe, successful and respected in the world].* Retrieved from http://www.mzz.gov.si/fileadmin/ pageuploads/Zakonodaja_in_dokumenti/dokumenti/ strategija_ZP.pdf.

82 | Vlada Republike Slovenije. (2017). *Premier dr. Cerar s kitajskim podpredsednikom vlade Gaolijem o nadaljnji krepitvi sodelovanja med državama [Prime Minister Dr. Cerar with Chinese Deputy Prime Minister Gaoli on further strengthening cooperation between the two countries].* Retrieved from http://www.vlada.si/predsednik_vlade/ sporocila_za_javnost/a/premier_dr_cerar_s_kitajskim_ podpredsednikom_vlade_gaolijem_o_nadaljnji_krepitvi_ sodelovanja_med_drzavama_828/.

83 | There are many cases of the youth succeeding in innovative technologies and business models creation, such as Zemanta, Kosei, Block, Bitstamp, Toshl etc. Furthermore, certainly the most successful story in Slovenia and at the same time one of the most successful start-ups in the world, created by two young Slovenes is Outfit7. It is best known for the creation of the Talking Tom app and it was acquired by the Chinese United Luck Consortium and later Chinese Zhejiang Jinke. All the mentioned young business holders place a greater value upon innovation, rather than earnings. See: Večer. (2017). *Prodan še en slovenski start-up – Zemanta.* Retrieved from https://www.vecer.com/prodan- se-en-slovenski-start-up-zemanta-6284255; Takahashi, Dean. (2017). Talking Tom maker Outfit7 confirms it has been sold to United Luck Consortium for $1 billion. *Venture Beat.* Retrieved from https://venturebeat.com/2017/01/20/ talking-tom-maker-outfit7-confirms-it-has-been-sold-to- united-luck-consortium-for-1-billion/.

economic opportunities for youth are infrastructure, logistics, hydroelectricity, tourism and agriculture[79]. The economic interests of Slovenia in these regards are more specifically closely connected to the Port of Koper,[80] Slovenian railways, the automobile industry, and by participating in the most technologically demanding fields, where Slovenia can offer productivity and creativity – youth being the main contributor.[81] Some of these fields are information technology, new materials, alternative energy sources, green connectivity, and good governance systems. The Chinese market undoubtedly represents great potential for the growth and development of Slovenian industries in many sectors, which also means greater potential for the employment of young workers.[82] Slovenian youth is well educated – it is a country with a skilled, productive, innovative but relatively cheap workforce – and the trend in education is going towards the areas of high-tech. Moreover, the number of start-up hubs and start-ups created by youth in Slovenia is increasing. BRI investments into innovative technology projects offer opportunities for those young people with skills, business sense and creative ideas.[83]

Better connections between the two countries also offer an initiative for improved youth mobility. This benefit is twofold: first, it offers learning mobility, and second, it provides for job mobility in the labour markets of Slovenia and China. Both facets of transnational mobility are necessary for the Slovenian youth to acquire new skills, and therefore, improve their employability and personal development. Studies confirm that learning mobility increases human capital, as the youth gains access to new knowledge, develops new language skills and intercultural competencies.

In addition, employers are aware of these benefits and appreciate them.[84] The experience of living in another country also impacts character traits of a young person – it increases independence and responsibility, helps to develop social skills, raises the level of solidarity and lowers the possibilities of prejudices that can lead to conflicts both between individuals and among larger groups, countries, nations. It indirectly diminishes various forms of intolerance, which would lower the chances of further collaboration between the countries, in our case Slovenia and China.[85] Therefore, the BRI is an opportunity for Slovenian youth to learn about both China and other countries. Participating within the Initiative offers them the possibility of gaining knowledge through cultural exchange, such as getting to know different political and social arrangements within countries, applying for scholarships, and allowing them to aspire to be a part of an expanding Slovenian diplomatic net.

One of the opportunities offered to the youth with the BRI is also to make use of the many institutions established to increase inter-cultural understanding between Slovenia and China. For example, the East Asia Resource Library (EARL) at the Faculty of Social Sciences, under the auspices of the latter faculty and the Faculty of Arts, University of Ljubljana, has the basic task of establishing a wide network of institutions from the East Asia region, which would contribute to scientific and professional literature and study materials, thus promoting understanding between the peoples, students and faculties of this region with Slovenes. The library is therefore a 'gateway' to the knowledge of East Asian culture, literature, languages, arts, geopolitics etc., and could gain more importance as the BRI, and its opportunities, gain visibility in Slovenia.[86]

4.3. The Importance of Youth in the Successful Implementation of the Initiative

The youth can be an important facilitator of the Belt and Road

84 | Movit. (2017). *Mobilnost mladih [Youth mobility]*. Retrieved from http://www.movit.si/info-servis-eurodesk/mobilnost-mladih/.

85 | Mladinski svet Slovenije. (2011). *Policy Paper on Youth Mobility*, Adopted at the 20th Regular Meeting of the MSS Assembly, on 08 December 2011. Retrieved from http://www.mss.si/datoteke/dokumenti/PP_youthmobility_web.pdf.

86 | Fakulteta za družbene vede. (2017). *Sedež EARL [EARL headquarters]*. Retrieved from https://www.fdv.uni-lj.si/knjiznica/earl/sedez-earl.

Initiative (BRI) in three ways in Slovenia. First, it can influence public opinion by improving the credibility of the project in the eyes of citizens and making visible the meaningful benefit of the Chinese presence in Slovenia. For instance, youths are strong supporters of Slovenian membership in the EU, due to several EU projects for young people (Erasmus+ being an example). The support of the youth tips the scale on the general public polls regarding Slovenia exiting or staying in the EU.[87]

Secondly, the youth is important as an innovative workforce, as a facilitator of the transfer of knowledge and skills, scientific progress, and as a generator of real community change. This chapter mentioned the problem of young people's trust in the social and political structures of the country, which is an indicator that youth could and should be included in the decision-making process regarding the project's effect on the whole community. Moreover, due to their skills – especially the ones that are characteristically youthful, such as mastery of IT – they are important for the successful realisation of the planning process, as well as the execution of the project. They can also influence the transfer of knowledge between Slovenia and China, as they reach higher levels of mobility in comparison to other generational groups.

And lastly, they can also influence the transfer of values, understanding and facilitate bridge-building between Slovenia and China. Young people are the decision-makers of the future; they should be educated to cooperate in a peaceful way, gaining enough understanding of the other nation's culture, languages and values. It is well known that personal ties can influence relations between countries, especially in economic terms.[88] Snyder *et. al* noted that 'developing trust through personal relations can have a profound effect on policy outcomes'.[89] Therefore, the participation of youth in various activities that can raise their intercultural understanding is of utmost importance in terms of *long-lasting* successful implementation of the Initiative in Slovenia.

87 | Mlad.si. (2016). *"Erasmus+ v Sloveniji – zgodba o uspehu!" Interview with the Director of MOVIT, National agency.* Retrieved from http://mlad.si/2016/05/erasmus-v-sloveniji-zgodba-o-uspehu/; European Commission. (2016). *Special Eurobarometer 451 Report: Future of Europe.* Retrieved from http://europe.vivianedebeaufort.fr/wp-content/uploads/2017/11/ebs_451_en.pdf.

88 | See for example Bandelj, N. (2002). Embedded Economies: Social Relations as Determinants of Foreign Direct Investment in Central and Eastern Europe. *Social Forces,* 81 (1), pp. 411–444.; University of Toronto. (2011). *Personal Relationships Key to Successful Diplomacy.* Retrieved from https://www.utoronto.ca/news/personal-relationships-key-successful-diplomacy-mulroney; Van Hoef, Y. (2014). Friendship in World Politics: Assessing the Personal Relationships Between Kohl and Mitterrand, and Bush and Gorbachev. *The Journal of Friendship Studies 2 (1): 62 – 82.* Retrieved from https://amityjournal.leeds.ac.uk/issues/volume-2/friendship-in-world-politics/.

89 | Snyder, R. C., Bruck, H. W., Sapin, B., Hudson, V., Chollet, D., & Goldgeier, J. (2002). *Foreign Policy Decision Making.* New York: Palgrave Macmillan.

三聯書店
http://jointpublishing.com

JPBooks.Plus
http://jpbooks.plus

EDITOR   Donal Scully
DESIGNER   Chloe Wong

TITLE   Young People and the Belt and Road –
              Opportunities and Challenges in Central and
              Eastern Europe

COMPILER   Young Belt and Road Series English Editorial Board

AUTHORS   Wang Chenxi, Jetnor H Kasmi, Altin Kukaj, Ajsela
              Toci, Toni Čerkez, Enna Zone Đonlić, Alexander
              Georgiev, Mark Georgiev, Nikolay Nedkov, Mihail
              Tsvetogorov, Anastasya Raditya Ležaić, Alexandr
              Lagazzi, Jaroslav Ton, Mikk Raud, Adrienn Lukács,
              Sintija Bērziņa, Ugnė Mikalajūnaitė, Ivana Jordanovska,
              Anastasija Milkovska, Slobodan Franeta, Luka Laković,
              Renata Ljuljdjurovic, Ivan Piper, Grzegorz Stec,
              Sebastian Sulkowski, Vladimir Vasile, Bianca Vlad,
              Đina Glavčić-Kostić, Sara Kljajic, Nikola Stojanovic,
              Lenka Čurillová, Jazmína Obedová, Nina Pejič,
              Urban Špital

SPONSOR   International Academy of the Belt and Road and
              Greater Bay Area & Belt and Road Center, Basic
              Law Foundation.

First published in January 2021

Published by Joint Publishing (H.K.) Co., Ltd.

20/F., North Point Industrial Building, 499 King's Road,
North Point, Hong Kong

Printed by Elegance Printing & Book Binding Co., Ltd.

Block A, 4/F, Hoi Bun Industrial Building, 6 Wing Yip Street,
Kwun Tong, Kowloon, Hong Kong

Distributed by SUP Publishing Logistics (H.K.) Ltd.

16/F., 220-248 Texaco Road, Tsuen Wan, N.T., Hong Kong